Seize the High Ground

The Army in Space and Missile Defense

NIKE - ZEUS
1957-1964

SAFEGUARD Systems Command
1969-1974

Space and Missile Defense Command
2001-Present

by
Dr. James Walker
Dr. Lewis Bernstein
Mrs. Sharon Lang

Historical Office
U.S. Army Space and Missile Defense Command
2003

CMH Pub 70-88-1

For sale by the Superintendent of Documents, U.S. Government Printing Office
Internet: bookstore.gpo.gov Phone: toll free (866) 512-1800; DC area (202) 512-1800
Fax: (202) 512-2250 Mail: Stop SSOP, Washington, DC 20402-0001

ISBN 0-16-072308-6

Table of Contents

List of Illustrations

List of Tables

The Commander's Introduction

Part of the U.S. Army's strength lies in its traditions. These traditions are epitomized in the framework of lineage and honors that link soldiers and their units. As the Army's newest major command, one might assume that the U.S. Army Space and Missile Defense Command (USASMDC) would not have a significant historical record. However, USASMDC and its predecessor organizations have spent many decades (since 1957) focusing on issues and experiments with missile defense, space-based communications, and sensor technologies. This focus can be seen as a natural outgrowth of the Army's continuing strategic defense mission: defending the U.S. homeland.

It is my pleasure to introduce this history of the U.S. Army's activities in space and missile defense. A glance through the pages of this survey will illustrate the importance of space and missile defense to America's military focus. As the command evolved from its beginnings in 1957 into its present shape, it retained a functional organizational structure that oversaw the development of various systems from the earliest developmental stages to operational use. The USASMDC was, and remains, an adaptable, technology-based organization, open to new ideas, innovations and forms of collaboration. But, at the same time, we have never lost sight of our primary goal, giving the individual soldier the best possible tools for finding and destroying the enemy. We support the warfighting combatant commanders, play an important role in the acquisition process, work to integrate space and missile defense solutions within the Army, and act as the service's advocate in Joint Warfighting forums.

In the continuing evolution of the USASMDC's missions, I thank the soldiers and civilians who have served and continue to serve with loyalty, courage, intelligence, initiative, ingenuity, and creativity. I have the honor to serve as the command's public face not only for the dedicated men and women who serve in the USASMDC today but for all those who have contributed to our accomplishments over the last 46 years. May all those who read this book learn from the achievements captured in its pages.

SECURE THE HIGH GROUND

Joseph M. Cosumano, Jr.
Lieutenant General, U.S. Army
Commanding

Preface and Acknowledgements

Seize the High Ground: The Army in Space and Missile Defense provides an overview of the Army's involvement in the development and use of space-based systems and missile defense to serve the nation. The Space and Missile Defense Command traces its origins to the founding of the nation, when strategic defense meant coastal fortifications. Through the 19th and 20th centuries, this concept expanded as new construction techniques were devised and coast artillery fire became more accurate. Since 1945 the concept of strategic defense has expanded beyond coastal fortifications and artillery to encompass outer space and missile defense. In order to meet these new challenges, the Army was specifically assigned to develop a system to detect, intercept and destroy enemy missiles. At the same time, the Army was intimately involved in the early days of space flight, building the missiles that launched the first American satellites and astronauts into orbit.

The Army's record of achievement in space and missile defense matters is a success story. Despite political controversies surrounding missile defense and conflicts over the Army's role in space, soldiers, scientists and technologists have been generally successful in devising ways to defend the nation from missile attack and in using space-based systems to increase the Army's combat power. Army operations since 1989 provide the historical evidence on which this judgment rests. Additionally, the functional task groupings the Army's space and missile defense units adopted to bring order to their activities may offer a template for future Army organization.

Although expanded and bearing a new title, *Seize the High Ground* builds upon the material presented in *Strategic Defense: Four Decades of Progress*, an earlier volume published by the USASMDC Historical Office in 1995. Sharon Lang and Lewis Bernstein, the authors of the present volume, are particularly indebted to Frances Martin, a coauthor of this 1995 volume. The authors also wish to thank Ms. Susan Gahagan of Sigma Services of America who designed the layout and Mr. Roy L. McCullough of Science Applications International Corporation (SAIC) for his comments and suggestions on a draft version of the manuscript.

James A. Walker, Ph.D.
Command Historian

Introduction
A Historical Perspective of Missile Defense and Space

In 1997, the Army established its newest major command, the U.S. Army Space and Missile Defense Command, to serve as its proponent for space and national missile defense and overall integrator for theater missile defense. The Army's interest in space and missile defense stems from its traditional mission of strategic defense that began in the 1790s, when it built and manned coastal forts to defend the nation against assault from the sea. In the early 20th century, soldiers continued to man coastal forts and by World War II, they operated anti-aircraft sites across the nation to defend against the threat of long-range bombers. By 1945, the Army had a secure grasp of sensor technology fundamentals that enabled it to take, process, and analyze millions of photographs for intelligence purposes; it had created and operated a large, secure, unified global communications system, and created the best code-breaking capability in the world. Additionally, along with its air arm, the Army was working on developing guided missiles.

At the end of World War II, the United States and the rest of the world were introduced to two new threats—ballistic missiles and nuclear weapons. Postwar assessments pointed the way toward new weapons systems and using rockets in space exploration. In the mid-1950s, Soviet intercontinental ballistic missiles threatened to destroy American cities with nuclear warheads. In 1955, the technological challenge posed by this new threat led the Army to begin studying the feasibility of creating a defense against ballistic missiles. In October 1957, the Army created the Redstone Anti-Missile Missile Systems Office in Huntsville, Alabama, initiating research that led to the NIKE-ZEUS anti-missile system. These efforts provided the foundation for the Army's space and missile defense program.

Through the late 1950s, the Army's efforts in rocketry, missile defense, and sensor technology were complementary—each capability worked to enhance the other. Although not seen at the time, they were inter-locking efforts. In fact, the Army built and launched the nation's first ballistic missile and earth orbiting satellite. The Mercury astronauts were placed in orbit by modified Army Redstone rockets, the Jupiter Cs. These feats, which challenged the U.S.S.R.'s first ventures into space, were the work of Dr. Wernher von Braun and his rocket team at the Army Ballistic Missile Agency at Redstone Arsenal, Alabama. The first communications and reconnaissance satellites were developed and launched through a partnership between private industry and government in which the Army played a prominent part. This link was temporarily broken by the Eisenhower Administration's decision to create the National Aeronautics and Space Administration (NASA) and redistribute space and missile roles and missions among the services. NASA received the Redstone program, the Explorer satellite program, and all the rocket and missile contracts the Army had with the California Institute of Technology's Jet Propulsion Lab, as well as responsibility for developing the 1.5-million-pound thrust Saturn rocket. The Army also transferred technical expertise from the Army Ballistic Missile Agency Development Operations Division to NASA.

The Army lost most because it had the most to lose, but it did retain responsibility for maintaining and improving worldwide communications and for missile defense. The former derived from its communications functions while the latter from its charge to defend American territory, in terms of coastal defense, then antiaircraft defense, and finally air defense. The Army's work served as a catalyst for a telecommunications revolution because satellites stitched the world together in a way very different from either wires or cables. As it temporarily lost part of its space functions, it concentrated on the air defense mission.

In 1962, the NIKE-ZEUS program made the first successful intercept of an intercontinental ballistic missile. In 1963, Project MUDFLAP achieved a satellite intercept. These successes generated controversy when Secretary of Defense Robert S. McNamara determined that the missile defense system was neither technologically feasible nor cost effective. He assumed the Soviets would overwhelm it by launching more missiles than could be intercepted. His own preferred solution was peace through terror or mutually assured destruction. In 1965, after China developed and tested nuclear weapons and missiles, strategists began to refer to an "Nth country" threat to stability. McNamara eventually agreed with the premise and announced a decision to deploy a system to protect against a possible Chinese attack, and development work proceeded. As arms control negotiations began with the Soviet Union, diplomats on both sides used their anti-missile systems as bargaining chips to obtain concessions from each other. By the 1970s, the Army activated the only missile defense system in the West, the Stanley R. Mickelsen SAFEGUARD Complex in North Dakota. The ratification of the SALT I agreement in 1972 limited deployment of ballistic missile defense systems. In 1975, as a cost-cutting measure, the United States deactivated its system.

Despite this setback, the Army's scientists and engineers continued to develop and test a missile defense system through the Ballistic Missile Defense Organization (BMDO). The period between 1974 and 1983 began with declining interest in Ballistic Missile Defense (BMD) initiatives, followed later by guidance to accelerate development of a defense for American ICBMs. The Systems Technology Radar, designed to provide data in terminal, low-altitude, and midcourse operations, was a major improvement over the SAFEGUARD Missile Site Radar. This unmanned system was capable of transmitting thousands of beams per second and used a versatile transmitted waveform combined with more advanced signal processors that permitted better target discrimination. With these advances over the SAFEGUARD radar, the Systems Technology Radar alone could serve as a radar system for defending Minuteman missiles. This radar system was also an important element in the underlay of the proposed layered defense concept.

There were also experiments with an airborne telescope. The Ballistic Missile Defense Organization's engineers recognized the limitations of ground-based radars and explored using airborne/spaceborne sensors to discriminate between targets in the Designating Optical Tracker Program. These experiments with a rocket launched infrared telescope demonstrated that a long-wave infrared sensor could discriminate between, designate, and track a reentry vehicle. Encouraged by these successes, the experimenting continued with an Airborne Optical Adjunct to investigate the technical feasibility of using airborne optical sensors to detect, track, and discriminate between ballistic missile reentry vehicles along with the ability to pass trajectory data to ground-based radars. Renamed the Airborne Surveillance Testbed, this optical sensor was the first Ballistic Missile Defense project incorporated in the next generation anti-ballistic missile initiative and played an important role in missile test programs and exercises.

During the 1970s, the Ballistic Missile Defense Advanced Technology Center explored military applications of neutral particle beams and high-energy lasers, two different directed energy technologies. The Advanced Research Projects Agency began initial research in the late 1950s and while Congressional parsimony and restrictions handicapped research, progress was made. The two primary efforts were the exoatmospheric neutral particle beam accelerator program and the collective ion accelerator experiment. By the late 1970s, Army researchers demonstrated that lasers could work with pointing and tracking devices to form an effective weapons system. After Congress began pressing Army officials to begin developing space-based laser weapons, the White Sands Missile Range, New Mexico, was designated as a suitable location for high-energy laser range testing. In 1980, following policy established by President Carter, Secretary of Defense Harold Brown directed the services to emphasize the use of lasers in space. BMDO focused on using lasers to destroy ballistic missiles in the boost or midcourse phase of their flights, before their reentry vehicles deployed.

To protect the Air Force's newest ICBM, the MX or Peacekeeper missile, the Army developed a low-altitude defense system composed of a series of radars, distributed data processors, and nuclear-tipped interceptors. Its size and design would complement any of the proposed MX ICBM deployments. In 1981, Secretary of Defense Casper Weinberger issued a Ballistic Missile Defense Program Directive to support all MX basing options. The directive also called for the development of a non-nuclear endoatmospheric weapon. With this guidance, the Ballistic Missile Defense Organization planned to convert the low-altitude defense system to a non-nuclear interceptor and renamed the program SENTRY. The next year, the Ballistic Missile Defense System Command terminated the SENTRY program.

The advances made in infrared sensor and computer technology encouraged scientists to experiment with hit-to-kill technology, that is, kinetic energy intercepts, leading to the first kinetic kill interception of a missile in space with the Homing Overlay Experiment. Launched by two Minuteman stages, the kill vehicle consisted of a computer, a long wavelength infrared optical sensor package for guidance, and a kill device. When the missile reached a point above the atmosphere, a sensor and computer on the launch rocket would locate and track the reentry vehicle and relay tracking data to the intercept vehicle. As the target neared, the kill vehicle would be launched and using its own infrared sensors and computer home in on the target. Right before intercept, the kill vehicle would unfurl the spokes of a 13-foot radial net that would capture the reentry vehicle.

In June 1984, the Homing Overlay Experiment successfully completed the first kinetic kill intercept. The kill vehicle intercepted a mock ICBM reentry vehicle more than 100 miles above the Pacific Ocean. In this test, the kill vehicle and the warhead closed at more than 15,000 feet per second and smashed into each other. The Homing Overlay Experiment was the first true revolution in ballistic missile defense since research began in the 1940s.

The Army's space and missile defense interests were revived by the internal debates over professionalism, equipment, and doctrine that occurred after the Vietnam War and President Reagan's 1983 Strategic Defense Initiative. However, this was foreshadowed by Army participation in the Tactical Exploitation of National Space Based Capabilities Program (TENCAP) beginning in 1973. By 1983, by virtue of its participation in TENCAP and its aggressive research program in missile defense, the Army found itself as the service with the most experience in dealing with technical missile defense problems as well as the biggest consumer of space products.

The Strategic Defense Initiative was buttressed by the Fletcher and Hoffman reports. The Hoffman group concluded that a missile defense could enhance deterrence and believed that an anti-tactical ballistic missile system could serve as a first step toward a national missile defense system. The Fletcher commission recommended a research blueprint for the Strategic Defense Initiative in the areas of Systems Concepts; Surveillance, Acquisition, and Tracking; Directed Energy Weapons; Conventional Weapons; Battle Management and Command, Control, and Communications; Survivability; Lethality and Threat Vulnerability; and Selected Support Systems. A somewhat constrained program based on this model became the guide for the Strategic Defense Initiative.

The Strategic Defense Initiative Organization was a multi-service group. However, the Army's anti-ballistic missile experience was its foundation, and the Army repeatedly took the lead in project development. This experience allowed the Strategic Defense Initiative Organization to protect the technology base, increase the emphasis on proof-of-feasibility experiments with greater investment in high risk-high payoff approaches, and continue examining multi-layered defense.

Researchers from the Strategic Defense Initiative Organization (SDIO), the Army, and the Air Force created a concept for tiered, or layered, defense against enemy missile systems to ease interception of an incoming missile during its three flight phases: boost, midcourse, and terminal. Each of the services was assigned elements designed to track or intercept during specific phases of the missile flight. The Strategic Defense Command and the Army assumed the lead in the effort.

The Strategic Defense Initiative concept for the boost phase incorporated the Boost Surveillance and Tracking System, the Space-Based Laser, and the Ground-Based Laser. The Strategic Defense Command shared responsibility for the Space-Based Laser with the Air Force, while it was assigned sole control over the Ground-Based Laser. In the midcourse phase, the system architecture envisioned a Space-Based Surveillance and Tracking System, a Space-Based Interceptor, a Neutral Particle Beam, and the Exoatmospheric Reentry-vehicle Interceptor Subsystem. The Air Force directed development of the Space-Based Surveillance and Tracking System and the Space-Based Interceptor while it shared responsibility with the Army and the Strategic Defense Command for the Neutral Particle Beam. The Army then directed the evolution of the Exoatmospheric Reentry-vehicle Interceptor Subsystem. The final defense layer, the terminal phase, employed the Airborne Optical Adjunct, the Ground-Based Radar, the Ground-Based Surveillance and Tracking System, and the High Endoatmospheric Defense

Interceptor. The Strategic Defense Command had the lead on all of these programs. All three primary elements, the Air Force, the Army, and SDIO, shared in the development of the Battle Management/Command, Control, and Communications systems. As these programs evolved to the demonstration stage, the command explored new areas in interceptor, sensor, and related technology. Advances have been made in optics, sensors, and data processing, which have subsequently been applied to existing and planned systems.

Existing in parallel and drawing on some of the strategic defense initiative's research was the anti-satellite program. In January 1989, the Defense Acquisition Board authorized developing an anti-satellite program to deploy in the mid-1990s. In March, the Army received the lead in this joint service effort, based on its record with ground-based interceptors. The program would counteract an already deployed Soviet anti-satellite system. The Defense Acquisition Board requirements included using both kinetic energy and directed energy approaches.

As funded, the anti-satellite program was distinct from strategic defense, but drew on the Strategic Defense Command's kinetic and directed energy research. Thus, Strategic Defense Initiative funding directly affected anti-satellite development. Although there were delays in the directed energy program, the kinetic energy program proceeded with only a few setbacks. The proposal for two versions of a kinetic energy weapon, one ground-launched, the other sea-launched, was reduced to a single system. In August 1990, the Rockwell International Corporation was awarded a contract to develop a ground-launched kinetic energy anti-satellite weapon. The first tests for this visual light sensor system were planned for January 1992. Following budget reductions and program restructuring, the Army recommended canceling both programs. Funding was restored after several senators wrote to President Bush to support the effort. In 1992, Congress directed that the program reflect the end of the Soviet threat and the proliferation of militarily significant space capabilities of a growing number of countries.

By June 1993, continued budget cuts forced the termination of the Kinetic Energy Anti-Satellite Joint Program Office. The Defense Authorization Act for 1994 directed its conversion to a command-managed technology program and progress continued at a slower rate. The work culminated in a September 1995 hotfire strapdown test that demonstrated the kill vehicle's ability to fly a predetermined simulated flight path by firing its divert/attitude control system thrusters. The system also successfully acquired and tracked a target with its onboard computers. Two years later, the prototype concluded a successful hover test, in which the sensor acquired and locked onto a simulated moving target.

The program experienced funding problems throughout its history, resulting in rescheduling and other setbacks. In 1998, the U.S. Space Command's Mission Needs Statement for Space Control included a requirement for an anti-satellite capability. In that same year, however, President Clinton used a line item veto to eliminate funding for the anti-satellite program as well as 42 other programs. This action was declared unconstitutional by the U.S. Supreme Court and the funding restored. Surviving on Congressional plus-ups, the program was transferred to the Army Aviation and Missile Command in October 2001.

The command also proceeded with laser experiments and created a dedicated test facility for them. In 1974, the Congress directed the Defense Department to create a military high-energy laser test facility to halt redundant development work at various government and contractor sites. In 1981, the Defense Department awarded a contract to construct a site at White Sands Missile Range, which was nearly complete by 1984. The High Energy Laser Systems Test Facility (HELSTF) would support Army and Department of Defense laser research and development, test, and evaluation, as well as integrate and operate lasers and related instruments, facilities, and support systems. HELSTF would also conduct and evaluate laser effects tests on materials, components, subsystems, weapons and systems. The facility became operational in September 1985 with an Air Force Lethality and Target Hardening program experiment for the Strategic Defense Initiatives Organization. In this test, a Mid-Infrared Advanced Chemical Laser destroyed a Titan booster rigged to simulate the conditions of a thrusting rocket booster.

In October 1989, the Secretary of the Army had the facility transferred from the Army Materiel Command to the Strategic Defense Command to centralize high-energy laser research. The actual transfer occurred in October 1990. HELSTF's mission expanded to include a full range of research, development, test, and engineering functions. These included test and evaluation, laser damage and vulnerability support, intelligence evaluation resources, advanced system integration center, range instrumentation, space surveillance, and anti-satellite contingency capability. The site has been active in the command's directed energy programs.

As the Army made progress in missile defense, experimented with anti-satellite weapons systems as well as laser and particle beam weapons, its long dormant interest in space began to revive. It was assisted by its own internal reformation and the announcement of President Reagan's National Space Policy in July 1982. The policy included commitments to explore and use space for peaceful purposes by all nations, pursue activities in space supporting the United States' right of self-defense, make space-based systems available to commercial and government users, and continue to study space arms control options that would limit testing and deploying specific weapons. In 1988, the policy was updated, reaffirming the national commitment to space exploration and addressing civil, military, and commercial space use. It called for American space policy to obtain scientific, technological, and economic benefits for the general population and to improve the quality of life on earth through space related activities, promote international cooperative activities while protecting American interests.

Because there was not a pre-existing critical mass of interest for space as there was for missile defense despite the Army's use of the medium, the way forward was more difficult. First, the Army had to reinvigorate its interest and begin to see space-based systems as force multipliers. While the National Space Policy indicated a broad interest, there was still no direct reason for the Army to become aware of its reliance on space-based systems. The event that drove this dependence home was Operation Urgent Fury, the invasion of Grenada in 1983.

In 1983, the Army Science Board's study *Army Utilization of Space Assets* concluded the Army was not using space systems to their full potential; to achieve better exploitation there must be a high-level commitment backed by sufficient resources. Operation Urgent Fury highlighted the services' scramble for access to limited space assets. Because it had used other services' systems too long, the Army was assigned the leftovers in a crisis. The subsequent Combined Arms Grenada Work Group recommended the Army develop, own, and control its own satellites to ensure critical communications in such operations. Later in 1983, an Army Space General Officer Working Group was founded to provide direction for Army space efforts. In 1984, the Army Science Board studied the Army's use of space to support its missions, concluding the Army made limited use of space assets and was neither active nor influential in designing and operating most of the space systems then in use. In August 1984, an Army Space Council was created to approve proposals and provide direction for the Army's involvement in and use of space.

In September 1984, the Vice Chief of Staff of the Army, General Maxwell Thurman, activated an Army Staff Field Element, the nascent form of the Army Space Command, at Air Force Space Command headquarters. By the end of 1984, the Army had four organizations to manage its space commitments. First, there was an Army Space Council chaired by the Vice Chief of Staff then the Army Space Working Group. The third organization was the Army Space Office, a focal point for space-related matters that served as a liaison to the Joint Staff and the Office of the Secretary of Defense, and finally the Army Staff Field Element of Air Force Space Command. The Space Office identified five high-priority space tasks: developing Army space policy; creating an Army space-related requirements and programs inventory; near-term enhancements to Army space involvement; developing Army space-related requirements based on an operational concept for space support to warfighting; and developing Army options to support a unified space command.

Space-related activity in the Army reached critical mass in 1985. That year, the Combined Army Combat Developments Activity created a Space Directorate, a Space Initiatives Study Group was formed to analyze the ways the Army should use space, and the Staff Element at the Air Force Space Command became the Army Space Planning Group—the Army element of the new U. S. Space Command. The following year, the Army Space Planning Group became the Army Space Agency. The Army Space Initiatives Study was published in December 1985. The study advocated making the Office of the Deputy Chief of Staff for Operations and Plans the senior Army staff proponent for space, and recommended that the Combined Arms Center at Fort Leavenworth become the Army proponent for space and that the Command and General Staff College become the lead Army school for space education. The study also urged forming an Army Space Command as the Army component of Space Command and advocated the Army integrate the use of space and space products into its doctrine. The report further called for establishing an Army Space Institute, the Army Space Technology Research Office, and the Army Space Agency. The Space Initiatives Study counseled that the Army train soldiers about space systems and create an additional specialty indicator to trace personnel with experience, education, and training in space systems.

The Army Space Institute (ASI) was founded in 1986 to serve as a clearinghouse for matters relating to the Army's use of space. Functioning as the Training and Doctrine Command (TRADOC) proponent for space and space systems, it was responsible for developing Army space concepts, doctrine, training, force structure, materiel requirements, techniques, and procedures that would apply space systems and technology to improve the execution of AirLand

Battle Doctrine and support the Strategic Defense Initiative. The ASI maintained a tactical focus throughout its existence, approaching its mission aggressively, and predicted that space systems would be available at the battalion and company levels. It also prepared for the Army Space Demonstration Program to show the ways current space-related products could support battlefield commanders and their units, down to the squad level. In 1986, shortly after the space activities skill code was established, ASI proposed to redefine it while realizing this did not address the basic need to build expertise. In 1987, a new Space Activities skill code definition was sent to the Vice Chief of Staff of the Army with specific qualifications in duty assignment, military training, and civilian schooling.

The Army had long had an interest in manned space flight. In January 1959, NASA dealt a blow to the Army's hopes for continued involvement in space exploration when it published the selection criteria for astronauts from the military services. One of the requirements—that an astronaut had to be an experienced jet aircraft pilot—eliminated Army personnel from consideration as astronaut candidates. In 1964, NASA dropped the requirement for pilot experience for crew members, but only in an effort to recruit "scientist-astronauts" to conduct research on space flights. Most of these candidates had doctoral degrees in the natural sciences, medicine, or engineering, or equivalent experience. Because few of its officers had advanced training in these fields, the Army was once again excluded from the manned space program. In January 1978, NASA announced the selection of 35 new astronaut candidates for the Space Shuttle Program, the first chosen since 1969. This group included the first women and racial minorities chosen; additionally, two new astronaut job titles were created, pilot and mission specialist. Both civilians and military officers were among the candidates; one of the latter was Major Robert L. Stewart, who became the Army's first astronaut.

In 1986, the Pentagon established the Military Man in Space program as part of Shuttle operations. The Air Force was the overall Executive Agent and the Office of the Deputy Chief of Operations and Plans, Department of the Army, became the Executive Agent for the Army program. The program was to evaluate, through experiments proposed by each uniformed service and approved by DoD, ways in which military operations on earth could be improved using space-related facilities and technologies. In 1987, the Army proposed three experiments it thought would improve its war-fighting capabilities, Terra View, Terra Scout, and Terra Geode. These three experiments played significant roles in the future of manned space flight. In 1987, as its participation in NASA burgeoned, the Army established an Army Astronaut Detachment at the Johnson Space Center. Between 1983 and 1989, interest in space was revived as the Army formed a space command and mounted an educational effort to show how space-based systems were invaluable to the warfighter.

Two events in the 1990s changed the way the Army looked at space and missile defense. First, the Cold War ended with the collapse of the Soviet Union and its Eastern European empire. While this ended the threat of nuclear confrontation, it ushered in an era of geopolitical instability. The proliferation of missile technology and weapons of mass destruction signaled a change in strategic defense. The emphasis shifted away from protecting the United States from wide-scale nuclear attack to protecting against limited attacks from hostile nations. The new emphasis was exemplified in President Bush's new approach to the problem of missile defense.

He ordered the Strategic Defense Initiative program to emphasize defense against limited attacks; the new system was called Global Protection Against Limited Strikes. It was designed to counteract strikes by various Third World countries developing ballistic missiles, or accidental or unauthorized launches from the Soviet Union. This reorganization unified the Army's space and strategic defense efforts in the new Army Space and Strategic Defense Command. Completing this unification was the transfer of the Army Space Technology Research Office from the Communications-Electronics Command in 1993 and the transfer of the Army Space Program Office in 1994.

The second event was the Gulf War of 1990-1991. Desert Storm was the nation's first space war and marked a change in military technology and tactics. During this war, the Army relied on space-based systems to provide soldiers with position and navigation information, multi-spectral imagery, satellite weather and communications, and ballistic missile warning. The war demonstrated the growing importance of space as a military medium.

The Gulf War demonstrated that space-related systems and products could successfully support the Army's operations. Units used the Global Positioning System (GPS) to navigate, control convoys and resupply operations, mark and breach minefields, and conduct artillery surveying and fire direction. Tactical units used weather receivers to obtain crucial weather information quickly. When weather information was combined with multi-spectral satellite imagery, maps using the latest intelligence were created and distributed in a timely manner. Tactical missile detection used space-based systems to warn units of incoming rocket attacks.

Commercial space systems played a large role in the Gulf War and had a large impact on the military. Although the military Defense Satellite Communications System carried about half of the communications traffic in the war, the INTELSAT system carried another quarter—the commercial system supplemented the military system. The WRAASE weather receiver was a commercial product and the topographical units' services expanded because of the commercial equipment and software bought during the war. Even the much-heralded GPS could not be distributed to the majority of units until the Army bought and sent commercial receivers to the Persian Gulf. A final enduring lesson from the Gulf War was the relatively short shelf life of combat experience. If the Army would retain its interest in space and space-based systems and products, the Army's space community must make a greater effort to normalize and operationalize space through the capture and dissemination of the lessons it learned from observations and historical study of training, exercises and combat operations.

It was also obvious that few commanders fully grasped the potential of the space-based systems to which they had access. Few understood how military space-related systems and their products could help them improve their tactical practices and their grasp of the operational art.

The Army found itself increasingly dependent upon space-based systems to conduct operations. The typical soldier relied on them to determine his position, locate the enemy, communicate with friendly forces, and fire "smart weapons." For the Army, space was becoming the new high ground, an important part of firepower and information dominance on the battlefield of the future. It became crucial for the Army to improve its space technology.

The end of the Cold War led to the reconfiguration of the Army in the 1990s. The Chief of Staff of the Army, General Gordon Sullivan, established a new vehicle to investigate and support necessary change, the Louisiana Maneuvers process. In 1996, the Army initiated the Army After Next Project to fashion requirements for the Army of the near future, concentrating on the possible shape of warfare between 2010 and 2025. The project would explore the nature of warfare thirty years in the future and help develop a long-term vision for the Army. In 1997 and 1998, a series of war games demonstrated how crucial space assets had become and would remain to modern land warfare. The Army After Next Space Game Two showed how space support could be integrated into a cohesive theater campaign. Its results gave the Army a better understanding of the ways in which space-based resources might affect military operations on the ground. The game also pointed out ways commercial space-based systems could amplify the commander's knowledge of the battlespace with improved position and navigation capabilities and imagery systems. Many of the Army's senior leaders identified space as the battlefield's new "high ground."

As the world political situation changed, the emphasis in missile defense changed from the strategic to the theater level. Although planning for theater missile defense began in the mid-1980s, the events of Operation Desert Storm proved the significance of theater missile defenses. In 1993, the Clinton Administration reemphasized theater missile defense efforts because the new threat was theater ballistic missiles controlled by Third World dictators. The first priority became deploying a theater missile defense system with space-based sensors. The second priority was deploying a national missile defense program to meet the threat posed by rogue nations. When it came to further research and development, follow-on technologies like directed energy efforts received the lowest priority rating. The Strategic Defense Initiatives Organization was reorganized as the Ballistic Missile Defense Organization, which reflected the new priorities and wider mission. With this shift to development and acquisition of systems, the new organization reported to the Under Secretary of Defense for Acquisition.

The 1993 Bottom-Up Review of the Military initiated by the Clinton Administration outlined the national security plans for 1995-1999. The goal was to field effective theater missile defense systems in the shortest time possible, while providing a basis for a speedy decision to deploy national missile defenses if a serious threat appeared. The review offered a three-tiered program that emphasized theater missile defense, with the bulk of funding going to create this tier. In contrast, national missile defense and the research for follow-on technologies and strategies would be funded at much lower levels.

The mid-1990s also saw modifications to the Anti-Ballistic Missile Treaty referring to theater missile defense systems, specifically theater high-altitude area defense. In 1995, the Clinton administration proposed that the boundary between tactical and strategic ballistic missiles be the ability to intercept a missile traveling at 5 kilometers per second, adding this should be based on demonstrated capability and not theoretical ability. Following two years of negotiations, officials agreed to the Russian proposal that theater missile defense systems with a demonstrated interceptor velocity of 3 kilometers per second would comply with the treaty. The proviso was that the systems would not be tested against target missiles having a range greater than 3,500 kilometers and a maximum flight velocity of no more than 5 kilometers per second. The governments of the United States, Russia, Belarus, Kazakhstan, and Ukraine signed the final agreement on 26 September 1997.

These critical developments encouraged the Army to reevaluate its goals for space and missile defense and led to creating a new command specifically to meet these ends. Shortly before its creation, these objectives were outlined in a Memorandum of Agreement with the Training and Doctrine Command. The agreement made the command the Army's proponent for space and national missile defense and the overall Army integrator for Theater Missile Defense. The command would determine space requirements for TRADOC approval and lead the integration of doctrine, training, leader development, organization, matériel, and soldier solutions across the Army and within appropriate joint agencies. The agreement also chartered the command to establish a battle lab to plan and conduct space and missile-defense warfighting experiments, the first outside TRADOC.

The new command organized itself to meet its new responsibilities and lead the way for Army space and missile defense. The goals the Space and Missile Defense Command (SMDC) set included integrating space support in full spectrum land operations, creating a global multi-element missile defense, cultivating space partnerships, and extending advanced space and missile defense technology for combat forces. The SMDC would be tailored to suit the Army's needs in the new century for an organization combining combat and materiel developments, acquisitions, and operations in one place. Integrating these functions in a single entity would save time, effort, and money by reducing the competition for space and missile defense resources within the Army, enabling it to better explore the global reach of the command's assets.

The Space and Missile Defense Battle Lab was founded to interact with the other TRADOC battle labs on space and missile defense models and simulations, experiments, and technology infusions. It was formed by combining the former Battle Integration Center and the Army Space Exploitation Demonstration Program (founded in the mid-1980s to acquaint the Army with new space products and to be an element of the Army Space Command). A Force Development and Integration Center was established as the Army's manager and developer for space and missile defense. It would develop, manage, and prioritize missile defense and space future operational capabilities across the entire spectrum of TRADOC responsibilities from personnel to combat developments activities. It would also advocate the Army's positions on space and missile defense to the joint technical development and operational communities.

One of the Battle Lab's goals was to develop a Synthetic Battlefield Environment that would link technology to the warfighter by providing weapons developers, battle planners, and commanders with interactive realistic scenarios. The Synthetic Battlefield Environment rested with the Extended Air Defense Testbed. Initiated in 1989, the testbed models air, land, sea, and space-based forces and their contribution to theater-level extended air defense, enabling the user to develop tailored simulations from the fire-unit up to the theater level for theater missile defense and the global level for national missile defense. The Battle Lab also established a synthetic environment that permitted simulated elements to be replaced with actual hardware. The Synthetic Battlefield Environment has continued to grow with the evaluation of new software and technologies that address aspects of space and missile defense. These include Project Stalker (which assists in locating, tracking, and destroying mobile transporter erector launchers) and the Battlefield Ordnance Awareness system (which collects and processes data on missile launches, artillery, and tank fire). These and other technological advances are brought to the soldier through traditional exercises, long-distance training, and the Space and Missile Defense War Game.

In addition, the Battle Lab brings products to the soldier through the Army Space Exploitation Demonstration Program. Many products could be used to illustrate the Battle Lab's successes with this program, including the Global Broadcast System–Joint In-Theater Injection, Joint Tactical Ground Station, and Force Protection Tactical Operations Center. In addition to the large systems, the demonstration program also developed technologies affecting communications available to the individual soldier or unit. These include the Iridium phone system, which provided a truly global phone for the soldier in the field; the Pager Alert Warning System, to notify troops in the expected impact zone of tactical ballistic missile attacks; and the Joint Expeditionary Digital Information program that combined these capabilities with a laser range finder, GPS satellite positioning, and text messaging to send and receive information via satellite.

The Space and Missile Defense Acquisition Center centralized materiel development and testing operations. As initially configured, its numerous elements included the Joint Land Attack Cruise Missile Defense Elevated Netted Sensors System Project Office, the Ballistic Missile Targets Joint Project Office, the Army Space Program Office, the High Energy Laser Systems Test Facility, and the U.S. Army Kwajalein Atoll. One of the more intriguing elements was the Joint Land Attack Cruise Missile Defense Elevated Netted Sensors System. Unmanned sensors, suspended in a compartment below the aerostat, would provide a 360° picture enhanced by Identification Friend or Foe systems. This data is relayed to a ground-processing center through a fiber optic tether, which would notify relevant interceptor systems. An aerostat can provide round the clock surveillance for up to thirty days. The system's primary focus was missile surveillance, tracking, and fire control for the various anti-missile systems using an aerostat, a tethered balloon designed with an inner ballonet.

The Missile Defense and Space Technology Center continued to support the Ballistic Missile Defense Organization and the Program Executive Office for Air and Missile Defense. It also established a space technology directorate to identify promising technology and set up a long-range research and development program. This organization's achievements are evident in the progress made by the variety of missile defense systems under development. While the technology associated with interceptor systems remains its primary focus, the center continued to explore innovations. Directed energy once again has its attention and the SMDC wrote the first Directed Energy Master Plan in 1999. Sensor technology has also advanced. One example sought to improve the interceptor systems' ability to interpret what they see, while another was designed to expand the area covered. The Technology Center's goal remains flexibility in order to respond rapidly to new programs and marketing opportunities.

The Army Space Command continued to support the warfighter with space products and capabilities, including communications through the Defense Satellite Communications System. It also managed the Army Astronaut Detachment at the Johnson Space Center.

The Army's attempt to normalize and operationalize its space assets were highlighted in the *Army Space Master Plan* and TRADOC Pamphlet 525-3-14, *Concept for Space Operations in Support of the Objective Force*. Both documents emphasized the individual unit's dependence on space-based assets to carry out the Objective Force's overarching mission. Space and missile defense emerged as key enablers to these missions in these documents. Space control's contribution to the Army's Objective Force and to the joint force commander cannot be overemphasized. The doctrinal and training implications of space control and missile defense technology hold the potential for changing warfare.

After the end of the Cold War, the fires that fed the missile defense debate appeared to have been banked by the disappearance of the Soviet threat. As new threats emerged, specifically the proliferation of both guided missiles and weapons of mass destruction, missile defense became a more urgent matter. When missile defense returned to the national stage, it seemed as though the debate had the same parameters that bound it when Robert McNamara thought it was both technologically and economically unfeasible. Partisans on both sides shed more heat than light on the subject, overshadowing the work of the Army's engineers and contractors, who performed amazing technological feats. As we conclude, missile and nuclear proliferation have compelled the United States to begin constructing a missile defense system.

The Army's earlier efforts in space and missile defense played out against a background of war and Cold War. The Army's earlier space and missile defense efforts operated in tandem through the late 1950s and produced breakthroughs in rocket and sensor technologies. After the Eisenhower Administration's forcible separation of the Army's space and missile defense programs, greater attention was paid to missile defense and communications. As the missile defense technologies matured and grew more sophisticated, the organizations supporting them became functionally organized. Army missile defense organizations developed to combine research and development, testing and evaluation, and acquisition functions in one place. They slowly evolved into centrally organized functions-based organizations. While not the result of an overt design, the change promoted collaboration and worked to short-circuit duplication of effort and pointless competition for scarce resources.

A series of events beginning with the end of the Army's Vietnam experience and the beginning of its participation in the TENCAP and ending with the publication of AirLand Battle Doctrine signaled a new interest in space and missile defense. The way missile defense and space-based systems were used in the Gulf War vindicated the Army senior leadership's decision to reenter space in order to influence the ways in which the systems it used would be developed. The centralization of the Army's space and missile defense programs that began in 1992 is creating a new organizational form. In 1997, the Army established its newest major command, the U.S. Army Space and Missile Defense Command, to serve as its proponent for space and national missile defense and overall integrator for theater missile defense. The new command would be tailored to suit the Army's needs in the new century for an entity that centralized combat and materiel developments, acquisitions, and operations. Integrating these functions in a single organization would save time, effort, and money by reducing the competition for space and missile defense resources within the Army, enabling it to better explore the global reach of the assets of the command.

Chapter 1
Aircraft, Rockets, Missiles and Radar, 1907-1961

Powered Flight, Radio, Rockets and Sensors

The desire to enhance the intelligence, command, control, and communications functions of armies has been a preoccupation of military commanders ever since the first armies took to the field. Prior to the eighteenth century, military commanders achieved such enhancement by establishing their military positions and observation posts on the highest elevations possible. In 1783, this concept of seizing the high ground took on brand new meaning with the launch of the first balloon. The balloon quickly gained popularity as a military observation platform and was used with varying degrees of success in a number of 18th and 19th century conflicts. The use of military balloons reached its apogee in World War I, when every major belligerent used tethered balloons to report enemy movements, direct artillery fire; track infantry progress in attacks, give details of obstacles and describe the effects of bombardment.[1]

The advent of the airplane ended the dominance of the military balloon. The Army became interested in powered aircraft shortly after the first flight of the Wright Brothers. On 1 August 1907, the Chief Signal Officer, Brigadier General James Allen, issued a memorandum creating an Aeronautical Division within his office that would have "charge of all matters pertaining to military ballooning, air machines and all kindred subjects."[2] Conventional wisdom suggested that aircraft be used to transmit messages and for reconnaissance. Consequently, in 1908, the Signal Corps assumed responsibility for the development and operation of military aircraft. This changed in the First World War.[3] In June 1917, General John J. Pershing created the Air Service of the American Expeditionary Forces, an organization that was independent of the Signal Corps. The creation of an autonomous Army Air Service in May 1918 further reduced Signal Corps involvement with the airplane. Thereafter, the Signal Corps' responsibility was limited to the development of airborne radios.

Along with powered flight, the early 20th century witnessed the advent of wireless radio, an invention that laid the foundation for electronic sensors. The Army's interest in new communications technologies can be linked to the variety of its missions that encompassed an enormous geographical expanse. These include the Army's coastal defense mission, the mission to defend Alaska and Hawaii, and the Army's responsibility for cable communications. Congressional parsimony reinforced the Army's desire to search for a new technological fix to keep communications and defenses operating properly throughout its enormous area of responsibility. New technologies such as the wireless radio enabled the Army to fulfill its varied missions and to become both more efficient and effective.[4]

World War I produced a sea change in the way American society viewed the relationships between business and government. The war strengthened the ties between the military and business, especially with respect to radio technology, the cutting-edge technology of the day. In the 1920s and 1930s, the Army experimented with mobile communications devices for airplanes and mechanized equipment, and with developing tools for signals intelligence.[5] The Army was also developing radio navigation aids, beacons, compasses, direction finders and electronic detection aids to enhance situational awareness. Throughout the 1920s and 1930s, public opinion envisioned that the Army and Navy should be used for narrowly defined defensive purposes. The Navy's experiments with aircraft carriers and the Army's justification for developing heavy bombers, radio detection, and ranging (radar) emphasized their potential contribution to coastal defense.[6] The defensive mindset of Army planning also determined priorities with respect to weapons development. At the same time, Congressional thriftiness also served to stifle technical innovation.[7]

Guided missiles and rockets, however, were weapons systems not readily attached to any defense-oriented Army doctrinal concepts. Nevertheless in 1936, in partnership with the California Institute of Technology (CIT or Cal Tech), the Army's Ordnance Department began basic research in rocket design and fuel and propulsion systems. On the eve of the Second World War, the Army was moving toward developing radar and rockets.[8] The war served as a forcing house for these innovations.

World War II: A New Threat Emerges

By 1940, German rocket scientists were using a supersonic wind tunnel to test design alternatives for radio-controlled bombs, rockets and flak shells, while United States Army scientists were still trying to obtain subsonic velocity data for rockets and bombs. A rocket program initiated in 1942 had, by 1944, evolved into a separate division of the Army Ordnance Department engaged in research on solid and liquid rocket fuel and rocket manufacture. In 1944, shortly after the German V-1 attacks began, the Army divided responsibility for guided missile development between the Army Air Forces (AAF) and the Ordnance Department to lessen secrecy and promote data sharing. The AAF was given "development responsibility…for all guided or homing missiles dropped or launched from aircraft…[or those] launched from the ground which depend for sustenance primarily on the lift of aerodynamic forces." The Ordnance Department would develop missiles "which depend for sustenance primarily on momentum of the missile." In January 1945, the General Staff made the Ordnance Department responsible for developing a missile suitable for antiaircraft use.[9]

Despite these promising developments, the U.S. Army's interest in long-range rockets and missiles remained theoretical. By contrast, the German Army introduced the Allies to the practical effects of long-range rocketry and missile technology when they launched the first ten V-1 rockets against the city of London on the evening of 12-13 June 1944.[10] The V-1, a precursor to the cruise missile, was a cheap and simple weapon to construct.[11] Beginning in 1942, the Germans produced approximately 30,000 of these weapons. In one nine-month period, approximately 10,000 V-1s or "buzz bombs" were launched at London from sites along the English Channel and from medium bombers. More than half of these reached the United

Kingdom, killing over 5,800 people and seriously injuring another 15,000.[12] These attacks forced the U.S. Army to grapple with the effects of this new weapon, a weapon that had no representation in U.S. Army doctrine.

Fig. 1-1. V-1 flying bomb over residential area of London.

Fig. 1-2. Launch of a V-2 rocket.

The Germans improved upon their rocket technology and, in September 1944, fielded the V-2. Unlike the V-1, the V-2 was a supersonic missile, with a top speed of 3300 mph and a range of 190-200 miles.[13] Although equipped with a three-axis gyro pilot, the V-2 remained an unreliable system. Studies suggest that only one in three reached their intended target. Nevertheless, it proved to be an effective psychological weapon. Its speed precluded any substantial advance warning and its impact velocity generated more extensive damage than the V-1. During the six-month attack on London, for example, over 1,000 V-2s reached Great Britain, killing 2,855 people and seriously injuring 6,268 others.[14]

The Allies implemented a number of defensive measures against these weapons. The first, and perhaps most successful, effort to defeat the V-weapons involved offensive bombing raids on production facilities, support facilities, and launch sites. This was supported by a layered defense system against the V-1 which "included an excellent detection and control system, high speed interceptors, radar-directed guns firing proximity fused shells, and barrage balloons."[15] The combined network destroyed 52.8 percent of the 7,488 V-1 missiles observed. The cost of such a network was staggering. According to a 1944 British Air Ministry report, the estimated cost of defending against the V-1 was £48 million.[16]

Attempts to intercept a V-2 were less successful. Initially, available technology was incapable of tracking or directing aerial defenses against the V-2.[17] As radars improved experts began to assess the feasibility of creating a missile barrier using a barrage of anti-aircraft artillery. This option was dismissed, however, when it was realized that "a barrage of 320,000 shells would be required to produce a likely kill. Of these shells, about 2 percent would be duds that would then fall on London, causing more damage and casualties than a V-2."[18] Thus, once a V-2 was launched the Allies had no means to intercept it.[19]

The Second World War proved to be a fruitful time for scientific research and development in the Army. By 1945, the Army had developed shoulder-fired rockets, truck- and tank-mounted rockets, and was working on larger guided missiles. The Army's role in shooting, processing, and analyzing millions of photographs for intelligence purposes caused a dramatic expansion in the field of aerial reconnaissance photography. Additionally, the Army's code breaking capacity enabled American policy and decision-makers to eavesdrop on enemies, allies and neutrals.[20] Based on prewar experiments and experience gained with a rudimentary long-haul communications and radio direction finding system, the Signal Corps managed to create and operate the largest, secure, unified, global military communications network in existence at that time. The Army also developed ground-based and airborne radars used in early warning systems and aerial bombardment.[21] By 1945, the Army was routinely using aerial and signals intelligence methods to gather, process, and disseminate information. The Army developed and operated a secure (simultaneously encrypting and decrypting), worldwide communications system, anti-aircraft and anti-missile early warning systems, and was developing solid-fueled rockets and liquid-fueled missiles.

As the war ended in May 1945, Colonel Holger N. Toftoy and Major James P. Hamill began contacting German rocket scientists and offering to transport them to the American Occupation Zone. Before the zonal boundaries solidified, they recruited more than 120 scientists and technicians, removed more than 100 assembled V-2 rockets, and transferred more than 300 freight cars full of documents and machinery to the American Zone. The U.S. Joint Intelligence Objectives Agency continued with Operation PAPERCLIP, recruiting German scientists and "inviting" them to the United Stated to continue their rocket developmental work in the postwar period.[22]

Postwar Assessments

Postwar investigations revealed that the Germans were far ahead of the United States in several scientific fields, including rocket and jet engine propulsion.[23] Captured documents revealed plans for a two-stage, intercontinental ballistic missile (ICBM) with a range of 3,500 miles, capable of reaching New York City. The United States was interested in German technical capabilities because they would benefit the American military and American industry On June 20, 1945, Secretary of State Cordell Hull approved the transfer of the German rocket specialists. Before the operation concluded, almost 500 rocket scientists and technicians were transported to the United States. Many of the German scientists were taken to Fort Bliss, Texas and given six-month contracts to work at the newly established Army Ordnance Research and Development Sub-office. While at Fort Bliss, the German scientists trained military, industry and university personnel in the intricacies of rocket and guided missile technology and helped refurbish, assemble and launch V-2s that had been shipped from Germany to the White Sands Proving Grounds in New Mexico.

In July 1945, in response to the experiments taking place at White Sands and elsewhere, a delegation of American officers recommended that the U.S. undertake a research and development program to develop a defense against these new weapons. A subsequent study by the Scientific Advisory Group of the Army Air Force, published in December 1945, took the issue further by exploring the use of missiles armed with nuclear warheads, and the use of an energy beam as a potential defense against missile attacks.

Dr. Vannevar Bush, the head of the War Department's Office of Scientific Research and Development, opposed the Army's case for missile development. He believed the Army overstated the benefits and advantages of missiles and satellites and argued that it would take many years to develop a reliable ICBM because of the relatively immature state of the technology. In the 1940s, first as Chairman of the Joint Research and Development Board of the War and Navy Departments, and later as Chairman of the Development Board of National Military Establishment, Bush challenged the Army and the Air Force to demonstrate that missiles and satellites could perform warfighting missions in a manner that was more cost effective than the available conventional means. In this manner, Bush successfully delayed large-scale research programs.[24]

On 29 May 1946, the War Department Equipment Board, headed by General Joseph W. Stillwell (the Stilwell Board), issued its report on the equipment needed by American ground forces in the postwar era. The Board recognized that "guided missiles, winged or non-winged, traveling at extreme altitudes and at velocities in excess of supersonic speed, are inevitable." The Stilwell Board went even further in their threat analysis by concluding that "intercontinental ranges of over 3,000 miles and payload[s] sufficient to carry atomic explosive[s] are to be expected." Based on this assessment, the board determined that no aircraft or missile could be allowed to gain access to areas deemed vital to the nation. Thus, future ground forces would require guided intercept missiles. Finally, the board advised that the development of defensive measures designed to counter atomic weapons should be "accorded priority over all other National Defense projects."[25]

Even before the Stilwell Board completed its report, the Army Air Forces had initiated research on the anti-missile concept. On 4 March 1946, the Army Air Force awarded two contracts for the study of anti-missile missiles. The University of Michigan received a contract for Project WIZARD (MX-794). General Electric received another for Project THUMPER (MX-795). The targets in both studies would be traveling at a rate of 4,000 mph at altitudes up to 500,000 feet. Project THUMPER went further, by specifically exploring "the interception of 'rocket-powered ballistic and glide missiles and supersonic ram-jets."[26] In 1947, the Air Force redefined these efforts as prolonged studies and, in March of the next year, canceled Project THUMPER. Project WIZARD meanwhile continued to develop new technologies until 1958, when it merged with the U.S. Army's NIKE-ZEUS project.

Early Division of Labor and Experiments

The first postwar years were a time of great turbulence. This brief period saw wartime demobilization, the beginnings of the Cold War and a massive military reorganization and unification effort that resulted, among other things, in the establishment of the Department of Defense. In 1947, Congress passed the National Security Act, unifying the armed services and establishing an independent Air Force. The next year, the Secretary of Defense, James V. Forrestal, negotiated specific roles and missions with each of the services. The Navy's primary responsibility remained sea operations and it retained the Marine Corps. The Army's primary responsibility remained land operations. In addition, the Army assumed responsibility for ground-based air defense of the continental United States as well as constabulary and occupation forces for Germany and Japan. The Air Force's primary responsibility lay in strategic air power, air transport and tactical air support of the Army's ground forces. The Air Force put forth a claim for jurisdiction over space and satellites, arguing they were a logical extension of strategic air power. The Army and Navy reluctantly conceded this point to the air arm.

After 1945, the Army remained interested in space age communications and missiles, despite the widespread belief that these devices were not military science but science fiction. On 10 January 1946, the Army demonstrated it could send radio waves anywhere when, as part of Project Diana, the "Evans Signal Laboratory succeeded in bouncing a radar signal off the moon." The signal was received seconds after it was transmitted. More than a technological trick, Project Diana's success showed that VHF radio signals could penetrate the electrically charged ionosphere around the earth. This was the beginning of space age signaling.[27]

Almost simultaneously, Operation PAPERCLIP yielded fruit at the White Sands Missile Range.[28] In mid-August 1945, 300 freight cars arrived in New Mexico. These carried 100 V-2 missiles and components and had been spirited out of the Russian Zone of Germany by Colonel Holgar Toftoy and Major James Hamill.

The rockets were refurbished, modified, and rebuilt for tests carried out between 1946 and 1952. Many of the parts brought from Germany were in poor condition or unusable. A total of 67 rockets were assembled and tested in this six-year period. These activities and tests provided the military with invaluable experience in missile assembly, static and pre-flight testing, as well as missile handling, fueling, launching and tracking. The project managers considered approximately two-thirds of the tests to be successful but even the failures yielded valuable information. In addition to the missile testing, scientific experiments conducted onboard the missiles produced new information about rocketry and the upper atmosphere.[29]

Rocket testing continued and the first multistage American rocket was the Bumper. The Bumper was a two-stage rocket comprising an Army Corporal rocket mounted inside a German V-2.[30] The Bumper Project began in June 1947 and the first Bumper flight was launched in May 1948.[31] The program sought to create a rocket that could reach higher altitudes at greater speed, to gain experience in launching two-stage missiles, and to investigate techniques for ensuring

Fig. 1-3. 120 mm gun in Chicago.

Fig. 1-4. The 75mm Skysweeper antiaircraft gun was the last conventional antiaircraft artillery weapon issued to ARADCOM. In this photo, soldiers practice with a Skysweeper at White Sands.

stage separation at high altitudes. The program also sought to investigate high altitude phenomena. Between May 1948 and July 1950 there were eight Bumper test flights. Six of these tests took place at White Sands, New Mexico and two occurred in Florida. The first fully loaded Bumper, launched in 1949, reached the greatest velocity and altitude ever attained by a man-made object up to that time, and measured temperatures at extreme altitudes. The WAC (Without Attitude Control) Corporal carried instruments that transmitted flight data to a ground station. This was the first time radio equipment operated at these extreme altitudes. The last two Bumper tests were conducted in Florida. One test ran into some difficulty because moisture had collected inside the missile. Despite errors in trajectory, however, the test missile flew at a speed of Mach 9, the highest sustained speed ever reached in the earth's atmosphere. The project demonstrated that adding multiple stages could increase a missile's speed. As a result of these tests, scientists were able to solve the problems of rocket motor ignition at high altitudes, as well as that of that attachment and separation of successive stages. These initial successes provided the necessary foundation for building larger and more complex missiles.

The Evolving Threat

These technological developments were made more significant by the beginnings of the Cold War in Europe and its extension to Asia. In 1947, the United States assumed Britain's role as guarantor of Greek independence and moved to assist both Greece and Turkey as part of the newly articulated Truman Doctrine. In 1948, the U.S. announced the Marshall Plan for European reconstruction. The subsequent two years witnessed a series of dramatic and troubling events. In August 1949, the Russians successfully exploded an atomic bomb. In October, the Communists achieved victory in China. In January 1950, the China and the Soviet Union signed the Sino-Soviet Alliance, forcing a critical reevaluation of American foreign policy as expressed in NSC-68 (April 1950).[32] In June 1950, the United States found itself desperately fighting the Soviet-backed North Korean invasion of South Korea. An initial retreat was followed by a successful U.N. offensive that advanced to the Chinese border. This provoked a Chinese Communist intervention (September-December 1950) that resulted in another retreat, stalemate, and eventual armistice (July 1953).

This sequence of events presented the Truman and Eisenhower Administrations with a world threatened by Soviet expansionism. The level of anxiety increased even further in 1953, when the Soviets detonated a hydrogen bomb.[33] At the same time, the Soviet Union experimented with long-range bombers and missile technology. During the 1954 May Day parade, the Soviets revealed the B-4 Bison, a long-range bomber. Finally, on 26 August 1957, the Soviet Union announced that it had successfully tested an ICBM – the SS-6. Concurrent with these offensive developments, the Soviet Generals had also initiated an anti-ballistic missile development program in September 1953.[34]

Nike II

As the race to develop powerful rockets proceeded, the Army had developed a missile for use against manned bombers. In 1945, the U.S. Army initiated Project NIKE to explore the use of missiles to counter the threat posed by supersonic aircraft.[35] In November 1951, the NIKE-AJAX missile intercepted an aircraft flying at a range of 15 miles, an altitude of 33,000 feet, and at a speed of 300 miles per hour.[36] Having successfully addressed the threat of a single long-range bomber, the Army began to focus on the threat posed by a mass aerial attack. This resulted in the development of the NIKE-HERCULES, a modified NIKE-AJAX.[37] The HERCULES intercepted its first drone aircraft in 1956. The next year, Operation Snodgrass demonstrated that the NIKE-HERCULES system could select a specific target within a formation of aircraft. Soon NIKE-HERCULES replaced AJAX in batteries across the country.[38]

Fig. 1-5. In June 1958, the first NIKE-HERCULES unit reached operational readiness status in Chicago. NIKE-HERCULES crew scrambles during exercises in Chicago.

In February 1955, the Army Ballistic Missile Agency (ABMA), located in Huntsville, Alabama, contracted with Western Electric Company and Bell Telephone Laboratories to conduct an 18-month study addressing means of countering the air defense threat of the future. Researchers were instructed to keep "in mind ballistic targets and the desire to defend against extremely difficult intercontinental ballistic missiles with a reasonable extension of current radar and missile technology."[39] As a result of intelligence data gathered on the Soviet long range missile program, however, later discussions placed greater importance on the threat posed by intercontinental ballistic missiles (ICBMs) and this became the primary focus of the NIKE II study.

In December 1955, Bell Labs presented the first full status report on the NIKE-II System to the Chief of Army Ordnance at Redstone Arsenal.[40] Bell Labs concluded that it was feasible to develop and deploy a missile defense system. Although many leading scientists scoffed at the concept of a missile intercept given the extreme high velocity (24,000 feet per second), Bell

Labs, using an analog computer, conducted a series of 50,000 computations simulating intercepts of ballistic missile targets. Armed with this data, Bell Labs concluded that it was possible to intercept an ICBM, or in other words, to "hit a missile with a missile."[41] Furthermore, they anticipated deploying such a system by late 1962.

The final concept proposed in October 1956 involved a common data gathering system used in conjunction with a missile equipped with interchangeable nose cones that could handle a full-range of potential threats. The missile would carry a 400-pound nuclear warhead, and be capable of executing 10-g maneuvers at 100,000 feet. Given the 95 to 100 percent attrition rate sought in an interceptor system, integrated multiple radars, high-speed communications, and data processing played key roles.[42] The ICBM's deceleration rate combined with the use of decoys would be countered with a series of three types of radars.[43] A series of forward acquisition radars and local acquisition radar would provide early acquisition data and relay data to target track radars. Finally, a missile tracking radar would guide the interceptor to its target. Researchers theorized that "a long-range, high-data-rate acquisition radar" combined with the NIKE-B/HERCULES could serve as an interim ICBM defense.[44] Although, "parts of the NIKE II system concept would be altered or discarded, the concept presented in 1956 defined ABM system technological requirements and its basing policy for the next 25 years."[45]

In November 1958, a NIKE-HERCULES intercepted a high altitude supersonic target missile. This feat was repeated in 1960, when the HERCULES shot down a CORPORAL ballistic missile and another HERCULES in tests at White Sands Missile Range.[46] These tests marked the first time a ballistic missile was destroyed by another missile.

Roles and Missions

The Truman-Eisenhower policy of forming a worldwide alliance system to contain the Soviet Union and China resulted in a shift in science and technology policy. In 1955, the Killian Committee (named after President Eisenhower's Chief Scientific Advisor Dr. James R. Killian) recommended the government continue developing intercontinental and intermediate range ballistic missiles (ICBMs and IRBMs), high altitude reconnaissance aircraft and reconnaissance satellites.[47] That same year, President Eisenhower proposed that the transit of satellites over the United States and the Soviet Union be unimpeded by either power.[48]

By May 1954, however, the *New York Times* reported that the Soviet Union might be gaining an advantage on the United States in rocket and missile development, to include the development of new supersonic missiles capable of intercontinental nuclear strikes.[49] The press dubbed these ICBMs the "ultimate weapons" for which there was no defense. On August 30, the National Security Council (NSC) recommended that the ICBM program be given the highest priority. President Dwight Eisenhower affirmed this measure on 13 September 1955, designating the ICBM program the nation's top research and development priority. Nevertheless, reports commissioned by the administration credited the Soviet Union with a substantial lead, talked of a "missile gap," and predicted that Soviet missiles would be able to overwhelm American retaliatory forces.[50] This worrisome scenario prompted greater attention to the need to develop a missile defense system.

In the 1950s, all three services were developing ballistic missiles of various ranges and exploring anti-missile systems. In August 1950, the Army and the Air Force signed the Vandenberg-Collins Agreement, establishing an integrated air defense effort incorporating antiaircraft artillery battalions and fighter squadrons.[51] As one political scientist noted:

> The fears of war breaking out at any moment, the perception of a hostile and potentially aggressive enemy capable of inducing heavy loss upon North America and the belief in the vast potency of a military technology capable of rendering obsolete whole weapons systems - all of these attitudes promoted duplication [of weapon systems] as a lesser evil to the possibility of unpreparedness.[52]

This attitude however began to change in 1956. In November, Secretary of Defense Charles Wilson, in a Memorandum to the Members of the Armed Forces Policy Council, further clarified Army and Air Force roles and missions.[53] He assigned to the Army responsibility "for the development, procurement and manning of land-based surface-to-air missile systems for point defense." Point or terminal defense focused on specific geographic areas, cities and vital military and industrial installations and addressed air targets at altitudes out to a horizontal range of 100 nautical miles. This assignment would be achieved with guided missiles, such as the NIKE I, NIKE B and land-based TALOS, and co-located radars.

In April 1957, a joint Army-Air Force committee, headed by Mr. Hector Skifter, reviewed the ballistic missile defense mission. They recommended "that the Army continue with the development of a terminal defense system and that the Air Force be given the responsibility for the early warning system against ballistic missiles." In addition to the missile system, the Army was responsible for developing Target Track Radar and Local Acquisition Radar. The new Secretary of Defense Neil McElroy affirmed this assessment with two memoranda issued on 16 January 1958. McElroy assigned to the Army primary responsibility for the BMD mission (missile, launch site, radars, and computer components). With its NIKE missile systems, the Army had progressed further than the Air Force, whose Project WIZARD had yet to produce a missile. However, the Air Force, based on their experience with early warning radars, continued to develop early warning radars, tracking and acquisition radars and communications links, ensuring that they were compatible with the NIKE-ZEUS system.[54]

Beginnings of Space Exploration

Initial American satellite launches were scheduled to be part of the International Geophysical Year (1957-1958) to foster multilateral exploration of the earth and its atmosphere.[55] The Soviets surprised the world by placing satellites in orbit in October and November 1957.[56] The public and congressional uproar resulted in the perception that the Soviets had surpassed the United

States scientifically. On 8 November 1957, President Eisenhower directed the Army to place a satellite in orbit by March 1958. On 31 January 1958, the Explorer I satellite was launched by an Army Redstone rocket. In addition to soothing the national pride, the instruments on the first Explorer satellite detected the Van Allen radiation belt circling the earth. It was not until 17 March 1958 that the Navy launched its first Vanguard satellite, which used solar cells (developed by the Army Signal Corps) to power its radio transmitters.[57]

Fig. 1-6. An Explorer satellite atop a Jupiter-C rocket.

Fig. 1-7. Major General John C. Medaris, one of the founders of the Army's missile program.

The revelations that followed the Sputnik launch reinforced the belief of American technological inferiority and raised fears in among the public that the realm of space was about to be dominated by an enemy power, the very situation the Stilwell Board had cautioned against in 1946. Sputnik had significant and far-reaching effects on American space, scientific and educational policies.[58]

Launching the Explorer I satellite with a Redstone rocket ended a process that began with the successful conclusion of the Project Bumper tests in 1950. That year the Army consolidated its missile programs, moving those projects and personnel at White Sands and other places to the Redstone Arsenal in Huntsville, Alabama. In response to a Chief of Ordnance directive, the new agency began work on a surface-to-surface multistage missile with a 500-mile range. In 1953, the Redstone missile was successfully tested at Cape Canaveral. In 1955, the Army recommended to the Department of Defense (DoD) that the Redstone missile be developed as the intermediate range missile recommended by the Killian committee. On 1 February 1956, the Army established the Army Ballistic Missile Agency (ABMA) at Redstone. Later that year the ABMA began a series of tests in Florida that launched a Redstone-C missile nose cone 682 miles into space and 3335 miles down range. In November 1956, Secretary of Defense Charles Wilson divided missile development responsibilities between the Army and the Air Force. The Army would be responsible for developing missiles with a range of less than 200 miles while the Air Force would be responsible for developing missiles with ranges of more than 200 miles. By 1958, the Army finished developing the Jupiter and handed responsibility for its operation and deployment to the Air Force.

Before the Secretary of Defense's division of labor between the Army and Air Force, the von Braun team at ABMA began to design a 12,000-pound booster rocket for space investigation, tentatively called Juno. The Army expanded the design effort to include a complete missile (later renamed Saturn). In December 1957, a Juno missile launched the first American lunar exploration mission.

Although there was relative unity of effort concerning ballistic missile defense, the national space effort was dissipated in a myriad of programs supervised by a plethora of civilian and military agencies. Without a supervisory, coordinating or directing body both military and civilian programs would duplicate each other in vain attempt to garner prestige. There was no non-military body to direct civilian space research and, much to the chagrin of the Air Force, the Army developed into the most successful and most experienced military space organization.[59] A general realignment of responsibilities transferred the Jupiter-C missile program from the Army to the Air Force. In a more far-reaching reorganization of the national space effort, Congress created a civilian space agency, the National Aeronautics and Space Administration (NASA) in 1958.[60]

Before NASA's creation, the Army built and launched the nation's first ballistic missile and earth orbiting satellite. The Mercury astronauts were placed in orbit by modified Army Redstone rockets. These feats, a response to the U.S.S.R.'s first ventures into space, were the work of Dr. Wernher von Braun and his rocket team at Redstone Arsenal, Huntsville, Alabama, many of whom had come to the United States as a result of Operation PAPERCLIP. Army research also contributed to the Apollo moon landings. Von Braun's team began working on a heavy rocket booster, the Saturn I, in the late 1950s and when it was transferred to NASA in 1959, the work continued and eventually resulted in the Saturn V rocket used to power the moon flights.

While building rockets to send satellites into orbit, the Army also sought to send soldiers into space. In 1958, the armed services were developing proposals for manned space flight. The Army's plan, Project Adam,[61] sought to launch a man in a sealed capsule on a sub-orbital trajectory using a modified Redstone rocket. The Army justified the project as the first step in improving troop transportation methods.[62] A Redstone rocket would carry a manned capsule to an altitude of 150 miles before splashing down in the Atlantic Ocean down range from Cape Canaveral. During the flight, the astronaut would perform psychological and physiological tests while undergoing acceleration and a brief period of weightlessness. The plan's elegant simplicity made it controversial and it was derided by many experts as the "shooting the lady out of a cannon" plan. Listening to the critics, the Secretary of Defense, Neil H. McElroy, ruled that the project needed further study and ABMA eventually abandoned it. However, with some minor changes, it became the basis for NASA's Project Mercury.

In the summer of 1959, ABMA made an even bolder proposal: plant a military colony on the moon. In Project Horizon, the Army planned, to land on the moon in 1965 and establish a 12-man outpost on the lunar surface in 1966. Providing the moon base with logistical support would require launching 64 Saturn rockets annually (one rocket every 5.7 days), with each rocket carrying more than 266,000 pounds of cargo. The Army expected the program to cost more than $6 billion. The Project Horizon cargo rockets would make a direct earth-to-moon trip, while the crew would first make a low-earth orbit, rendezvous with a space station, and only then fly to the moon. The space station would be manned by a crew of 10 men who would be rotated every few months, with some of them rotating to the moon. The lunar base would be constructed underground and include living quarters, storage areas, nuclear reactors, laboratories, a hospital, a communications center, dining rooms, and recreation rooms. Like Project Adam, Project Horizon never got off the ground. In 1959, the Eisenhower Administration decided to promote the civilian use of space, created NASA and transferred the von Braun team to the new agency. The Army gave Project Horizon to NASA, which shelved the plan.[63]

Between 1958 and 1961, the Army transferred most of its space programs to the new agency. NASA inherited not only the missile programs but also the Redstone Arsenal missile development facilities, renamed the Marshall Space Flight Center. NASA received the Redstone program, the Explorer satellite program, and the entire complement of rocket and missile contracts the Army had with the California Institute of Technology Jet Propulsion Laboratory. NASA also became responsible for developing the 1.5 million-pound thrust Saturn rocket. The Army also transferred technical expertise (approximately 6500 people) from the ABMA Development Operations Division to the new agency.

End Notes

[1]Lee Kennett, *The History of Strategic Bombing* (New York: Scribners, 1982), pp. 24-25, examines these issues as does Eileen Lebow, *A Grandstand Seat: The American Balloon Service in World War I* (Westport: Praeger, 1998) from a U.S. Army vantage point. A contemporary account of the American effort may be found in Spalding W. Ovitt and L.G. Bowers (ed.), *The Balloon Section of the American Expeditionary Forces* (New Haven: Tuttle, Morehouse and Taylor, 1919).

[2]The memorandum may be found in Charles DeForest Chandler and Frank P. Lahm, *How Our Army Grew Wings: Airmen and Aircraft before 1914* (New York: The Ronald Press Company, 1943), p. 80, note 6. The official birthday of the United States Air Force is 1 August 1907.

[3]Unmanned balloons as well as rigid and non-rigid airships were used for reconnaissance, bombing and transport in the First World War by France, Italy, Germany, Britain, and the United States. Easily accessible sources include Lord Ventry and Eugene M. Kolesnik, *Jane's Pocket Book of Airships* (New York: Collier Books, 1976), pp. 81, 96; Robert Jackson, *Airships* (Garden City, N.Y.: Doubleday and Co., 1973), pp. 119-120, 125-126; Douglas Botting, *The Epic of Flight: The Giant Airships* (Alexandria: Time-Life Books, 1981), pp. 51-73 and Kennett, *Strategic Bombing*, pp. 5-6. During World War II, the Japanese used unmanned balloons armed with incendiary bombs to attack North America unsuccessfully. The most comprehensive technical account is Robert C. Mikesh, *Japan's World War II Balloon Bomb Attacks on North America* (Smithsonian Annals of Flight #9, Washington, D.C.: Smithsonian Institution Press, 1973, reprinted 1990).

[4]See Emanuel Raymond Lewis, *Seacoast Fortifications of the United States: An Introductory History* (Washington, D.C.: Smithsonian Institution Press, 1970), Russell J. Parkinson, Politics, Patents and Planes: Military Aeronautics in the United States, 1863-1907 (Ph.D. dissertation, Duke University, 1963) and Hugh G.J. Aitken, *Syntony and Spark—The Origins of Radio* (Princeton: Princeton University Press, 1985).

[5]Initial experiments using AM (amplitude modulated) transceivers revealed that internal combustion engines caused radio interference. Early portable sets were relatively heavy and although field tests were conducted, they were not produced in quantity until the beginning of the Second World War in Europe. The FM (frequency modulated) radio was developed, tested and found to eliminate noise and interference, but a combination of factors, including the Army's technological conservatism and the communications industry's fear of technological obsolescence meant that this innovation was not adopted until 1940. See Robert A. Miller, The United States Army during the 1930s (Ph.D. dissertation, Princeton University, 1973), John W. Killigrew, The Impact of the Great Depression on the Army, 1929-1936 (Ph.D. dissertation, Indiana University, 1960), Tom Lewis, *Empire of the Air: The Men Who Made Radio* (New York: Edwin Burlingame Books, 1991) and Rebecca Robbins Raines, *Getting the Message Through: A Branch History of the U.S. Army Signal Corps* (Washington, D.C.: U.S. Army Center of Military History, 1996), pp. 229-233. Several of the more accessible works on signals intelligence include Ronald Clark, *The Man Who Broke Purple: The Life of Colonel William F. Friedman, Who Deciphered the Japanese Code in World War II* (Boston: Little,, Brown and Co., 1977), David Kahn, *The Codebreakers: The Story of Secret Writing* (New York: Macmillan Company, 1967) and Ronald Lewin, *ULTRA Goes to War* (New York: McGraw-Hill Book Co., 1978).

[6]The Army's sensor research dates back to the early 1920s and conducted the development work in its own laboratory at Fort Monmouth in the 1930s. The moving force behind the Army's effort was Colonel William R. Blair, whose interest in direction finding dated back to his service as the head of the Signal Corps Meteorological Section of the AEF during World War I. In 1936, the first field tests were held at Newark Airport and in 1937 the Secretary of War and the assistant chief of the air corps saw radar demonstrated at Fort Monmouth. See the biographical entry on "Blair, William S." in the *Dictionary of American Biography* (New York: Scribner, 1943-) supplement VII. Blair was awarded a patent for his radar work in 1937. A brief, accessible history of the Army's prewar interest in radar and its simultaneous development in the United States, Great Britain, France, Germany and Japan may be found in Dulany Terrett, The *Signal Corps: The Emergency (to December 1941) (United States Army in World War II, The Technical Services*) (Washington, D.C.: Center of Military History, 1956; reprint ed., 1986), pp. 35-48. Also see David K. Allison, *New Eye for the Navy: The Origin of Radar at the Naval Research Laboratory* (Washington, D.C.: Naval Research Laboratory, 1981), Robert Buderi, *The Invention that Changed the*

World: How a Small Group of Radar Pioneers Won the Second World War and Launched a Technological Revolution (New York: Simon and Schuster, 1996).

[7]The next paragraphs are based on Constance McLaughlin Green, Harry C. Thomson and Peter C. Roots, *The Ordnance Department: Planning Munitions for War* (*United States Army in World War II: The Technical Services*) (Washington, D.C.: Center of Military History, 1955; reprinted 1990), *passim* and LTC Eddie Mitchell, Apogee, Perigee and Recovery: Chronology of Army Exploration of Space (RAND Corporation Working Draft, May 1989). RAND published LTC Mitchell's chronology in 1991. In it, Mitchell gathered together many streams of information and his indefatigable research provided the basis for the chronological structure of this piece which was supplemented by the material in the SMDC archives as well as other primary and secondary sources. However, while chronology is useful for examining the manner in which events unfolded and were inter-connected, once the Army began to use space as a force multiplier and enhancer, the chronological approach must be combined with an analytical framework to understand the way this new field was perceived.

[8]In an interesting aside, Dr. Robert Goddard offered the Army the fruits of his research into rockets in 1918. At the Aberdeen Proving Grounds that November he test fired a "recoilless gun" or "rocket gun" for the Army. The test "showed potential" and the Ordnance Department recommended further development as a possible antitank weapon shortly after the Armistice. The problems that needed to be solved to perfect this weapon included creating a suitable explosive charge for the shell and perfecting the electrical firing mechanism. These difficulties, along with the advent of peace, demobilization and austere budgets caused the Army to shelve the project. It was not taken up again until 1940. See Green, et al., p. 356.

[9]Green, et al., p. 234. While the AAF was responsible for developing missiles, Ordnance was responsible for developing and supplying warheads and "destructor sets" for these missiles.

[10]V-1 originally stood for Versuchmuster, meaning experimental model, but later meant Vergeltungswaffe, vengeance weapon.

[11]Constructed in 550 man-hours, the rocket was 26 feet long with a wingspan of 19 feet. The V-1 ran on a pulse-jet motor, which used low-grade petroleum. Flying at a speed of 350-400 mph, the V-1 with its 1870 pound warhead had a range of 250km/160 miles. Centre for Defense and International Security Studies (CDISS), *Cruise Missiles: A Brief History: 1900-1945*, http://www.cdiss.org/cmhist.htm.

[12]Figures cited in Forrest Pogue, *The Supreme Command* (*United States Army in World War II, European Theater of Operations*) (Washington DC: Center of Military History, 1954, 1989) p. 252. Great Britain was not the only target for V-1 attacks. Liege and Antwerp, for example, also suffered large casualties from this weapon.

[13]CDISS, *Ballistic Missiles: The German V-2 Campaign, 1944-1945*, http://www.cdiss.org/v2.htm. Also known as the Aggregat-4 (A-4), the V-2 measured 46 feet in length and five in diameter and carried a 2,201 pound warhead.

[14]Pogue, p. 252.

[15]An interesting discussion of the layered defense used against the V-1s and the difficulties encountered can be found in Kenneth P. Werrell, *Archie, Flak, AAA, and SAM: A Short Operational History of Ground-based Air Defense* (Maxwell AFB, AL: Air University Press, 1988) and at http://www.strandlab.com/buzzbombs/index.html.

[16]CDISS, Cruise Missiles.

[17] CDISS, *Countering the V-1 & V-2 in WWII*, http://www.cdiss.org/scdnt2.htm.

[18]William S. Mark, Jr., Joseph P. D'Arezzo, R.A. Ranson, and G.D. Bagley "Detection and Plotting of the V-2 (Big Ben) Missile as Developed in ETO," 4 July 1945, Document 142.0423-16 Jul-Sep 1945, cited by Donald R. Baucom. *The Origins of SDI, 1944-1983* (University Press of Kansas, 1992), p. 4.

[19]In some respects the only defense was what the British called a "Bob Hope" - bob down and hope for the best. Jim Garamone, "SECDEF outlines need for national missile defense," American Forces Press Service, in *Kwajalein Hourglass* August 21, 2001: 5.

[20]See Kahn, *Codebreakers* and Lewin, *ULTRA*, David Kahn, "Cryptology and the Origins of Spread Spectrum," *IEEE Spectrum* 21 (September 1984): 70-80, Ronald Lewin, *The American Magic: Codes, Ciphers and the Defeat of Japan* (New York: Farrar, Straus and Giroux, 1982) as well as Edward J. Drea, *MacArthur's Ultra: Codebreaking and the War Against Japan, 1942-1945* (Lawrence: University Press of Kansas, 1992).

[21]The Army's experience with radar illustrates the fallacy of thinking that technology is an end to itself as it included the failure to warn against the Pearl Harbor attack and provided the ability to track V-2 rockets but not the ability to destroy them in flight. As the war went on, skepticism about the efficacy of the technology was replaced by growing sophistication in its use.

[22]The dossiers of those "invited" were marked by paperclips.
[23]Baucom, *Origins*, p. 4.
[24]Vannevar Bush was one of the most effective public figures American science produced in the 20th century. He was a Professor of Electrical Engineering, Dean of the School of Engineering, Vice President and President of MIT. During World War I, he worked for the Navy devising ways to detect submarines. Between 1928 and 1935, he was part of the team that developed a "network analyzer" and a "differential analyzer," the beginnings of digital computing. An academic inventor and entrepreneur, he was one of the founders of Raytheon. As Chairman of the President's National Defense Research Committee, President of the Carnegie Institution and Director of the Office of Scientific Research and Development, he was instrumental in developing the atomic bomb. He also recommended forming the National Science Foundation in his 1945 report, *Science—The Endless Frontier*. The standard biography is G. Pascal Zachary, *Endless Frontier: Vannevar Bush, Engineer of the American Century* (New York: Free Press, 1997). Monographs that attest to his varied influence include, James M. Nyce and Paul Kahn (eds.), *From Memex to Hypertext: Vannevar Bush and the Mind's Machine* (Boston: Academic Press, 1991), Colin B. Burke, *Information and Secrecy: Vannevar Bush, Ultra and the Other Memex* (Metuchen, NJ: Scarecrow Press, 1994) and Montgomery C. Meigs, Managing Uncertainty: Vannevar Bush, James B. Conant and the Development of the Atomic Bomb, 1940-1945 (Ph.D. dissertation, University of Wisconsin—Madison, 1982). He was the co-author of one of the standard texts in electrical engineering and wrote *Operational Circuit Analysis* (New York: John Wiley and Sons, Inc., 1929), an early exploration of digital computing. As a public figure, he was the author of the 1945 report, *Science—The Endless Frontier: A Report to the President on a Program for Postwar Scientific Research* (reprint ed., Washington, D.C.: National Science Foundation, 1980) as well as *Endless Horizons* (Washington, D.C.: Public Affairs Press, 1946) and *Modern Arms and Free Men: A Discussion of the Role of Science in Preserving Democracy* (New York: Simon and Schuster, 1949). In a 1945 article, he predicted ways of augmenting human thought to make knowledge more accessible through "associative trails," thus predicting both hypertext and the internet fifty years before they took shape. See Vannevar Bush, "As We May Think," *The Atlantic Monthly* 176.1 (July 1945):101-108.
[25]Excerpted from the Appendix of *History of the Plato Antimissile Missile System: 1952-1960* by Ruth Jarrell and Mary T. Cagle (Redstone Arsenal, AL: U.S. Army Ordnance Missile Command, 1961), quoted in Baucom, *Origins*, p. 6.
[26]Baucom, *Origins*, p. 6.
[27]Raines, p. 329 and "Army's Role in Space, FY 86," chart 25, Project Diana. The project was named after the Roman goddess of the moon. In fact, the idea for geosynchronous satellites relaying radio and television signals was first developed in Arthur C. Clarke, "Extra-Terrestrial Relays: Can Rocket Stations Give World-Wide Radio Coverage?," *Wireless World* 51 (October 1945): 305-308. The technology was maturing and was seriously considered in John R. Pierce, "Orbital Radio Relays," *Jet Propulsion* 25 (April 1955): 153-157 and the first satellite was launched in 1962.
[28]For web pages associated with this project, see http://www.wsmr.army.mil/paopage/Pages/V-2.htm, accessed on 30 January 2003. The paragraphs on the V-2 experiments are based on the material found here.
[29]The V-2 tests went to higher altitudes (70 miles on 10 May 1946), saw the first separation of a nose cone (30 July 1946), took and returned the first motion pictures that showed the earth's curvature (24 October 1946), successfully demonstrated the use of an auto pilot system (23 January 1947) and mapped 800,000 square miles of the Earth's surface (26 July 26).
[30]The Corporal was the first operational American ballistic missile. Its roots lay in the experiments carried on by the Ordnance Department during World War II resulting in the Private rocket. They were tested in 1944 (Camp Irwin) and 1945 (Fort Bliss). See Green, et al., *passim*. The Corporal was its lineal descendent and both were the fruits of the ORDCIT program that began in 1936. Despite its initial unreliability, it was deployed and its accuracy improved with further modifications. With a 75-mile range and the capability to carry either nuclear or conventional explosives, the Corporal remained in service between 1955 and 1963. See http://www.wsmr.army.mil/paopage/Pages/Corppr.htm, accessed 31 January 2003.
[31]Material on the Bumper Project may be found at http://www.wsmr.army.mil/paopage/Pages/bump.htm, accessed on 30 January 2003.

[32]NSC-68 was the sixty-eighth National Security Council Memorandum. The NSC was one of the bodies established in 1947 in a reorganization of the defense establishment. This policy memo was prepared in April and signed by President Truman in September 1950. It was not declassified until 1975.
[33]The United States detonated its first hydrogen bomb in February 1954.
[34]Donald Baucom, MDA Historian, "Missile Defense Milestones 1944-2000," http://www.acq.osd.mil/bmdo/bmdolink/html/milstone.html.
[35]Named for Nike the Greek Goddess of Victory.
[36]The NIKE-AJAX, also known as NIKE-I, measured 34 feet in length and 12 inches in diameter and weighed 2,455 pounds. It had a range of 25-30 miles and up to 70,000 feet altitude. The first NIKE-AJAX battalion deployed at Fort Meade, Maryland, in March 1954. John Lonnquest and David Winkler, *To Defend and Deter: The Legacy of the United States Cold War Missile Program* (Rock Island, IL: U.S. Army Construction Engineering Research Laboratories, 1996).
[37]The NIKE-HERCULES, originally called NIKE-B, measured 41 feet in length and 31.5 inches in diameter, weighed 10,710 pounds, and had a range of 75miles. The Hercules could reach altitudes up to 15,000 feet. Lonnquest and Winkler, *To Defend and Deter*.
[38]The Army deactivated its last Nike-Hercules batteries in July 1979, but they remain operational elsewhere in the world. Christina M. Carlson and Robert Lyon provide an interesting overview of the NIKE deployment and life at NIKE batteries in *Last Line of Defense: NIKE Missile Sites in Illinois* (Denver, CO: National Park Service, 1996).
[39]Bell Laboratories, *ABM Research and Development at Bell Laboratories: Project History* (Whippany, NJ: Bell Labs for the U.S. Army Ballistic Missile Defense Systems Command, 1975), p. I-1.
[40]Bell Labs, *Project History*, pp. I-2-I-3.
[41]A second study conducted by the Atomic Energy Commission and Department of Defense concluded, in their September 1957 report, that it was feasible to develop a warhead capable of destroying an ICBM warhead for the NIKE II. They added that this warhead could be available by 1961.
[42]*Ibid*, pp. I-1-I-15. World War II air defense objectives found a 10-15 percent attrition rate acceptable.
[43]*Ibid*, p. I-4. Simulation calculations found that an ICBM would decelerate at a rate up to 60g's, based on the shape of the nosecone, thus easing interceptor maneuverability requirements.
[44]*Ibid*, p. I-5.
[45]James A. Walker, Frances Martin, and Sharon S. Watkins, *Strategic Defense: Four Decades of Progress* (Huntsville: U.S. Army Space and Strategic Defense Command, 1995), p. 10.
[46]The CORPORAL intercept took place on 3 June 1960 and the HERCULES in September 1960.
[47]Paul B. Stares, *The Militarization of Space: U.S. Policy, 1945-1984* (Ithaca: Cornell University Press, 1985), p. 31 and William E. Burrows, *Deep Black: Space Espionage and National Security* (New York: Random House, 1986), p. 71. Dr. Killian was the President of M.I.T. and also recommended establishing a civilian agency to handle research and development and have civilian scientists guide the space program.
[48]The Eisenhower Administration's proposal was imbedded in the Soviet and American dilemma over arms control, where much of the working vocabulary is borrowed from economics and game theory. The object of arms control is to allow each party to reach a state of normative parity in order to participate in the game, but not an overwhelming advantage to win. If either party achieves a destabilizing advantage, the incentive for surprise attack is increased. Allowing Soviet and American satellites free passage over each other's territory would be tantamount to giving each side access to the other's secrets in a limited fashion. Doing so would mean both parties would have their national security enhanced. This factor introduces a Byzantine complexity to the already complicated world of Cold War diplomacy and espionage. See also R. Cargill Hall, "The Origins of U.S. Space Policy: Eisenhower, Open Skies and Freedom of Space," *Colloquy* 14 (December 1993):5-6, 19-24.
[49]*New York Times*, 5 May 1954, quoted in Bell Laboratories, *ABM Research and Development at Bell Laboratories: Kwajalein Field Station* (Whippany, NJ: Bell Laboratories for U.S. Army Ballistic Missile Defense Systems Command, 1975), p. 23.
[50]The 1957 Gaither Report, for example, is often cited as a key document in this discussion.
[51]Document 14 - Vandenberg-Collins Agreement, 1 August 1950 in Richard Wolf, The *United States Air Force Basic Documents on Roles and Missions*, (Washington, DC: Office of Air Force History, 1987).

[52]Ernest Yanarella, *The Missile Defense Controversy - Strategy, Technology, and Politics, 1955-1972* (Lexington: University Press of Kentucky, 1977), p.19, quoted in Christina M. Carlson and Robert Lyon, *Last Line of Defense: NIKE Missile Sites in Illinois* (Denver, CO: National Park Service, 1996), p. 23.
[53]Document 21 Secretary of Defense Wilson's Memorandum, 26 November 1956 in Wolf, *Air Force Documents.*
[54]The memoranda only address the development of the system. There is no mention of organizational control over a deployed system.
[55]The International Geophysical Year programs are described in J. Tuzo Wilson, *IGY: The Year of the New Moons* (New York: Knopf, 1961).
[56]The Russian accomplishment should not have been a surprise. Russian scientists and engineers had been active in producing literature about space travel. Konstantin Eduardovich Tsiolkovsky theorized about rockets flying through space as early as 1883 and had fully developed a theory of rocket flights and space travel by 1903. The Red Army captured Peenemünde and part of the German design team in May 1945; moving them to the Soviet Union in 1946. Instead of trying to improve the V-2 design, the Russians embarked on a new design tack. By 1951, Russian scientists were convinced on the basis of their calculations and experiments that space flight was possible and began developing medium and intercontinental range ballistic missiles in 1953. In 1956, they announced they would launch a satellite during the International Geophysical Year the following year. See "Article on Space History." In his first chapter, McDougall points out that the Soviets had invested considerable resources in their rocket programs since the 1930s.
[57]Briefing, "Army's Role in Space FY 86."
[58]Paul Dickson, *Sputnik: The Shock of the Century* (New York: Walker and Co., 2001) is a recreation of Sputnik's effect on American public opinion and popular culture. Three of its immediate effects were an emphasis on language and science education and their funding in American schools. Contemporary critics recognized that Sputnik delivered its biggest blow to American complacency. Some called the Russian feat a gimmick rather than a real technological feat and merely a showcase for big booster rockets. They also pointed out that American scientists had achieved radical size and weight reductions of thermonuclear weapons as well as rocket and telemetry components and, given the American alliance system, big booster rockets were irrelevant as either weapons delivery or satellite orbiting systems. Their voices were discounted in the general rush to judgment. Senator Lyndon B. Johnson, Chairman of the Senate Armed Services Committee, called the launching of Sputnik I "a technological Pearl Harbor." See McDougall, p. 152.
[59]See Jane van Nimmen, Leonard C. Bruno, Linda N. Ezell, *NASA Historical Data Book* (4 vols., Washington, D.C.: Scientific and Technical Information Division, National Aeronautics and Space Administration, 1988-1994). The most pertinent volumes for this section are the first two, NASA *Resources, 1958-1968* and Programs *and Projects, 1958-1968.*
[60]See McDougall, chapters 6 and 7 and Roger E. Bilstein, *Orders of Magnitude: A History of NACA and NASA, 1915-1990* (Washington, D.C.: National Aeronautics and Space Administration, Office of Management, Scientific and Technical Division, 1989), pp. 47-48. The NACA was the National Advisory Committee on Aeronautics.
[61]For an overview, see Anthony M. Springer, "Project Adam: the Army's Man in Space Program," *Quest* Summer/Fall 1994:46-47 and Chapter 4 of Loyd S. Swensen, Jr., James M. Grimwood, Charles C. Alexander, *This New Ocean: A History of Project Mercury*, (Washington, D.C.: National Aeronautics and Space Administration, NASA History Office, Office of Policy and Plans, 1998) (NASA Special Publication-4201, NASA History Series, 1998).
[62]Army Ballistic Missile Agency, Redstone Arsenal, Alabama, "Development Proposal for Project Adam," ABMA Report No. D-TR-1-58, 17 April 1958, p. 2.
[63]For greater detail, see Anthony Springer, "Securing the High Ground: The Army's Quest for the Moon," *Quest* 7.2:32-38 and Frederick I. Ordway, Mitchell R. Sharpe and Ronald C. Wakeford, "Project Horizon: An Early Study of a Lunar Outpost," *Acta Astronautica* 17. 10:1105-1121.

Chapter 2
Rockets, Communications and Deploying Ballistic Missile Defense, 1958-1975

Satellites and Communications

The decisions taken in 1958 diminished the Army's overt space role but reaffirmed its missile defense role. While the Army gave NASA control of all aspects of the Army Launch Vehicle Program and the Air Force gained control of the Jupiter-C IRBM program and responsibility for developing an ICBM, the Army continued to bear the primary responsibility for Ballistic Missile Defense. The Navy transferred the Vanguard Program and part of the Naval Research Laboratory to NASA while retaining proponency for sea-launched missiles. Through 1960, the Army continued to contribute to communications, meteorological, reconnaissance, research and exploration satellites.[1] The first Vanguard satellite, with its instruments transmitting data using batteries designed by the Army Signal Corps, continued to send data to earth until 1964. This satellite remains in orbit today and will stay in orbit for another 2000 years. The Vanguard I launch was quickly followed by Explorer III (26 March 1958), the first satellite to carry an on-board tape recorder to store data to be transmitted to a ground station when it came in range. Explorer IV (26 July 1958) was placed in an elliptical inclined orbit by a Juno I rocket. It gathered data from a high altitude nuclear explosion and measured solar radiation for three months.

In December 1958, the Army and Air Force put the first communications satellite in orbit. Called Project SCORE (Signal Communications by Orbiting Relay Equipment), the Army Signal Research and Development Laboratory began working on the satellite that June. Since the Air Force proposed placing the entire Atlas rocket into orbit, the communications equipment was integrated into the rocket's fairing pods. The rocket was placed in a low earth orbit and the satellite's life expectancy was approximately two weeks. The satellite could receive, store and send voice and coded signals-one voice channel or seven teletype channels on two tape recorders. Shortly before the satellite was launched, a tape-recorded message from President Eisenhower was placed aboard. The system worked and the president's message was broadcast on short wave radio. In addition, the satellite responded to real-time and store-forward voice and teletype transmissions.[2]

When Vanguard II was launched into low earth orbit (17 February 1959) it carried an Army developed cloud imaging sensor. However, stability problems (the satellite wobbled in orbit) precluded imaging efforts. That same year, the Army Signal Corps, the Weather Bureau, RCA and several other organizations collaborated to develop a weather satellite. For the Army, this built on the work done on the first two Vanguards and on an experiment placing a television camera in orbit to take pictures of the earth. This was expanded to create the first weather satellites, TIROS I and II (Television Infrared Observation Satellite), which NASA launched in April and November 1960. The satellites used multiple television cameras to transmit pictures of cloud patterns to ground stations in New Jersey and Hawaii. The program proved so successful that eight satellites were put into orbit between 1960 and 1965.[3]

The Army was also involved in communications satellites. The Courier was the first to transmit ultra-high frequency (UHF) radio waves. The attraction of this portion of the electromagnetic spectrum was that it was relatively unused and free from man-made or atmospheric interference. In 1958, as the SCORE project started, the Army began work on the ADVENT and COURIER programs. ADVENT, put into operation in 1960, was "a twenty four hour, equatorial synchronous, military satellite communication program" established at the Advanced Research Project Agency's (ARPA) direction.[4] COURIER was more advanced and could simultaneously transmit and receive "about 68,000 words per minute while traveling through space at 16,000 miles per hour and send and receive facsimile photographs."[5] Data could also be stored for later transmission. It was the first satellite powered by long life solar cells that recharged nickel cadmium storage batteries. Although the COURIER's effective life was only seventeen days, it proved that all types of messages and data could be received and transmitted by satellite.[6]

The Army's work served as a catalyst for a telecommunications revolution. Satellites stitched the world together in a way very different from either wires or cables. In 1962, NASA and AT&T joined to launch Telstar, the world's first active communications satellite, picking up, amplifying and re-broadcasting signals from one point on earth to another. Telstar broadcast the first live television pictures between continents, illustrating this new technology's potential. Later that same year, Congress passed the Communications Satellite Act of 1962 establishing the quasi-governmental Communications Satellite Corporation (COMSAT). This body managed an international syndicate, INTELSAT, whose members shared access to a global telecommunications satellite system. This system increased the number of transoceanic telephone circuits and made live television coverage possible anyplace on the globe.[7]

NASA continued to use Army developed Jupiter/Juno missiles for space probes to the moon and the sun. Pioneer II was launched in December 1958 and traveled 63,580 miles on a voyage to the moon. In March 1959, an Army rocket launched Pioneer IV on a voyage to the moon. Pioneer IV passed within 37,300 miles of the moon before going into permanent solar orbit. However, the Soviet Luna I probe that passed within 3,700 miles of the lunar surface overshadowed this achievement. A Juno II also successfully launched the Explorer VII satellite in October 1959. The Explorer VII carried a scientific package for detecting micrometeors, measuring the earth's radiation balance, and conducting other experiments.[8] It was the last of the satellites launched as a part of the IGY and its scientific instrument package began a new era in weather forecasting and meteorology.

NIKE-ZEUS Testing

Fig. 2-1. Emblem of the NIKE-ZEUS Project Office.

The division of responsibility between the Army and the Air Force over missiles began in 1944, shortly after the German V-1 attacks began. The Army divided guided missile development responsibility between the Army Air Forces (AAF) and the Ordnance Department, intending to lessen secrecy and to promote data sharing. The AAF was given "development responsibility…for all guided or homing missiles dropped or launched from aircraft…[or those] launched from the ground which depend for sustenance primarily on the lift of aerodynamic forces." The Ordnance Department would develop missiles "which depend for sustenance primarily on momentum of the missile." In January 1945, the General Staff made the Ordnance Department responsible for developing a missile suitable for antiaircraft use.[9] This division of labor continued through the postwar era, with the Air Force claiming responsibility for intercontinental ballistic missiles as the logical extension of long-range bombers, while the Army viewed missiles as very long-range artillery weapons. In January 1958, Secretary of Defense Neil McElroy assigned the Army primary responsibility for developing all aspects of ballistic missile defense, including missiles, launch sites, radars and computer components. At the same time, the Air Force continued its responsibility for developing early warning radars, tracking and acquisition radars and communications links to ballistic defense installations.

Fig. 2-2. Brigadier General Ivey Drewry became the first NIKE-ZEUS Project Manager in August 1962. Brigadier General Drewry led the Army's missile defense program until his retirement in November 1967.

Fig. 2-3. The NIKE Family of missiles. From back to front: NIKE-AJAX, NIKE-HERCULES and NIKE-ZEUS.

The importance of the BMD program became apparent in August 1957 when the Soviet Union announced a successful ICBM test flight using its SS-6 ICBM.[10] In January 1958, the National Security Council assigned the highest national priority ("S-Priority") to the NIKE-ZEUS antimissile missile development program. One problem remained however – locating a site appropriate for field testing, which presented a new set of obstacles.

The test range should be located far enough away to allow ICBM testing in an uncluttered area that could be secured from "curious adversaries"[11] White Sands had been the desired location. Distances, however, would not allow the interceptor to be tested to its full capability.

Range restrictions, which forced the premature destruction of some shots, eventually eliminated a second choice, Point Mugu, California. The Atlantic Range, stretching south from Cape Canaveral, was eliminated because of the area's high population density and the absence of suitable American territory for testing and tracking facilities.

Following a requirements review, the Army Rocket Guided Missile Agency (ARGMA) decided that Kwajalein Atoll in the Marshall Islands presented the most logical solution because it was part of the Pacific Islands Trust Territories, served as an American naval base and had an already existing logistical structure. In addition, it was geographically perfect, within a day's flying distance of Hawaii, lying 4,800 miles from the United States, which made it ideal for testing interceptors against testing vehicles launched from Vandenberg Air Force Base, California .

In 1959, the DoD's Ballistic Missile Committee approved the test program, which began at White Sands Missile Range (WSMR), New Mexico; Kwajalein Atoll became the down range test site. The Kwajalein Test Center was officially established on 1 October 1960. Ultimately, the significance of the Kwajalein site to the Army's interceptor test program resulted in the transfer of the site from the U.S. Navy to the U.S. Army on 1 July 1964.[12]

Fig. 2-4. Testing began at the Kwajalein Missile Range in December 1961. This aerial view shows the island in January 1962.

On 26 August 1959, the NIKE-ZEUS flight test program began with the launch of the first NIKE-ZEUS missile at WSMR. Although deemed a partial success, the missile broke up shortly before sustainer-booster separation. This missile and two others fired in 1959 were designed for uncontrolled flights, constructed with fixed-fins and a dummy nose instead of the thrust vectoring nose. It was not until the fourth test, conducted on 3 February 1960, that a NIKE-ZEUS test flight completely at met all objectives.

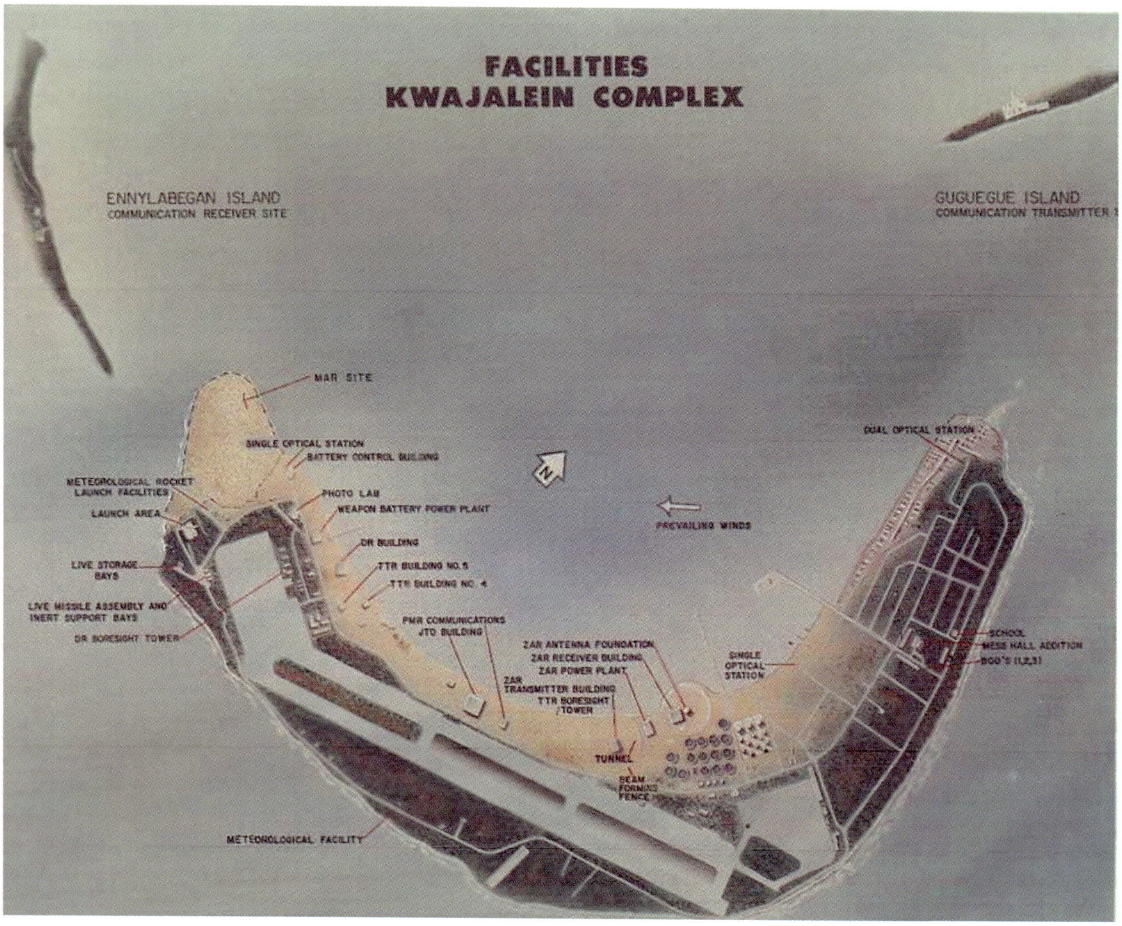

Fig. 2-5. A map of the facilities at the Kwajalein Complex in the 1960s.

By the end of its first year, the NIKE-ZEUS test program had completed five successful tests using two different versions of the missile.[13] Nevertheless, the argument against ABM development and in favor of the production of offensive systems remained strong. On 18 October 1960, the President's Science Advisory Committee concluded, "There has been very considerable progress in the ZEUS program within the last year. This does not, however, appear to be any reason for changing the major conclusion we drew last year… that with respect to defense of population against a major attack, fallout shelters should have priority over extensive ZEUS deployment." They recommended instead continued research and development in conjunction with full testing at the Kwajalein site followed by "very limited deployment in the near future."[14]

In contrast, the 21 November 1960 report to the Chief of Staff by the NIKE-ZEUS Ad Hoc Advisory Committee provided three recommendations: (1) That the production program for the NIKE-ZEUS batteries begin immediately at the rate of four per year; (2) The units be deployed in the defense of the North America in support of the antimissile defense plans established by the

North American Air Defense Command; and (3) The present NIKE-ZEUS research and development program be continued with the primary objectives of determining the system effectiveness against various types of threats and of improving this effectiveness consistent with the state of the art.

While it was testing the ZEUS, the Army pursued funding for a production program that aimed for operational status in 1962.[15] Secretary of Defense McElroy favored a continued research and development effort and no funds were granted in fiscal year (FY) 1959. Congress reversed itself and provided $137 million in production funds for FY 1960, but the Eisenhower administration refused to spend the funds.

Fig. 2-6. Pictured are the many different radars of the NIKE-ZEUS system. The photo was taken at Kwajalein, 20 July 1963.

In January 1961, the ARGMA submitted a revised "NIKE-ZEUS Defense Production Plan" which called for producing and deploying 70 batteries for 29 defense centers at a cost of $8 billion over eight years.[16] At the same time, a new administration had entered the White House and President John Kennedy ordered a review of the BMD program.[17] Robert McNamara, the new Secretary of Defense, determined that deploying the NIKE-ZEUS system was neither technologically feasible nor cost effective. Unimpressed by the challenges that the system would pose to Soviet planners, he argued that the Soviets could simply counter by increasing the number of their offensive missiles and overwhelming the system. Nevertheless, he approved significant funding for a continued research and development effort that would keep the program at a "top priority" level.

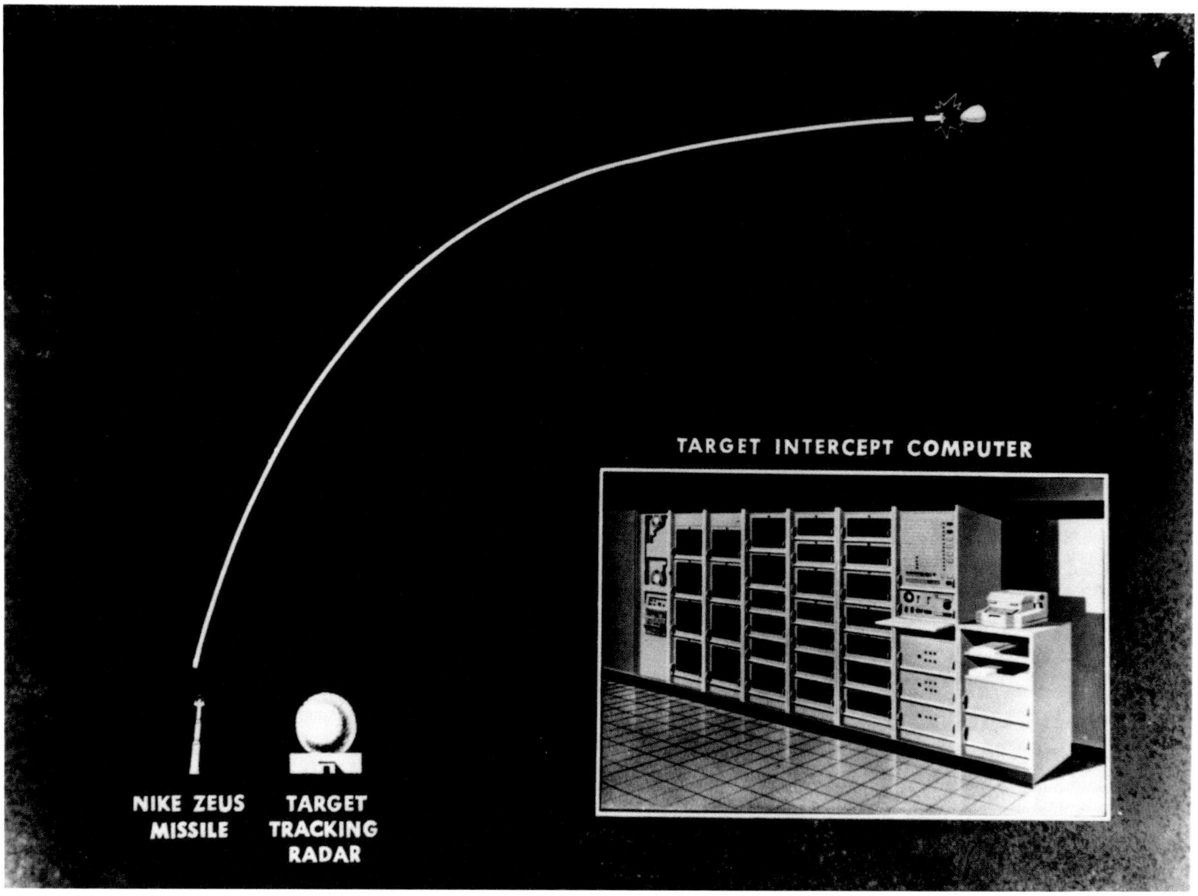

Fig. 2-7. Artist's conception of the target track radar and the target intercept computer.

The First Intercept

The research and development program proceeded even as the political developments unfolded. In May 1961, an advanced ZEUS missile was successfully fired at White Sands. Systems demonstrations began in November 1961. On 14 December, the NIKE-ZEUS system intercepted a NIKE-HERCULES missile in the first integrated system test.[18] The ZEUS passed within 100 feet of the HERCULES missile, well within the distance defined for a successful nuclear intercept.

Fig. 2-8. On 12 December 1962, the NIKE-ZEUS Project Office achieved the first fully successful intercept of an ICBM, seen in the horizon over the ZEUS Acquisition Radar.

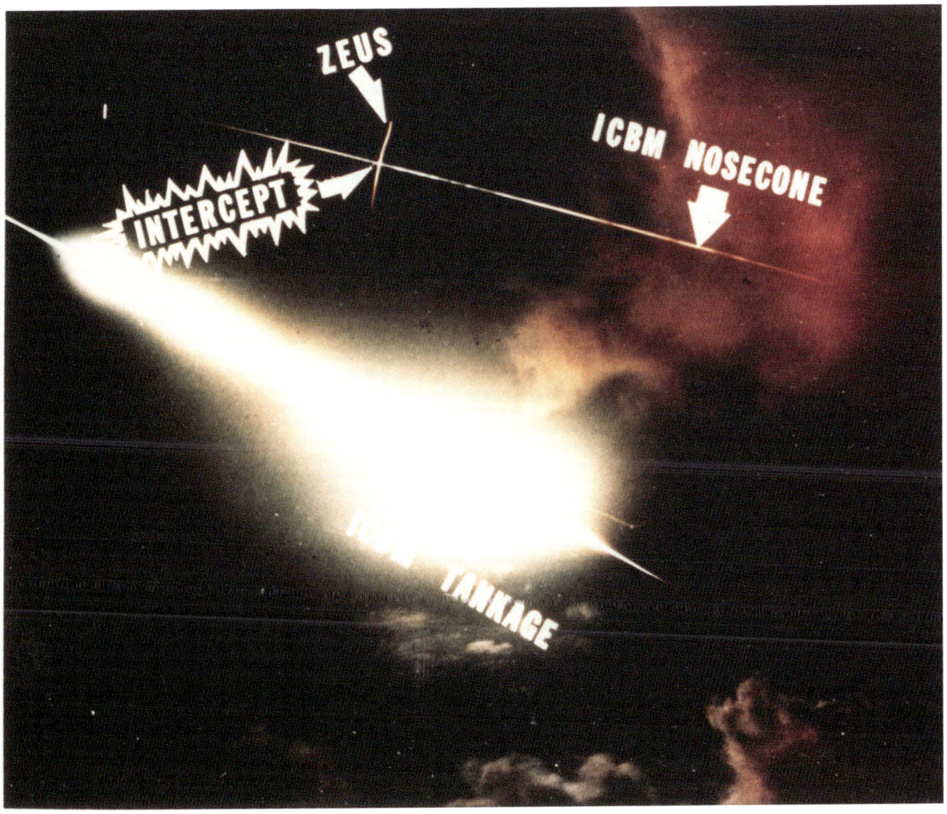

Fig. 2-9. An annotated photograph illustrating the Army's successful ICBM intercept.

The most important demonstration took place on 26 June 1962 with the first attempted intercept of an ICBM fired from Vandenberg AFB to Kwajalein, a distance of 4,500 miles. Unfortunately, the radar malfunctioned and the interception attempt failed. A second attempt on 19 July 1962 intercepted an Atlas D nose cone traveling 16,000 mph. One wire service release declared the intercept a "majestic bull's-eye, comparable some have said, to a bullet hitting a bullet." Project Office officials declared the test only partially successful.[19] The U.S. Army made history on 12 December 1962 when the NIKE-ZEUS Project office made a fully successful intercept of an ICBM nose cone, passing well within the acceptable limits for a simulated nuclear warhead.

Fig. 2-10. A NIKE-ZEUS missile on a launcher during testing at White Sands Missile Range.

A New Mission: Project MUDFLAP

As related above, after the Soviet Union launched the first man-made satellite Sputnik I, American interests focused upon matching and surpassing this feat. By 1959, however, new concerns had arisen. In June of that year, Dr. Walter Dornberger, a former German general who helped develop the V-2 and worked as an Air Force and NASA consultant in 1950s, warned the audience at the Dr. Robert H. Goddard Memorial Dinner that the United States should prepare a defense against nuclear bombing from earth-orbiting satellites.

On 27 April 1962, Secretary of Defense McNamara announced a new requirement for the NIKE-ZEUS system. By 1 May 1963, the NIKE-ZEUS Project was to provide the capability for a satellite interception demonstration at Kwajalein, known as Project MUDFLAP.[20] Bell Labs modified a ZEUS missile and began testing at White Sands in December 1962. Their missile ultimately reached an altitude of 151 nautical miles. In March 1963, testing transferred to Kwajalein and, on 23 May 1963, a NIKE-ZEUS missile successfully intercepted an AGENA D earth satellite. From this moment forward, the missiles and personnel at Kwajalein were maintained in a state of readiness to launch a ZEUS in an anti-satellite mode. Training and test launches continued in 1964, until officials terminated the "ready requirement."

After deciding not to deploy the NIKE-ZEUS system, no further live ICBM target tests were conducted.[21] With simulated targets and other programs, the test program continued until 9 December 1964 at White Sands Missile Range and May 1966 at Kwajalein.[22] Ultimately, the NIKE-ZEUS test program conducted 79 developmental and 68 systems tests, 147 firings altogether. Of the developmental firings – 56 at White Sands and Point Mugu and 23 at Kwajalein – 22 were failures, 12 were partial successes and 45 were full successes. Similarly, the Systems Tests conducted at White Sands and Kwajalein recorded 7 no tests, 15 failures, 7 partial successes and 39 successes.

Project Defender

Even as Secretary McElroy defined the specific ballistic missile defense (BMD) responsibilities of the services in January 1958, he assigned direction of the development effort to ARPA, in order "to make [the] most effective use of our overall national capability."[23] Project Defender was the ARPA-directed BMD effort. Scientists at ARPA explored broad concepts in missile defense.[24] With the boost missile boost intercept or BAMBI program, for example, they looked at satellite tracking to launch ground-based hit-to kill interceptors, which would intercept Soviet missiles over the Arctic. The Guidelines Identification Program for Antimissile Research (GILPAR) study examined lasers, neutral particle beams and "tailored-effects" produced by nuclear devices. As the Department of Defense leader of the BMD program, ARPA worked with the Army's NIKE-ZEUS program, providing funding and facilities for advanced testing. It also developed a new ground-based phased array radar system, which was incorporated into the Army's subsequent BMD initiatives.[25]

Continued Interest in Satellite Communications

In the early 1960s, the Army bowed out of its role in space exploration although it retained a role in satellite communications as well as its vitally important role in missile defense. The

Army was forced out by several developments, including the establishment of NASA and the subsequent demilitarization of many space missions, the DoD giving new space roles and missions to the other services, treaties, the centralized management, development and operation of long range military communications systems, and the distractions of the Vietnam Conflict.[26]

Vietnam skewed Army thinking away from space and using space-based instruments as a force multiplier. Between 1961 and 1973, the Army was slowly committed and subsequently withdrawn from Vietnam. An Army theater command was established and approximately two-thirds of the troops in-country were soldiers. The Army committed two corps headquarters, seven divisions, two separate infantry brigades, one airborne brigade and one air cavalry regiment to the theater. The war was the Army's major focus while American soldiers were involved in combat through 1973 as well as during the subsequent support effort. During its involvement, surviving and winning the conflict was the primary focus of the Army's efforts. Space-based satellites did not offer any direct tactical aid to the soldier on the ground. Satellite-assisted communication was the only way space-based assets influenced the ground fighting. The conflict in Vietnam demanded the Army's full attention, to the detriment of many other research and development initiatives.

Instead of thinking about space-based assets as force multipliers or used strategically to shape future wars, there was an understandable, natural desire to field robust, effective tactical weapons systems that troops could put to use immediately. Instead of thinking about the future of space-based systems, the Army concentrated on developing and fielding small, accurate battlefield missiles for ground support aircraft and the infantry. At the same time, however, the Army made substantial contributions to developing a worldwide communications network for the DoD, directly contributing to the design of the first geosynchronous satellite, SYNCOM, and to the Initial Defense Satellite Communications System. The Army also set up and managed the global network of ground stations that provided reliable communications to Army theater commands. In addition, the Army established the Strategic Communications Command to manage and operate the Army component of the Defense Communication System. In 1970, the Secretary of Defense modified an earlier directive and allowed each service to conduct research and develop programs that would serve its unique needs for battlefield surveillance, communication, navigation, mapping and charting. However, the Army was not able to take advantage of this opportunity until it began to think about the future of warfare and its place on the battlefield.

In a sense freed from concerns about the use of space, the Army concentrated its technological efforts on communications and on improving ballistic missile defense. The NIKE-ZEUS project continued with its successful test program but certain weaknesses proved troubling, such as the difficulty in differentiating between warheads and decoys.[27] Officials also believed that a saturation attack would overwhelm the capabilities of the discrimination radar as the Target Track and Missile Track Radars could only focus on one target or interceptor at a time. In addition, scientists had little data on reentry phenomena or the effect of the ZEUS warhead on other components. As a result, Department of Defense officials began to look to the needs of the future, while continuing with research and defense.

NIKE-X: A New Organization

Fig. 2-11. Emblem of the NIKE-X Project Office, which replaced NIKE-ZEUS in February 1964.

On 5 January 1963, Secretary of Defense McNamara directed that the development of an ABM be a priority program, and one that incorporated the most technologically advanced components and techniques available. This program, known as NIKE-X, incorporated a variety of studies and initiatives designed to develop the next generation ABM system intended to counter the ICBM threat envisioned for the 1970s. The NIKE-X system would be composed of higher speed, higher capacity radars and computers, and an interceptor missile fast enough to be launched and to intercept an enemy warhead after it reentered the atmosphere. Combined with the existing ZEUS long-range missile, the NIKE-X would provide a layered defensive network.

Oversight of the NIKE-X program was assigned to the NIKE-ZEUS Project Manager in 1963. One year later, on 1 February 1964, the Army officially changed the name of the project office from NIKE-ZEUS to NIKE-X.[28] The Army also assumed responsibility for the Kwajalein Test Range, a logical transfer of authority given the role of Kwajalein in the Army's ABM research and development effort.[29]

On 5 June 1965, anticipating a production order for the NIKE-X, the Secretary of the Army approved a centralization of all facets of NIKE-X and established the NIKE-X System Manager at the Department of the Army level.[30] Under the centralized arrangement, the System Manager oversaw all elements of research and development, testing, production, training and deployment. The concept was implemented one year later. General Harold Johnson, Chief of Staff of the Army, identified the program for "exceptional management" based on the "scope and importance to the national defense of the NIKE-X Ballistic Missile Defense System,"[31] At that time, he appointed the Chief of Research and Development, Lieutenant General Austin Betts, to serve as the NIKE-X System Manager. The NIKE-X Project Office and the NIKE-X Engineering/Service Test Organization were subsequently placed under the operational control of the System Manager, who in turn would report to the Army Chief of Staff.[32]

Within a year of his appointment, the System Manager had contacted the various army commands and agencies that would have a role in a future deployment. One of these letters of instruction assigned the Corps of Engineers the task of designing and constructing the nuclear-hardened tactical facilities and support structures that would be required in the event the system was deployed. In response, the Corps established a special NIKE-X Division in October 1967.[33] Its sole mission was "to develop criteria, design, and construct developmental, training, support, and tactical facilities" for the planned ABM deployment. In 1968, as a result of a Memorandum of Agreement with the Corps of Engineers, the Huntsville Division also came under the operational control of the NIKE-X System Manager, further centralizing the ABM program.

Fig. 2-12. Dual salvo launch seen near the headquarters of the NIKE-X Project Kwajalein Test Site.

Debating Deployment Options

In 1963, Secretary McNamara ordered a new study of the ABM initiative. His focus was the impact caused by an ABM system on deterrence and relations between the United States and the Soviet Union. The Commission, headed by Lieutenant General Austin Betts, supported the missile defense program. The Betts Report concluded:

> (1) offensive technology had not hopelessly outstripped defensive technology – rather, the two technologies were roughly equal; (2) a BMD system would limit damage in case of a nuclear attack, with the amount of limitation dependent on the scenario; and (3) BMD would not disrupt the balance of mutual nuclear deterrence.[34]

Despite these findings, no deployment decision was forthcoming.[35] Instead, McNamara informed the Senate, "without question, offensive capability or what I will call the capability for assuring the destruction of the Soviet Union is far and away the most important requirements [sic] we have to meet."[36] This argument coupled with the estimated $16 billion cost for a deployment and the growing opposition of the scientific community influenced the Secretary's cautious approach toward missile defense.

By the mid-1960s, some scientists had concluded that it was unrealistic to deploy an ABM system. Tests conducted by the Reentry Body Identification Group in 1958 revealed that multiple warheads could overwhelm the NIKE-ZEUS system.[37] A similar study conducted in 1959 by the President's Science Advisory Group produced comparable conclusions. Thus, a number of the government's leading scientists opposed an ABM deployment. In 1961, ARPA Director Dr. Jack Ruina testified, "that he felt a 'great deal of pessimism about ever developing a complete and adequate umbrella against ICBM attack.'"[38]

The basis of their opposition was that the ABM program undermined nuclear deterrence.[39] Opponents believed that it would be impossible to build an airtight defense, and the other side would simply build more and better ICBMs. Wrapped around the arguments were the on-going negotiations for a nuclear test ban treaty, an agreement that would always be in jeopardy if the United States continued to develop ABMs that required nuclear testing. The conflict within the government spilled into the public arena when two senior scientists published an article in the October 1964 issue of *Scientific American* on the futility of pursuing an ABM program. Herbert York, Department of Defense Director of Research and Engineering, and Jerome Wiesner, head of the President's Scientific Advisory Committee, argued that a deployment would not only prove ineffective but would lead to a new type of arms race. This arms race would focus on the development of improved warheads and penetration aids. They concluded, "It is our considered professional judgment that this dilemma has no technical solution."

The Evolving Threat: The Nth Country

In the 1960s, new international developments began to determine the progress and priorities of ABM research. In October 1964, the nuclear club expanded to include China. While the

Chinese had exploded a nuclear device, they did not yet have a delivery system. Within a year, however, this situation changed when they completed a device that could be delivered by bomber.[40] On 27 October 1966, China announced it had successfully test-flown a guided missile carrying a live nuclear warhead. In that same year, the Chinese deployed the Dong Feng-2 (also known as the CSS-1), an intermediate range ballistic missile with a range of over 1200 kilometers that could threaten American military bases in Japan.[41]

With these developments in China and continued advances in technology, the Secretary of Defense ordered new studies to reassess the development and the feasibility of ABM deployment. Beginning in 1965, strategists began to look at the possibility of limited strikes by nations other than the Soviet Union - the so-called "Nth country" threat. The team theorized, "such an attack would probably consist of a limited number of unsophisticated inaccurate ICBMs, designed to terrorize rather than neutralize strategic targets."[42] The dangers presented by Nth country threat scenarios somewhat lessened concerns over destabilization between the Soviet Union and the United States.

In February 1965, Bell Labs began to investigate modifications to the NIKE-X aimed at achieving effective "high altitude defense against relatively unsophisticated attacks with deployment growth to meet sophisticated threats."[43] The result was an M-Multifunction Array Radar supplemented by an off-the-shelf VHF radar to provide long-range detection of sneak attacks. The Missile Site Radar (MSR) and SPRINT remained the key elements to the system for missile guidance and short-range intercepts. Following this presentation, Bell Labs received authorization to revise the NIKE-X requirements, "providing a more cost-effective defense against a possible Nth-country threat, in addition to the more sophisticated Soviet-type threats."[44] The modifications would enable the system to provide "a general defense" for the entire continental United States.

Tasked by the Director of Defense Research and Engineering in October 1965, the Army and Bell Labs designed a system, which would provide a defense against a "simple" first-generation Nth country threat. The recommended deployment (DEPEX-II) employed a minimum amount of hardware: 4 VHF radars and 12 Missile Site Radars, with 20 modified ZEUS missiles at each site. Recognizing the limitations of such a deployment, in November 1965 three teams began research on active defense of hardened sites. During this phase, engineers developed the concept of "pitch and catch" for the missile launch phase, increasing the potential flight time for the SPRINT missile.[45] The advances made from these studies were significant. As a result, the Office of Deputy Chief of Staff for Operations concluded in "NIKE-X Studies for 1966 (X-66), Report to the SECDEF," that the likely effectiveness of NIKE-X validated the cost of deployment at DEPEX-II. The study also found that "NIKE-X would add to U.S. deterrence and provide significant reduction in fatalities in the event deterrence fails." Despite these assessments, Secretary McNamara continued to oppose any deployment options. A series of events in 1967, however, brought the ABM issue to the forefront.

1967: A Turning Point

In November 1966, Secretary McNamara announced that the Soviet Union had deployed an ABM system around Moscow. Sixty-four launchers surrounded Moscow, equipped with the Galosh, a nuclear-tipped interceptor with an estimated range of 200 miles.[46] With this announcement, McNamara hoped to undercut arguments for the deployment of an American ABM system and to gain support for increasing the deployment of offensive weapons to offset the Soviet defenses. Meanwhile, President Lyndon Johnson expressed growing concerns on this subject. Given the situations in China and the Soviet Union, and considering the Joint Chiefs of Staff recommendation in favor of ABM deployment, President Johnson was inclined to favor a deployment decision. Instead, McNamara proposed that the President tie ABM deployment funds to arms control talks with the Soviets. An ABM system need only be deployed if talks with the Soviets failed.

At the June 1967 Glassboro Summit, President Johnson tried to convince Soviet Premier Alexsei N. Kosygin to abandon Soviet missile defense efforts. Johnson argued that continued deployment would lead to another arms race. Without this agreement, the U.S. "would be compelled to increase the number of warheads in its ICBM arsenal to overwhelm any defenses."[47] Kosygin had already made his position known. In a February 1967 press conference, he observed, "a defensive system, which prevents attack, is not a cause of the arms race but represents a factor preventing the death of people." Kosygin countered the arguments of Johnson and McNamara at Glassboro by arguing that "Defense is moral; offense is immoral."[48]

Deployment Options: The I-67 Studies

In December 1966, the Department of Defense tasked Western Electric and Bell Laboratories to construct a NIKE-X deployment model that would combine both area defense and hardsite defense capabilities. The plan, officially designated "Plan I-67 Area/Hardsite Defense," had two primary objectives: "defense against a deliberate Chinese People's Republic industrial/urban attack (countervalue); and defense against a deliberate high-level ICBM attack from the U.S.S.R. (counterforce) aimed at U.S. strategic locations."[49] In subsequent meetings with Secretary McNamara, Bell Labs officials were directed "to minimize the cost of an ABM deployment, while providing a system of high reliability."

On 5 July 1967, after a six-month study period, officials briefed McNamara on the I-67 concept. Study results were based upon three conditions: "(1) specific design threat, (2) total investment cost not to exceed 5 billion dollars, and (3) initial operating capability within 54 months of deployment decision, thereby limiting choice of equipment to NIKE-X elements." The recommended deployment consisted of: 6 Perimeter Acquisition Radars (PAR), including one in Alaska, 17 Missile Site Radars (MSR) including one in Alaska and Hawaii, 480 Spartan interceptors and 455 Sprint interceptors. Of the Sprint deployment, 325 would be allocated for the defense of Minuteman sites. At that meeting, McNamara ordered a 30-day study of the evolving threat posed by China and the ability of the system to grow accordingly. The Montgomery Committee found that the NIKE-X DEMOD I-67 "constituted an adequate base for proceeding."

Arguments favoring the deployment of an American ABM system had continued in 1967 as the Chinese threat was renewed. In June 1967, the Chinese exploded their first thermonuclear device. This achievement was followed in October, by the successful launch of a nuclear-tipped missile that struck its intended target. In December, the Chinese conducted another nuclear test.

NIKE-X Becomes SENTINEL

Fig. 2-13. The Sentinel became the new symbol in 1967.

Although still opposed to the concept, in an 18 September 1967 speech to the UPI Editors and Publishers in San Francisco, Secretary of Defense McNamara announced the government's decision to deploy a light ABM system composed of NIKE-X components.[50] This system, identified as SENTINEL, would provide protection for urban/industrial areas against possible ICBM attacks by China.[51] It would also provide a defense in the event of an accidental launch by any power. Finally, the plan included an option to defend the Air Force's MINUTEMAN missile sites.

Deployment preparations began almost immediately. The Army had 54 months to reorient the program from research and development to production and deployment. The initial deployment consisted of 6 PARs, 17 MSRs, 480 Spartans, and 220 Sprints. On 1 November 1967, the Department of Defense announced the locations of the first ten SENTINEL sites: Boston, Chicago, Grand Forks Air Force Base, North Dakota, Salt Lake City, Detroit, Seattle, Hawaii, New York, and Albany, Georgia.[52] Two weeks later, the Secretary nominated the SENTINEL System production program to the S category of the master urgency list.

Two days after the deployment announcement, the Secretary of the Army signed the SENTINEL System Manager charter. The SENTINEL System Manager reported directly of the Chief of Staff of the Army and functioned as an element of the Office of the Chief of Staff. His mission, as stated in the charter, was to "develop and, when so directed, assure the timely, effective deployment of the SENTINEL System, and provide a single point of contact within the Department of the Army for the coordination and direction of all activities pertaining to the SENTINEL System."[53] Organized in the centralized manner devised by the NIKE-X Project, the SENTINEL System Manager headed the SENTINEL System Organization that was composed of the SENTINEL System Office in Washington, D.C., the SENTINEL System Command in Huntsville, Alabama, and the Sentinel System Evaluation Agency in White Sands Missile Range, New Mexico. Since the primary focus for this new organization was to be on systems/operations of the SENTINEL system, a parallel command was established to address further R&D efforts: the Ballistic Missile Defense Research Office.[54] As part of this reorganization, the Assistant Secretary of the Army for Research and Development recommended the transfer of ARPA's

ABM research, and, in March 1968, ARPA's Project Defender transferred to the Army and the Ballistic Missile Defense Research Office.

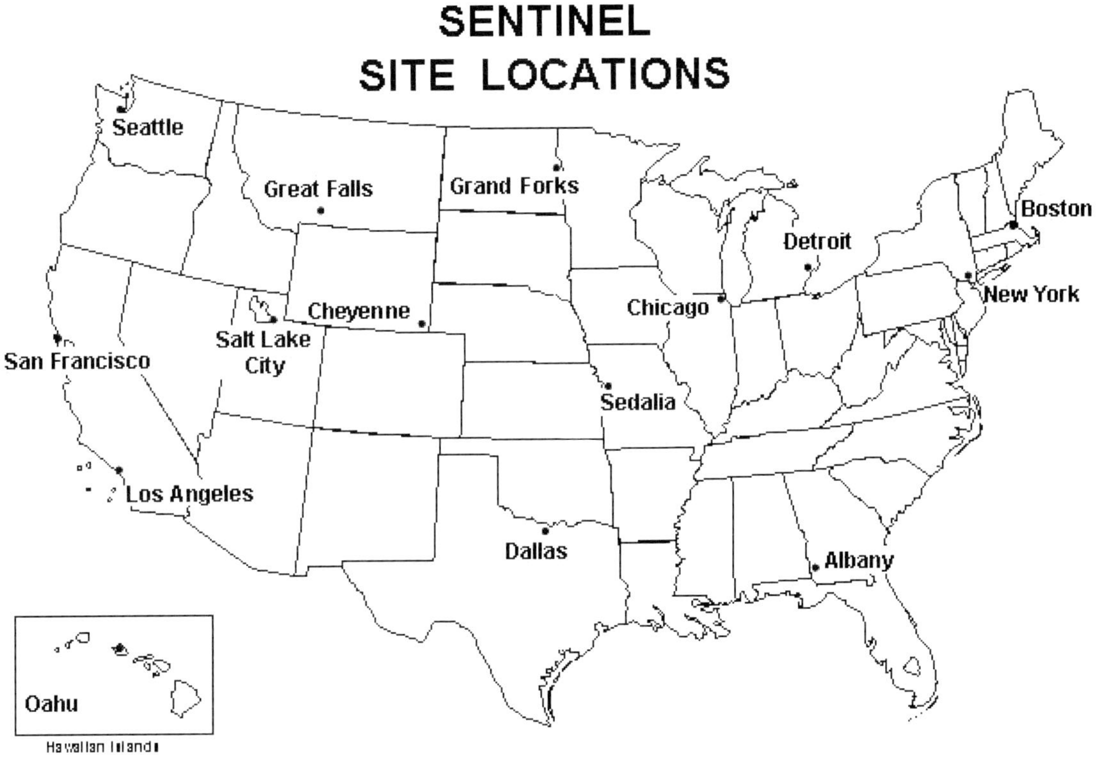

Fig. 2-14. SENTINEL sites were established to defend urban and industrial areas. The map does not show the sites in Washington, D.C. and Fairbanks, Alaska that were never publicly announced.

Even as the SENTINEL organization geared up for production in 1968, the attention of the Army and the nation was diverted.[55] The Army had an increasing role in the war in Vietnam and was less inclined to support funding for more than a thin ABM system. The public was also focused on Vietnam and becoming increasingly anti-military. Secretary McNamara and Assistant Secretary of Defense Paul Warnke announced that the Chinese program currently lagged a year behind expectations, which suggested that the need for a deployment was less urgent. Finally, the Johnson Administration signed the Treaty on the Nonproliferation of Nuclear Weapons and agreed to begin strategic arms limitation talks with the Soviet Union to limit both offensive and defensive nuclear weapons.

SENTINEL Deployment Suspended

Congress approved land acquisition near Boston for construction of the first SENTINEL site on 13 September 1968. Opposition, however, grew, and the SENTINEL sites served as rallying points for protesters. Scientists and residents raised safety concerns with the deployment of nuclear weapons near urban centers. Others argued that an ABM site in their neighborhood would make their city a target rather than protect it from attack.

The controversy continued unabated with the inauguration of the new administration. As a result, on 20 January 1969, President Richard M. Nixon took office and initiated a Department of Defense review of strategic offensive and defensive priorities. In conjunction with this order, the new Secretary of Defense Melvin Laird ordered a temporary halt to the SENTINEL deployment pending the results of this review.

NIKE-X/SENTINEL Components

On 25 September 1964, the Army Materiel Command awarded what was then the largest single contract in Army history. Western Electric Company received a $309,664,200 contract to fund research and development work and testing on the NIKE-X from October 1964 through September 1965. Although no deployment decision had been made at that time, this contract represented a definite commitment to BMD research and development. The primary focus of this initiative was on the Multifunction Array Radar, the Missile Site Radar, the Sprint missile and the Zeus/Spartan missile.

ZEUS DM/SPARTAN

The oldest component of the SENTINEL deployment was the SPARTAN. In June 1965, Deputy Director for Research and Engineering, Dr. Harold Brown, directed the Army to prepare a proposal to use a modified ZEUS missile in the barrage defense role. This research produced the SPARTAN (originally known as the ZEUS DM 15X-2).[56] Launched from an underground cell, the SPARTAN was a three-stage interceptor armed with a high-yield nuclear warhead designed to destroy ICBMs in the exoatmosphere.[57]

Building upon the knowledge gained in the ZEUS testing and incorporating the projected tactical design into the first flight missile, the development test program was comparatively short, comprising just 15 missile flights. The SPARTAN test program began on 30 March 1968 with the first launch of a SPARTAN missile from Kwajalein. The program concluded seven years later on 17 April 1975. In addition to the flight tests, twenty missiles were fired as part of the SAFEGUARD System Test Program and five production missiles were used in the Product Assurance Verification Test. These tests, conducted against intermediate range ballistic missiles, intercontinental ballistic missiles, space points, and simulated targets, demonstrated the versatility of the SPARTAN interceptor with regard to range, altitude, dynamic pressure, and third-stage ignition. The SPARTAN program made many contributions to interceptor technology. Among the innovations found in this missile are nuclear hardening technology, an ablator to protect the missile from extreme heats, a missile guidance set to "[ensure] proper operation during severe shock, vibration and noise," and a fluid-sphere gyro that increased reliability over conventional gyros.

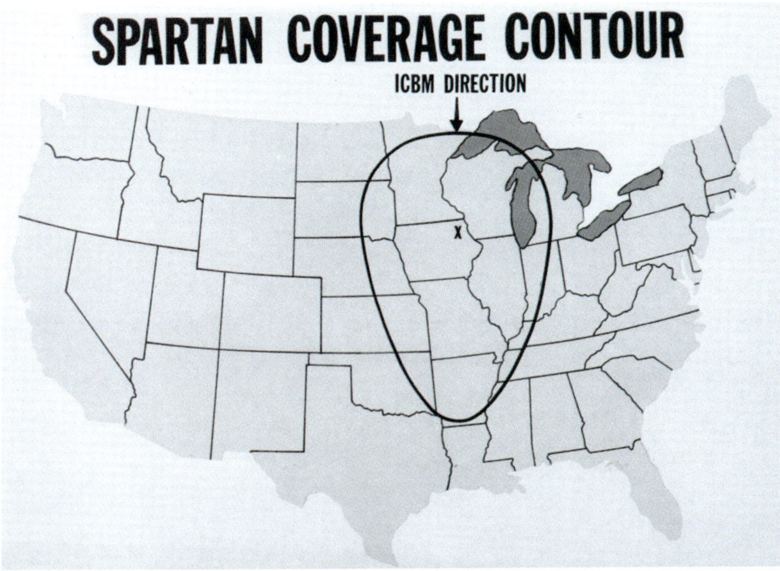

Fig. 2-15. Elliptical footprint of the area covered by a SPARTAN missile system from a hypothetical base in Iowa.

2-16. The control and guidance sections of the SPARTAN missile are loaded into a launch cell on Meck Island.

SPRINT

Developed as part of a 1962 study, the second interceptor was the two-stage, short-range SPRINT.[58] Armed with a low-yield nuclear warhead, the SPRINT was designed to maneuver within the atmosphere to intercept warheads that had survived the area defense provided by SPARTAN. This maneuverability maximized the time available for discrimination of warheads and decoys.

In order to meet anticipated deployment deadlines, the SPRINT test program began almost immediately. The 1963 SQUIRT flight tests, for example, looked at heat shield materials for the SPRINT. The first test of the SPRINT itself came in November 1965 at White Sands Missile Range. The developmental test program ended on 12 August 1970 following the forty-second SPRINT launch. In the next phase, the SPRINT was integrated with the other components of the system. In 34 tests at Kwajalein Missile Range, the SPRINT successfully intercepted 32 targets - IRBMs, ICBMs, space points, and simulated targets, well within the required miss distances.[59] The SPRINT test program concluded on 30 April 1975 with the intercept of a short-range low altitude space point.

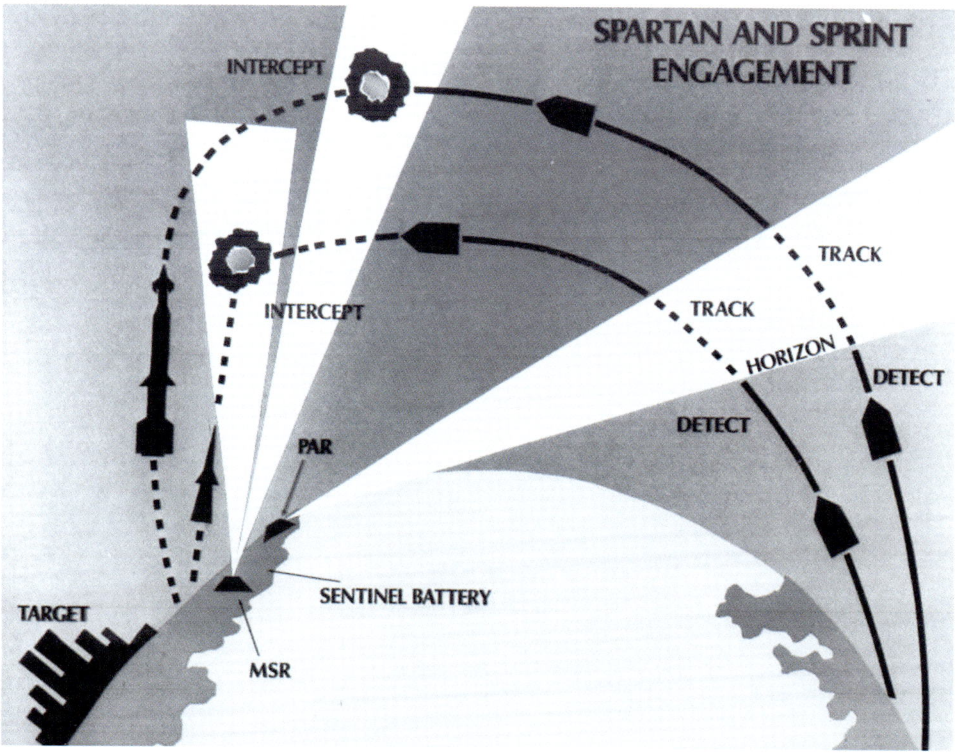

Fig 2-17. Artist's conception of the SPRINT and SPARTAN engagement concept from a coastal SENTINEL battery.

2-18. SPRINT test vehicle #2 on 4 June 1965 at White Sands Missile Range.

By all accounts, the SPRINT was an engineering marvel. Flying at tremendous speeds, "the missile's skin became hotter than the interior of its rocket motor and glowed incandescently." Its ability to accelerate to extreme velocities and maneuver within the atmosphere represented significant advances in missile technology. Among the many innovations attributed to this project were new valves to control airflow for high speed acceleration, a new high-burn rate propellant, special heat shield coatings that enabled radar tracking, missile communications that could be maintained through an ion layer, shock proofing and nuclear hardening technologies.

Fig. 2-19. SPRINT missile in launch cell for FLA-1 on 16 November 1965.

Multifunction Array Radar (MAR)/Perimeter Acquisition Radar (PAR)

In 1960, Bell Labs and the Army began to explore phase controlled scanning antenna radars.[60] Benefits from this type of system were many. These antennas had increased blast resistance capability, greater power handling capability, flexibility of beam adjustment, and the ability to combine multiple functions in one radar. In addition, the inertia-less beams in the phased array system could more easily support a "high-traffic-level-threat." The ARGMA granted authorization to develop a prototype Multifunction Array Radar (MAR), to be constructed at White Sands in June 1961. Testing of the MAR revealed the need to conduct exhaustive tests on such elements as the Traveling Wave Tube, which was incorporated into full radar in the thousands. Nevertheless, the White Sands experiments demonstrated the stability and accuracy of the system as well as the broad frequency bandwidth capability and microsecond switching. The experiments also illustrated the significant role of the centralized digital computer, which controlled all radar functions and executed large-scale, real-time data processing. The phased array radar was able to steer its beam electronically in a few millionths of a second.

Fig. 2-20. The Multifunction Array Radar was constructed at White Sands Missile Range.

Fig. 2-21. Artist's concept of the Perimeter Acquisition Radar.

As part of the proposed NIKE-X deployment, the MAR was defined as an L-band, high-power, phased array radar. Serving as the primary sensor in the system, the MAR had four functions: (1) search and verification, (2) threat evaluation, (3) target track, and (4) missile track. The long-range tracking and discrimination requirements dictated a high power requirement for the MAR and thus a separate transmitter and receiver array faces.[61] The systems costs associated with MAR and the revised deployment requirements from the I-67 study resulted in the final deployment of the less expensive Perimeter Acquisition Radar (PAR).

The configuration planned for the PAR was comparable to that of the MAR-I system tested at White Sands. This factor combined with the availability of UHF components produced a PAR design that was considered to be off-the-shelf technology. Therefore, a complete prototype of the system was never constructed. The one and only PAR system, is located near Grand Forks, North Dakota.

The PAR was a nuclear-hardened, electronically steered, phased array radar operating at ultrahigh frequency (UHF). Initially designed as an early warning system, the final version of the PAR provided tracking for the SPARTAN intercept. With a detection range of 1,000 to 2,000 miles, its primary role was to provide long-range surveillance to detect enemy missiles. In a secondary function, the PAR provided data on satellites for the North American Air (now Aerospace) Defense Command. Although other components of the I-67 deployment were terminated later, the PAR still operates as part of the Air Force's early warning system.

Missile Site Radar

In the initial design concept, the Missile Site Radar (MSR) would provide multiple tracking of defensive missiles and short-range target tracking.[62] Subsequent studies, addressing the defense of smaller cities, produced changes in the MSR design. In 1965, the role of the MSR increased to include search, acquisition and tracking of incoming targets. To achieve this mission, each MSR would be equipped with its own data processing and command and control center allowing independent operations.

The MSR was an S-band phased-array radar. Unlike the PAR, the MSR was designed to have one, two or four antenna faces, each equipped for both transmitting and receiving. A prototype MSR, constructed on Meck Island in the Kwajalein Atoll, began operations in January 1969 and participated in the full series of tests of the SPRINT and SPARTAN missiles. Designed for continuous operation, the MSR operated at a higher average power than any other radar in its frequency. In conjunction with its Missile Site Data Processor, the fully operational MSR processed "its own autonomous target data as well as data from the … PAR, discriminates between warheads and other objects, and launches and guides interceptor missiles on appropriate trajectories via an RF command guidance link to the SPARTAN and SPRINT missile farms."[63]

Fig. 2-22. Cut-away drawing of the tactical Missile Site Radar.

A New Direction: SAFEGUARD

Fig. 2-23. The Institute of Heraldry issued this shoulder sleeve insignia on 8 May 1969. It remained a symbol of this organization until the mid-1990s.

As a result of the controversy over the proposed SENTINEL deployment, the Nixon administration ordered a review of strategic offensive and defensive priorities. Secretary of Defense Laird instructed Deputy Secretary of Defense David Packard to conduct a review of the Pentagon budget and the U.S. strategic force structure. The scope of both studies encompassed the SENTINEL anti-ballistic missile system. On 20 February 1969, one month after beginning his study, Packard presented his findings to the President. Packard's presentation included four options. The first called for the deployment of a "thick" ABM system that incorporated long- and short-range missiles to protect the 25 largest cities in the nation. The second proposal was the continuation of the SENTINEL system as defined during the Johnson administration. The third option, known as I-69, called for the deployment of the SENTINEL system to protect ICBM fields rather than cities. The fourth and final proposal was not to construct an ABM system.

Directed to conduct a thorough study of all four options, Packard returned with a recommendation for the I-69 deployment. Unanimously endorsed by the Joint Chiefs of Staff, I-69 was a phased deployment plan with 12 sites across the nation. The first phase would provide protection for some of the Minuteman sites and the second would complete Minuteman coverage and "cope with more sophisticated threats."[64] An initial deployment at two sites would save $500 million in the first year, while still supporting R&D and only delaying full operating capability by 9-12 months.

When compared to the original SENTINEL system, the I-69 Modified SENTINEL would provide increased coverage of the National Command Authority, with the addition of 20 SPARTAN and 50 SPRINT missiles to protect Washington from a Soviet attack.[65] Although fewer SPRINT missiles would be deployed, they would be better distributed thus protection of the MINUTEMAN sites was virtually unchanged. The primary change came in area defense. By relocating and orienting radars to look in directions other than North, the new system would be able to provide protection against Soviet submarine-launched or fractional orbital space bombardment missiles. The reduction in radars, however, eliminated the defense of Hawaii and Alaska and resulted in some gaps in the continental United States. The modified SENTINEL found support in two distinct camps. The first held that the "deployment [filled] important gaps in the protection of our deterrent and [provided] options for meeting possible new threats ... that have not yet appeared, such as accurate Soviet MIRVs." The second saw the deployment "primarily as a useful first step toward obtaining a major damage limiting capability against the Soviet Union as well as a necessary step in maintaining an invulnerable deterrent."

Given the build-up in the Soviet offensive forces and their deployment of an ABM system, President Nixon favored the deployment of an American defensive system. On 14 March 1969, President Richard Nixon officially redirected the BMD program – creating the SAFEGUARD program. In his speech, Nixon specified three defense objectives. The first priority was "protection of our land-based retaliatory forces against a direct attack by the Soviet Union." The second was to provide a "defense of the American people against the kind of nuclear attack which Communist China is likely to be able to mount within the decade." And, the third sought to provide "Protection against the possibility of accidental attacks from any source." Nixon declared that the purpose of SAFEGUARD was "... to deny other countries the ability to impose their will on the United States and its allies under the weight of military superiority."

Components remained unchanged from the SENTINEL system but deployment concepts were redrawn. This new SAFEGUARD System was to be a phased deployment, rather than the SENTINEL's fixed deployment schedule. Construction would begin with two Phase I sites – Malmstrom Air Force Base, Montana, and Grand Forks Air Force Base, North Dakota, and a Ballistic Missile Defense Center (BMDC) at Cheyenne Mountain, Colorado. Annual reviews, by the President and the Foreign Intelligence Advisory Board, would assess the need to construct the other ten sites, based upon technical developments, threat and diplomatic context.

The nation and the Senate, however, remained divided on the ABM issue.[66] Throughout the spring and summer, opponents published reports that the SAFEGUARD system was "neither

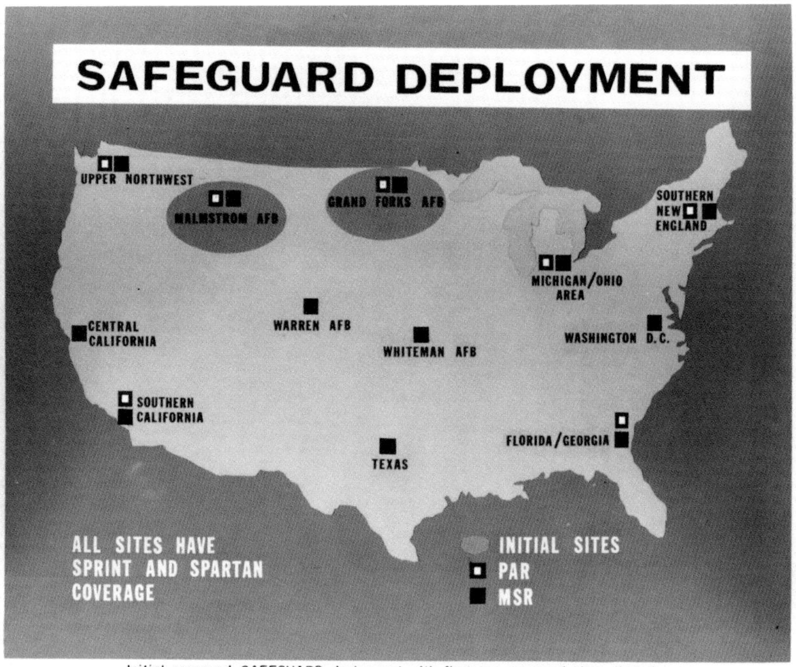

Initial proposed SAFEGUARD deployment with first two approved sites circled.

Fig. 2-24. The initial SAFEGUARD deployment, proposed in 1969, with the first two approved sites circled.

feasible nor desirable." The division in the Senate remained very close. Then on 14 July, Senator Winston Prouty (R-VT), broke with his state's leadership and spoke in favor of missile defense. As he stated, in the event of a missile attack "I discovered that there are now two grim alternatives – do nothing or push the button that unleashes our devastating nuclear fury ... SAFEGUARD provides an additional alternative, an extra button."[67] The debate would continue, culminating in the 6 August 1969 vote on the authorization bill. Although three separate amendments sought to restrict ABM to an R&D effort, the SAFEGUARD deployment authorization passed.[68]

Reorganization: SENSCOM to SAFSCOM

With the president's announcement, the Army organization established with the SENTINEL system deployment was redesignated SAFEGUARD.[69] The SAFEGUARD System Manager's mission was unchanged. As defined in the 1969 Charter, the new system manager's duties were to "develop and assure the timely, effective deployment of the SAFEGUARD Ballistic Missile Defense System, and provide a single point of contact within the Department of the Army for the coordination and direction of all activities pertaining to the SAFEGUARD BMD System."[70] The first site was to be operational within 54 months.

In order to achieve this mission, the Army began to establish supporting commands dedicated to the SAFEGUARD mission. The SENTINEL Logistics Command, a major subordinate command of the Army Materiel Command became the SAFSLOG. The U.S. Army Strategic Communications Command organized the SAFEGUARD Communications Agency at Fort Huachuca, Arizona.[71] These were followed by SAFEGUARD System Site Activation Commands at Grand Forks and Malmstrom created in 1970, and in Colorado Springs for the Ballistic Missile Defense Center and Fort Bliss, Texas for the Central Training Facility established in 1971. In April 1971, the SAFSLOG established the U.S. Army SAFEGUARD Depot Activity at Redstone Arsenal, Alabama. Following training, the depot cadre was assigned to Glasgow Air Base, Montana.[72] Other participating agencies were the U.S. Continental Army Command, U.S. Army Air Defense Command, Office of the Chief of Engineers, Office of the Surgeon General, U.S. Army Combat Developments Command, U.S. Army Security Agency and the U.S. Army Intelligence Command.

Deployment-Phase II

On 30 January 1970, President Nixon announced his decision to extend the deployment of SAFEGUARD to include a third site - Whiteman Air Force Base, Missouri. At the same time, advance preparation was to begin at five additional sites - Washington, D.C. and Warren Air Force Base, Wyoming, and unnamed sites in the Northeast, Northwest, and the Michigan-Ohio area. When submitting the proposal to Congress, Secretary Laird included additional SPRINT missiles at the first two sites. Although Laird described this proposal as "the minimum we can and must do both in cost and in system development, to fulfill the President's national security objectives," the Senate Armed Forces Committee did not approve the entire package.[73] In October 1970, funds were granted for the Whiteman site and advance preparation at Warren, but no monies were allocated for the other four sites. With this decision, the Whiteman site was designated the Fire Control Center, an intermediate command center reporting to BMDC, with Malmstrom to serve as the alternate.

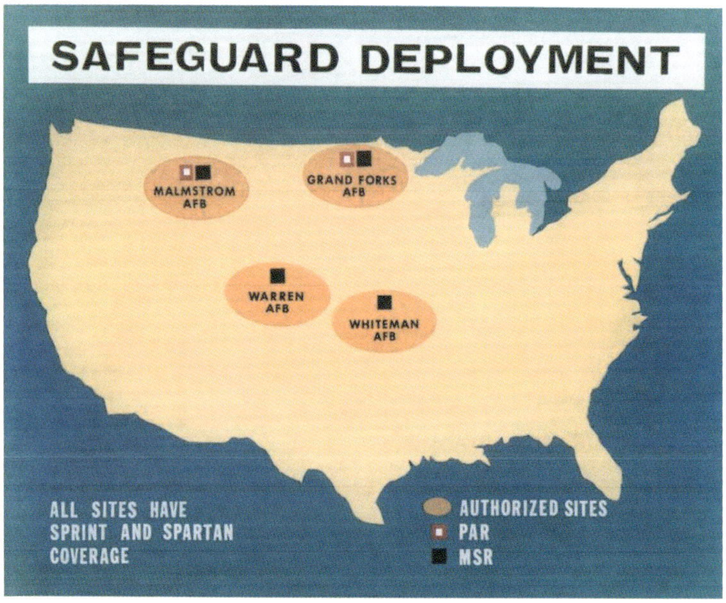

Fig. 2-25. Phase Two of the SAFEGUARD deployment with the four authorized sites circled.

After a thorough review of the SAFEGUARD program, Secretary of the Army Stanley Resor presented the President's request for FY 1972. The plan called for continued construction at the Grand Forks and Malmstrom sites, initial construction at the Whiteman site and "steps toward deployment of a fourth site at either Warren AFB or in the Washington, D.C. area."[74] Lieutenant General Alfred Starbird explained to Congress that while "a full light area defense deployment of the entire U.S. [continued] to be a desirable objective," this plan enables the Army to be responsive to the threat. The addition of Warren AFB "would allow timely deployment of additional MINUTEMAN defense and light defense of some inland strategic bomber bases and command and control center at Omaha and Colorado Springs." Meanwhile, the National Command Authority was deemed vulnerable to attack by both ICBMs and SLBMs and officials believed that a Washington deployment would add credibility to the deterrent. Before any decision was made on this controversial proposal, however, the United States and the Soviet Union signed the Anti-Ballistic Missile (ABM) Treaty, which imposed limits on the nation's ABM program.

The Anti-Ballistic Missile Treaty

Even as the Senate debated the Phase I deployment of SAFEGUARD in the summer of 1969, officials observed the system's potential use in arms negotiations. Senator Henry Jackson (D-WA) stated, "anyone who wants a successful negotiation with the Soviets to halt the further evolution of dangerous strategic armaments should be a strong proponent of the SAFEGUARD ABM." He added "the chance is promising that we could come to an agreement with the Soviet Union for a limited ABM defense on both sides ... provided that we do not foolishly throw that chance away by now scuttling our own program."[75]

Negotiations with the Soviets soon began. In November 1969, the United States and the Soviet Union initiated the Strategic Arms Limitation Talks (SALT I) to place limits on both ABM defensive systems and strategic nuclear offensive systems. Secretary of Defense Melvin Laird opposed cuts to the SAFEGUARD program or a halt to the deployment plans arguing that these would damage the American position in these talks. A new role was thus attributed to the SAFEGUARD System: that of a bargaining chip in the SALT talks.

Following two and a half years of meetings and back channel discussions, the two nations came to an agreement on ABM systems. On 26 May 1972, President Nixon and Soviet General Secretary Leonid Brezhnev signed the Anti-ballistic Missile (ABM) Treaty.[76] Ratified by the U.S. Senate on 3 August 1972, the treaty went into effect on 3 October 1972.

The ABM Treaty limited both nations to two ABM sites: one near the National Command authority, and the other near an ICBM complex. Each site could be equipped with 100 interceptors and launchers, with an additional 15 launchers located at test sites. The treaty also specified the number and type of radars that could be constructed at the different sites. While deployed systems could be upgraded and modernized both nations agreed "not to develop, test, or deploy ABM systems or components which are sea-based, air-based, space-based, or mobile land-based."[77]

Further restrictions were placed on the ABM program on 3 July 1974, when President Nixon and General Secretary Brezhnev signed a protocol to the 1972 ABM Treaty. The protocol limited each country to one ABM site, located at either the National Command Authority or an ICBM complex. With the reduction in sites, the number of interceptors and launchers permitted was also reduced from 200 to 100. This agreement went into force on 24 May 1976.

SAFEGUARD Deployed - Site Activation

Groundbreaking for the Phase I SAFEGUARD sites began with the PAR site, near Concrete, North Dakota, in April 1970. This was followed in June 1970 with site preparation at the second SAFEGUARD site at Malmstrom AFB, Montana. Construction began on the Ballistic Missile Defense Center in Cheyenne Mountain one year later in December 1971.

Fig. 2-26. Tying rebar-construction began on the nuclear hardened facilities in 1970.

Fig. 2-27. Despite weather conditions, which ranged from -40°F to 100°F, the Top-Out Pour for the Missile Site Control Building took place on 12 October 1971.

With the signing of the ABM Treaty, however, Secretary Laird advised that several actions be implemented immediately.[78] The SAFEGUARD deployment in North Dakota would remain unchanged. The Army was, however, to suspend (1) construction of the SAFEGUARD site at Malmstrom AFB, Montana, (2) all future work at the other sites, and (3) all R&D programs which are prohibited by the treaty. At the same time, Laird recommended preparing for the dismantling of the Malmstrom site, which would begin with the ratification of the treaty.[79] Finally, the Army and the SAFEGUARD System Organization were to initiate planning to cancel the 12-site deployment, but were to address the deployment of an ABM site near Washington, D.C. "on the fastest reasonable schedule." Any planning for a Washington ABM site ended with the 1974 protocol. At that time, the United States elected to maintain an ABM facility at an ICBM complex, while the Soviets continued to operate their Galosh system around Moscow.

The Stanley R. Mickelsen SAFEGUARD Complex

As mentioned above, the components of the SENTINEL and SAFEGUARD systems were identical - the Perimeter Acquisition Radar (PAR), the Missile Site Radar (MSR), SPRINT and SPARTAN missiles. The deployed system included all of these elements in various configurations near Grand Forks, North Dakota. The word complex was chosen to reflect the geographically dispersed organization. The PAR site is located near Concrete, the MSR near Nekoma, while the four Remote SPRINT Launch (RSL) sites can be found near Hampden, Dresden, Concrete, and Fairdale. On 21 June 1974, Army officially designated the SAFEGUARD tactical facilities in North Dakota the Stanley R. Mickelsen SAFEGUARD Complex (SRMSC).[80]

Fig. 2-28. Lieutenant General Stanley R. Mickelsen (1895-1966) recognized for his support of the Ballistic Missile Defense Program.

On 1 October 1974, the SRMSC achieved its equipment readiness date, with the completion of the construction and equipment installation phase. The Army officially accepted and dedicated the complex, the first new military installation since World War II. Delivery of missile warheads began in February 1975 after SRMSC received certification for its nuclear mission. The SAFEGUARD system achieved initial operating capability on 1 April 1975. On this date, operational control of 28 Sprint and 8 Spartan missiles and the "fully netted" system was turned over to the commander of the Continental Air Defense Command.

On 28 September 1975, three days ahead of schedule, the Stanley R. Mickelsen SAFEGUARD Complex reached full operational capability and became the first and only ABM system in the western world. In addition to the radars, the fully operational system included a total of 30 SPARTAN and 70 SPRINT missiles. As directed by the Secretary of Defense, SAFEGUARD was used as an educational source for the development and deployment of an ABM system.

Command and Control

The 1969 SAFEGUARD System Charter assigned to the SAFEGUARD Organization oversight of the research, development and deployment of the American ABM system. That same document specified that the ultimate user of this system would be the Army Air Defense Command (ARADCOM). Preparations began in October 1971, when the ARADCOM issued General Orders creating the first two units to man the SAFEGUARD sites - the U.S. Army SAFEGUARD Command, Grand Forks and the U.S. Army Surveillance Battalion, Grand Forks. With an authorized strength of 684, the mission of the SAFEGUARD Command was to "defend the Continental United States from a ballistic missile attack; specifically, to establish an area defense for existing retaliatory missile sites." Stationed at the PAR site, the 400 personnel of the Surveillance Battalion were "to provide long range surveillance and early warning of a ballistic missile attack against the Continental United States."[81] Both units had an organizational date of 1 September 1973.

In 1974, the Department of Defense deactivated ARADCOM. The SAFEGUARD Command, Surveillance Battalion, and Ballistic Missile Defense Center subsequently transferred to the Ballistic Missile Defense Organization effective 3 September 1974.[82] The SAFEGUARD mission was essentially distributed between the Ballistic Missile Defense Program Manager (administrative) and the Continental Air Defense Command (operational).[83] The CONAD subsequently assumed operational control of the SRMSC when it reached initial operating capability. The CONAD itself was inactivated on 30 June 1975. Responsibility for an operational SAFEGUARD subsequently rested with the Aerospace Defense Command, whose duties included U.S. air defense and aerospace surveillance.

Fig. 2-29 and 2-30. As construction progressed in North Dakota, missile testing continued on Kwajalein. The SPARTAN missile is launched from Mount Olympus (right). The final SPARTAN launch (M2-25) occurred on 17April 1975 (below).

Fig. 2-31 and 2-32. During flight, the SPRINT missile achieved such speeds that it would become incandescent. SAFEGUARD tests included salvo tests for both missiles. A SPRINT salvo launch. (right)

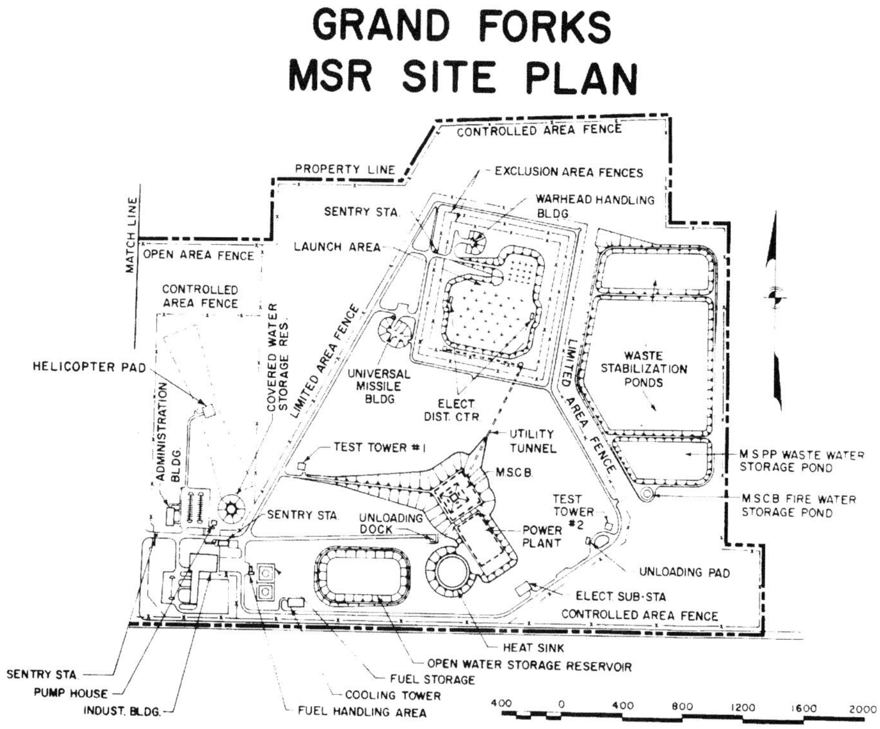

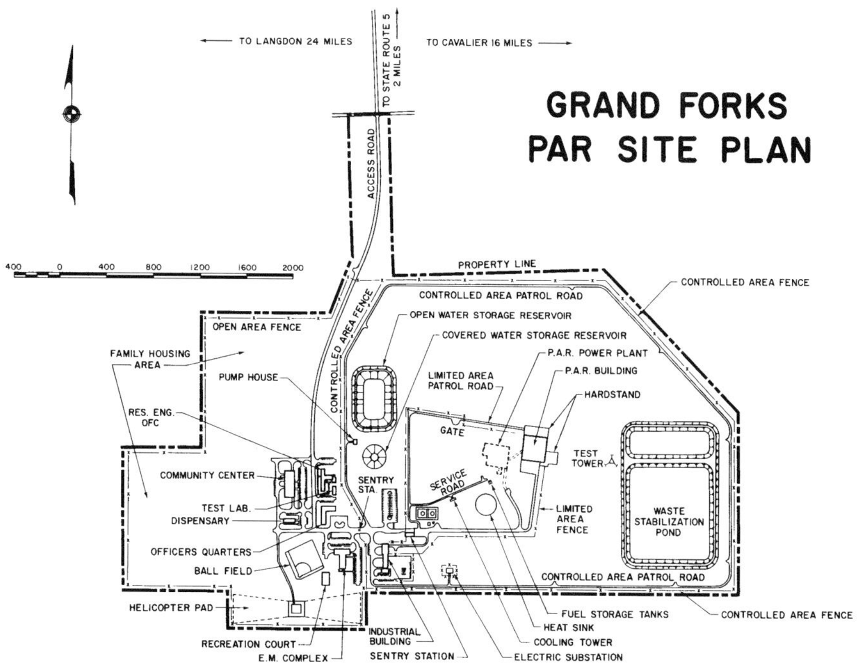

Fig. 2-33 and 2-34. Site plans for the MSR and the PAR illustrate the vastness of the complex. The MSR complex at 433 acres was much larger than the PAR, which encompassed 279 acres.

Fig. 2-35. The PAR building, the largest radar facility of its kind measures one acre at its base and has the equivalent height of a 12-story building. The PAR is now operated by the U.S. Air Force.

Fig. 2-36. Four remote SPRINT launch sites equipped with 12-16 SPRINT launch stations provided additional protection.

Fig. 2-37. The SAFEGUARD site became the western world's only operational ABM site. The MSR and its missile silos stand ready to protect the nation.

Deactivation

On the same day that the SRMSC reached full operating capability, the House Appropriations Committee recommended the deactivation of the SAFEGUARD site by the end of the fiscal year. They reasoned that the costs of operating such a system, combined with the limitations imposed by the ABM Treaty and the development of MIRVed missiles by the Soviet Union, would render the benefits from the system negligible.[84] The rest of the House concurred on 2 October 1975.

In response to the House action, Secretary of Defense James Schlesinger submitted a request to the Senate Appropriations Committee that the SAFEGUARD remain operational.[85] In his letter, Schlesinger emphasized the valuable experience to be gained from operating such a complex system. He added that the PAR system could provide a supplement to the nations' early warning system as it could detect missiles over the Arctic region. More importantly, Schlesinger argued, the United States should not terminate its ABM system without gaining some concessions from the Soviets.

While the Senate Appropriations Committee concurred with Schlesinger's arguments, the Senate as a whole, in a series of "relatively close votes," opted to discontinue operation of the SAFEGUARD complex. A strong factor in this development was Senator Edward Kennedy's amendment to the FY 1976/7T Appropriations Bill introduced on 18 November 1975. The amendment read, "Provided further that funds provided in this act for the Operation and Maintenance of the ABM Facility (other than funds provided for operation and maintenance of the PAR) may be used only for the purpose of the expeditious termination and deactivation of all operation of that facility." The amendment was incorporated into the final legislation signed on 9 February 1976.[86]

In December 1975, the Joint Chiefs of Staff ordered the SAFEGUARD Command to terminate the BMD mission. As directed by Congress, the Joint Chiefs of Staff ordered the deactivation of SAFEGUARD on 10 February 1976. At that time transmission for the Missile Site Radar and the missile launch capability were terminated. The removal of missiles and warheads, begun in December 1975, was completed in September 1976. Contractors salvaged materials from the MSR and RSL sites and later sealed the structures and silos. With the completion of this process in September 1977, the SRMSC entered caretaker status.

With the completed link to the North American Air Defense Command (NORAD) Combat Operations Computer, on 3 January 1977 the PAR became a part of the NORAD Early Warning Sensor system. The Air Force assumed tactical responsibility for the site on 22 August. The entire PAR complex was subsequently transferred to the Air Force on 1 October 1977 as the PAR Attack Characterization System. Operated by the Aerospace Defense Command, at Peterson AFB, Colorado, the PAR's space track capability became operational in December of that year.[87] The Air Force continues to operate the PAR radar system to this date. Its missions are "to provide detection and warning of a ballistic missile attack against the U.S. and Canada" and "to track thousands of man-made objects orbiting ... the earth."[88] Ultimately, the Stanley R. Mickelsen SAFEGUARD Complex remained in operation for 136 days. To date (2002), no other ABM system has been deployed by the nations of the western world.

A Tentative Return to Space

As the ballistic missile defense program became more technologically sophisticated, it appeared to be operating in many ways apart from an Army severely traumatized by its Vietnam experience. As the Army retreated into itself, making an inventory of its problems and challenges, it reverted to a more traditional form of existence. After the almost simultaneous ends of the Vietnam commitment and conscription, the Army was free to re-concentrate its efforts on becoming a professional all-volunteer force trained and prepared to fight a conventional war against a conventional enemy. Many date the Army's rebirth after Vietnam to the DePuy reorganization and formation of Forces Command and Training and Doctrine Command. This was followed by doctrinal debates between 1975 and 1982 over the significance of the Army's 1944-1945 experience in Europe and lessons derived from the 1973 Yom Kippur War as well as the DePuy-Gorman Training Revolution that created the Combat Training

Centers. The first stirrings of this revival may also be seen in 1973 with the formation of the Army Space Program Office (ASPO). ASPO was designed to carry out the Army Tactical Exploitation of National Capabilities Program (TENCAP) by serving as a liaison to other national program offices. ASPO uses the TENCAP to find ways the Army can exploit the current and future tactical potential of national intelligence programs by integrating them and their products into its tactical military decision making process as rapidly as possible. The TENCAP marshals data from various intelligence and electronic warfare communications and processing systems and integrates them to provide theater commanders and tactical units with timely targeting, battle planning and battle damage assessment information. The TENCAP systems provide for receiving, processing, exploiting, storing and disseminating combat intelligence data from national and selected theater collectors. The TENCAP owes its strengths to ASPO's relatively flat organizational structure and its adoption of the "80 percent solution;" field the product to the ultimate users and gain feedback from them. It is an activist program that helps reduce risk and cost. Engaging in an active dialogue with the end-users of its products ensures that problems are quickly identified, workable solutions are developed rapidly and the end user is always aware of product improvements. This method enabled ASPO to field a family of TENCAP systems.[89] The Army's operation of TENCAP has served as a model to the sister services.

End Notes

[1]"Major General Robert O. Hammond and Captain Frank B. Bragg, Army Space Program Fact Sheet." In addition, the Army Map Service made the maps of the moon used by the Apollo astronauts and the Army Corps of Engineers built most of NASA's launch, test and research facilities.
[2]Raines, pp. 331-332; McDougall, p. 190.
[3]McDougall, p. 221; Charles C. Bates and John F. Fuller, *America's Weather Warriors, 1814-1985* (College Station: Texas A&M University Press, 1986), pp. 174-176; Raines, p. 333; Mitchell, p. 24. See also, Army Space Reference Text, Chapter 2: Army Space History, compiled at TRADOC and available at http://www.fas.org/spp/military/docops/army/ref_text/, accessed on 4 February 2003.
[4]Mitchell, p. 24.
[5]Raines, p. 332.
[6]See Army Space Reference Text, Chapter 2: Army Space History and Mitchell, p. 24 and Raines, p. 332.
[7]Raines, p. 333; MacDougall, pp. 353-360 and 431; Bilstein, p. 56. Also, see Andrew J. Butrica (ed*.), Beyond The Ionosphere: Fifty Years of Satellite Communication* (Washington, D.C.: National Aeronautics and Space Administration, NASA History Office, 1997). Pertinent contributions to this volume include Daniel R. Headrick, "Radio Versus Cable: International Telecommunications Before Satellites," pp. 3-7, Donald C. Elder, "Something of Value: Echo and the Beginnings of Satellite Communications," pp. 33-40, Donald R. Glover, "NASA Experimental Communications Satellites, 1958-1995," pp. 51-64 and David J. Whalen, "Billion Dollar Technology: A Short Historical Overview of the Origins of Communications Satellite Technology, 1945-1965," pp. 95-127.
[8]See http://history.msfc.nasa.gov/milestones/chpt4.pdf accessed 7 February 2003.
[9]Green, et al., p. 234. While the AAF was responsible for developing missiles, Ordnance was responsible for developing and supplying warheads and "destructor sets" for these missiles.
[10]The SS-6 was described as "a single stage missile with clustered engines that developed twice the power of the American Atlas or Titan ICBMs."
[11]Walker, et.al. *Four Decades*, p. 15.
[12]The test site transferred from the U.S. Navy to the Army's NIKE-X Project Office under AMC General Orders 47, dated 19 June 1964.
[13]Beginning with the ninth test, a new advanced ("wingless") missile design was introduced.
[14]"Memorandum for Record, Subject: NIKE-ZEUS, 18 October 1960, The President's Advisory Council."
[15]Baucom, *Origins*, p. 14, 17 and Walker, et al, *Four Decades*, pp. 18-19.
[16]To fully equip 70 batteries required the production of 3,160 missiles.
[17] Baucom, *Origins*, p. 14, 17 and Walker, et al, *Four Decades*, pp. 18-19.
[18]Two other firsts were attained on 14 December 1961. A NIKE-ZEUS fired from Pt Mugu made the longest and highest flight to date and the Kwajalein Test Facility achieved its first test firing of a NIKE-ZEUS, in this case against a space point.
[19]Due to a roll-over anomaly in the last ten seconds of flight the ZEUS passed within 2 kilometers of the target. Quoted in Bell Labs, *Kwajalein Field Station*, p. 55.
[20]Bell Labs, *Program History*, pp. I-31-I-32 and Bell Labs, *ABM Research and Development at Bell Laboratories: Kwajalein Field Station* (Whippany, NJ: Bell Labs for the U.S. Army Ballistic Missile Defense Systems Command, 1975), p. 56.
[21]The thirteenth and final live fire test took place on 14 November 1963 with the successful intercept of a TITAN I boosted ICBM.
[22]Target Track Radar-4 continued to operate on Kwajalein, in support of various test programs, until February 1974.
[23]The ARPA already oversaw advanced military research on space and missiles. President Eisenhower had created ARPA in response to the Soviet's 1957 Sputnik launch.
[24]Baucom, *Origins*, pp. 15-17.
[25]*Ibid*, p. 17. With this phased array radar, ARPA "demonstrated the ability to electronically steer a radar beam in two dimensions using computes to control the beam."
[26]In 1961, the DoD gave the Air Force the mission of operating military satellites and space vehicles. The Defense Communications Agency assumed most of the Army's role in developing the communication payloads for satellite

systems. Today, the U.S. Army Satellite Communications Agency, created at Fort Monmouth in 1962, is responsible for managing ground terminals and ground support for space systems. In 1964, the SYNCOM III (Synchronized Communications Network) satellite was placed in geosynchronous orbit over the Pacific Ocean and served as a communications satellite. In 1966, the first eight satellites of the Initial Defense Communications System were launched. They were dispersed around the equator and eventually the system consisted of twenty-six of these satellites. Each satellite was able to place two points 10,000 miles apart in continuous communication. Each satellite could handle two high quality and five tactical quality voice circuits between two ground stations equipped with Army-developed steerable antennas. By 1967 the system was relaying photographs and other data from Vietnam to Hawaii and Washington. See http://www.fas.org/spp/military/docops/army/ref_text/, accessed on 4 February 2003.

[27]Baucom, *Origins*, p. 19.

[28]Army Materiel Command General Orders 4, dated 30 January 1964.

[29]The Kwajalein Test Range transferred from the control of the U.S. Navy to the Army on 19 June 1964.

[30]NIKE-X Project Office, Organization Plan for U.S. Army support of NIKE-X Deployment, dated 20 January 1965 (revised 19 March 1965). The centralized format became the basis for the execution of the SENTINEL and later the SAFEGUARD systems.

[31]General Harold K. Johnson, Memorandum to the Secretary of the Army, Subject: Revised Charter, NIKE-X Weapons System, dated 1 October 1967.

[32]Department of the Army General Orders 39, dated 14 October 1966 and DA General Orders 44, dated 10 November 1966. The NIKE-X System Manager continued to serve in the dual capacity as the Army's Chief of R&D.

[33]Unlike other COE divisions, the NIKE-X had neither civil works responsibilities nor geographical boundaries. Office of the Chief of Engineers General Orders 17 dated 9 October 1967. James Kitchens III, A History of the Huntsville Division – 15 October 1967 – 31 December 1976 (Huntsville, AL: Corps of Engineers, 1978), p. vii.

[34]Baucom, *Origins*, pp. 22-23.

[35]The focus at this time was on a terminal defense system operating at a moderate range. In this design "a battery at the most could defend one city; in the larger cities, more than one battery was required." "Stock Speech – Sentinel Anti-Ballistic Missile System" Fiscal Year 1968 Supporting Documents.

[36]Quoted in *Ibid.* This was from Senate testimony in 1965.

[37]Engineers addressed this and other criticism, that the ZEUS system could not distinguish between targets and decoys, in initiatives such as the 1967 Reentry Measurements Program

[38]"U.S. Expert Doubts Full ICBM Defense," *New York Times*, 17 February 1961, p. 9 cited in *Ibid.*

[39]Reprinted in part from Walker, *Four Decades*, p. 29.

[40]Michael Tronolone, Jr. (CPT), "More Than 50 Years of Nuclear Terror: A History of the Ballistic Missile Threat," *ADA Magazine* at http://147.71.210.21/adamag/August%202000/TBMHidst.html.

[41]*National Security Archive Electronic Briefing Book No. 26*: "The Chinese Nuclear Weapons Program: Problems of Intelligence Collection and Analysis, 1964-1972" edited by William Burr (31 March 2000) at http://www.gwu.edu/~nsarchiv/NSAEBB/NSAEBB26/index.html.

[42]Walker, *Four Decades*, p. 29.

[43]Bell Labs, *Project History*, p. I-41. Furthermore, with the addition of a larger nuclear warhead and a modified ZEUS missile, fired in a barrage mode the system "could provide large volume kill capability" and thus, "the NIKE-X was expanded to provide capability for a broad general defense of the whole continental United States."

[44]This umbrella was achieved in part by increasing the size of the nuclear warhead on a modified Zeus missile. M.D. Fagan, editor, *A History of Engineering and Science in the Bell System: National Service in War and Peace (1925-1975)* (Whippany, NJ: Bell Telephone Laboratories, 1978), p. 435.

[45]Bell Labs, *Project History*, p. I-43. Prior to this period and the development of the MSR, it was believed that the radar had to be locked onto the missile prior to launch.

[46]Soviet Defense Minister Malinovskii had proclaimed to the 22d Congress of the Communist Party in October 1961 that "the problem of destroying rockets in flight has been successfully solved." *New York Times*, 24 October 1961, p. 1 quoted in Daniel S. Papp, "From Project Thumper to SDI: The Role of Ballistic Missile Defense in U.S. Security Policy" *Aerospace Power Journal*, Winter 1987-1988, at http://www.airpower.maxwell.af.mil/airchronicles/apj/apj87/papp.html.

[47]Quoted in http://www.basicint.org/NMDhistory60.htm. See also *Foreign Relations of the United States 1964-1968*, Volume XIV – *Soviet Union*, edited by David C. Humphrey and Charles S. Sampson General Editor: David S. Patterson, Department of State Publication 10779 at http://www.state.gov/r/pa/ho/frus/johnsonlb/xiv/1400.htm.

[48]Aleksei Kosygin quoted in Baucom, *Origins*, pp. 34-35. Kosygin goes on to state "Some persons reason thus: Which is cheaper, to have offensive weapons that destroy cities and entire states or to have defensive weapons that can prevent this destruction? At present the theory is current in some places that one should develop whichever system is cheaper."

[49]Bell Labs, *Project History*, I-44.

[50]"Remarks by Secretary of Defense Robert S. McNamara Before United Press International Editors and Publishers San Francisco, California, September 18, 1967," reprinted in Annual Historical Review FY68.

[51]It was argued, in part, that "by deterring Communist China from nuclear blackmail, we hope to discourage nuclear weapon proliferation among the present non-nuclear nations of Asia." Excerpt from FY68 "Stock Speech."

[52]Three additional sites were added in May 1968 - San Francisco, Los Angeles, and Sedalia, Missouri. Two others were announced on 13 November 1968 - Warren AFB, Wyoming, and Malmstrom AFB, Montana. Two additional sites, Washington, D.C. and Fairbanks, Alaska, were never publicly announced. This deployment ensured that the combined footprint of the various batteries covered the entire nation.

[53]Charter of the SENTINEL System Manager, effective date 15 November 1967.

[54]Department of the Army General Orders 48, dated 15 November 1967. Under this GO, the Kwajalein Test Site transferred to the U.S. Army SENTINEL System Command, as a subordinate activity. The Research Office, which reported to the Army's Chief of R&D, was subsequently renamed the Advanced Ballistic Missile Defense Agency (ABMDA). The two organizations SENTINEL and ABMDA were collocated and coordinated both in Washington and in Huntsville.

[55]Baucom, *Origins*, pp. 38-39.

[56] The name was changed in January 1967 to avoid any confusion with the former NIKE-ZEUS initiatives.

[57]Bell Labs, *Project History*, Chapter 10 - "Spartan Missile Subsystem". The SPARTAN missile was 55 feet 2 inches in length with a total diameter of 43.1 inches.

[58]Bell Labs, *Project History*, Chapter 9. The SPRINT missile weighed 7600 pounds at launch and measured 27 feet in length and 4 feet 5 inches in diameter at the base.

[59]On 5 October 1973 during the first intercept test of a production Sprint test data showed that the interceptor actually hit the target.

[60]Bell Labs, *Project History*, pp. I-33-36, I-39-40, 2-16-24, and Chapter 8. Originally known as the Zeus Multifunction Array Radar (ZMAR) the name was changed to MAR with end of the ZEUS program. Compared to the mechanically steered and dish type radars used in NIKE-ZEUS, the phased array system operated at a higher speed and could be housed in nuclear hardened facilities.

[61]The MAR operated at 100MW at its peak and 2 to 3 MW average per transmitter face. Ibid, I-39.

[62]Ibid, 2-3, Chapter 7.

[63]Ibid, p. 7-5.

[64]Bell Labs, *Project History*, I-46.

[65]Memorandum from Henry Kissinger to President Nixon, 5 March 1969, Subject: Modified Sentinel System with attachment "Issues Concerning ABM Deployment." Declassified on 7 July 1998 by the National Archives. Located at http://www.gwu.edu/~nsarchiv/NSAEBB/NSAEBB36/18-04.htm.

[66]Baucom, *Origins*, pp. 43-50. Most members of the House of Representatives supported the program.

[67]Senator Winston Prouty quoted in Ibid, pp. 48-49 from the *Congressional Record*, 91st Congress, 1st Session, 14 July 1969, 115: 19420-23.

[68]Vice President Spiro Agnew cast the tie breaking vote.

[69]General Orders 18, dated 25 March 1969. These were the SAFEGUARD System Office, the SAFEGUARD Systems Command, and the SAFEGUARD System Evaluation Agency. The Kwajalein Missile Range was part of the SAFSCOM

[70]SAFEGUARD System Charter, dated 20 June 1969.

[71]USASTRATCOM General Orders 115, dated 16 June 1969.

[72]Army Materiel Command General Orders 71, dated 15 April 1971 and SAFLOG General Orders 5, dated 21 July 1971. The SAFEGUARD Logistics Command and the SAFEGUARD Depot Activity merged with the SAFEGUARD Systems Command in January 1973.
[73]Statement of Secretary of Defense Melvin R. Laird on FY71 Modified Phase II SAFEGUARD Program, 24 February 1970 located in U.S. Army SAFEGUARD System Command *Annual Historical Summary 1 July 1969 to 30 June 1970*, Volume II: *Supporting Documents* (Huntsville, AL: SAFSCOM, n.d.).
[74] Statement by Lieutenant General Alfred D. Starbird, SAFEGUARD System Manager, Office of the Chief of Staff of the Army, Department of the Army before the Department of Defense Subcommittee, Committee on Appropriations, United States Senate, First Session, 92d Congress, dated 23 April 1971 located in U.S. Army SAFEGUARD System Command *Annual Historical Summary 1 July 1970 to 30 June 1971.*, Volume II: *Supporting Documents* (Huntsville, AL: SAFSCOM, n.d.).
[75]Congressional Record, 91st Congress, 1st Session, 9 July 1969, 115; 18895 quoted in Baucom, *Origins*, p. 46.
[76]Copies of the ABM Treaty and its protocols can be found at the State Department web page at http://www.state.gov/www/global/arms/treaties/abmpage.html
[77]A 1 November 1978, Agreed Statement from the Standing Consultative Commission (SCC) sought to establish rules for the use of air defense radars at ABM test ranges and to clarify the meaning of the term "tested in an ABM mode."
[78]Memorandum from Secretary of Defense Laird to Secretary of the Army, dated 26 May 1972.
[79]Hampered by financial issues, labor disputes, and weather conditions, the Malmstrom site was not as developed as that at Grand Forks. At termination, construction was approximately eight percent complete. By July 1974, most of the Montana site had been removed or was buried beneath the earth. James Kitchens, *A History of the Huntsville Division, U.S. Army Corps of Engineers, 1967-1976* (Huntsville: U.S. Army Corps of Engineers, 1978).
[80]General Order 21, dated 21 June 1974. Lieutenant General R. Mickelsen, former Commanding General of the U.S. Army Air Defense Command, had in 1955 urged the Army "to make the ABM weapon a 'firm Army requirement' and assure its development at an optimum rate." Stephen Moeller, "Vigilant and Invincible: United States Army Air Defense Command," *Air Defense Artillery* May-June 1995:33.
[81]ARADCOM General Orders 354 and 355, dated 22 October 1971. The Command's 684 authorized employees were to be 62 officers, 22 warrant officers, 432 enlisted and 168 civilians. The 400 PAR personnel were to be as follows 41 officers, 14 warrant officers, 109 enlisted and 136 civilians.
[82]General Orders 30, dated 29 August 1974. In a separate reorganization, the SAFEGUARD System Organization became the Ballistic Missile Defense Organization in 1974.
[83]Operational control did not encompass such matters as administration, logistics, discipline, internal organization or unit training. These remained with the Ballistic Missile Defense Program Manager as the component commander for BMD forces.
[84]While multiple warheads were not a new phenomenon, MIRVed missiles were equipped with multiple independently targetable reentry vehicles.
[85]Baucom, *Origins*, p. 96.
[86]Public Law 94-212, dated 9 February 1976.
[87]Now known as Cavalier Air Force Station, the PAR was part of the Air Force's Aerospace Defense Command (1 May 1977 to 30 November 1979); the Strategic Air Command (1 December 1979 – 30 April 1983) and the Space Command (1 May 1983 - the present).
[88]"Cavalier Air Station Gets New Name," *Cavalier Chronicle*, 2 March 2000 electronic copy located at http://srmsc.org/ext0020.html.
[89]Sharon Carvalho, "The '80 percent solution' Concept Gets It to Field, Gets Users' Feedback," *The Eagle* March 1998:2.

Chapter 3
Communications, Sensors, Maintaining Interest in Missile Defense, and the Strategic Defense Initiative, 1970-1989

The Slow Revival of Interest in Space

Although handicapped by the policy changes of the late 1950s and early 1960s that centralized control of space, intelligence and communications programs, and wracked by the consequences of the Vietnam War, the Army maintained an interest in space and increased its stake in ballistic missile defense. Since the Army was the service most advanced in the use of space at the time, it lost the most during the reallocation of roles and missions. These institutional changes affected the ways the Army exploited space. One of the most dramatic changes occurred in 1961, when Secretary of Defense Robert S. McNamara's DoD Directive 5160.32 Development of Space Systems removed the Army from the business of launching satellites and conducting DoD satellite reconnaissance efforts. While the directive centralized control, supervision and coordination of satellite development and operations, it allowed the Army to continue its work on communications satellites and ground stations. Through the 1980s, the Army used space to provide theater commanders with long-haul communications systems.

Change in the Army's interest in space began when Secretary of Defense Melvin H. Laird modified McNamara's management and decision-making practices. The Nixon Administration appointed a Blue Ribbon Defense Panel that made more than 100 recommendations about the department's organization and functions in a 1970 report. A number of the proposals were implemented while Laird was Defense Secretary. He did not completely end McNamara's system, but described his policy as "participatory management." While retaining policymaking decision authority for himself and his deputy secretary, the Joint Chiefs of Staff and the Services became responsible for detailed force planning, while the individual military departments gained more responsibility for managing their own development and procurement programs. The policy gained the senior military leadership's cooperation in reducing the defense budget and the size of the military establishment. The Army saw immediate advantage to this new system when the secretary revised DoD Directive 5160.32 in September 1970, changing the division of DoD satellite development responsibilities three ways. First, each service conducted research and received approval to develop "unique battlefield and ocean surveillance, communication, navigation, meteorological mapping, charting and geodesy satellites." The Air Force still performed research and development and produced systems for launch support, launch vehicles, warning and surveillance satellites to detect enemy nuclear capabilities, and orbital support operations. Finally, the DoD Director of Research and Development became the focal point for space technology and systems to prevent duplication, minimize technical risk and cost, and ensure multiple service needs were met. This new policy allowed the Army to slowly return to space.

SAFEGUARD-The Next Generation: Hardsite Defense

In the post-Vietnam period, the Army experienced a renewed emphasis on professionalism and modernization. As part of this renewal, the Army continued to concentrate on ballistic missile defense. Beginning in 1969, as the Army pursued deployment of the SAFEGUARD System, the SAFEGUARD Systems Command (SAFSCOM) received orders from the Deputy Secretary of Defense to address the next generation of BMD development. In February 1971, SAFSCOM established the Hardsite Defense (HSD) Project Office, a prototype demonstration program.[1] As described by Secretary of Defense James Schlesinger, Site Defense "would give us the option to defend our Minuteman force against a Soviet ballistic missile attack ... or in the event that an acceptable ... limitation of strategic offensive arms cannot be achieved ... it would give us the option to deploy a more advanced ABM system."[2]

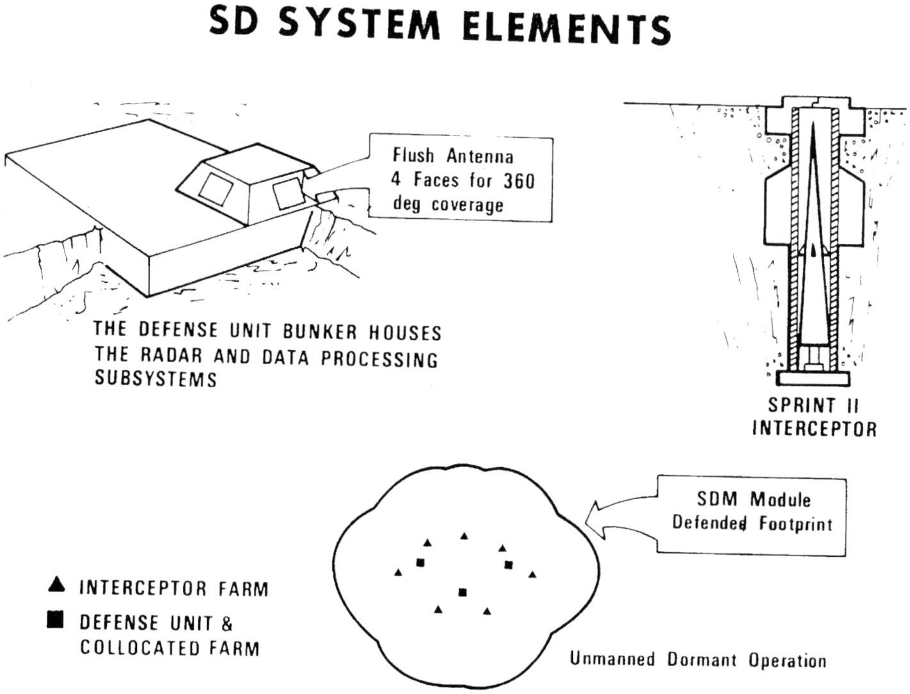

Fig. 3-1. The artist's drawing illustrates the system elements of the Site Defense concept and its proposed deployment within the defense unit.

Under the revised charter, the SAFEGUARD System Manager had two distinct missions, to "develop and assure timely, effective deployment of the SAFEGUARD Ballistic Missile Defense system and [to] plan and carry out a Hardsite prototype demonstration program."[3] The program included a deployment option, but no decisions were taken at this point. The resulting concept called for phased array radar, an interceptor, and commercial data processing equipment to be deployed in groups to protect MINUTEMAN sites and each other.[4] The new radar, smaller and built to a greater degree of nuclear hardening, would be more resistant to nuclear effects. The new interceptor, the SPRINT II boasted greater accuracy and maneuverability and improved silo hardening. With these innovations, the HSD-augmented SAFEGUARD would be capable of handling a larger, more sophisticated threat than SAFEGUARD. In February 1972, the Secretary of the Army announced the award of the Site Defense Prototype Demonstration Contract.[5] A demonstration program for the prototype was planned for Kwajalein in 1976.

Fig. 3-2. Computer systems play vital roles in missile defense. One such system for Site Defense was this CDC-76-computer, which was operated and maintained by the control consoles in the foreground-May 1974.

Everything changed in 1974, when a congressional ban on prototyping limited site defense to research and development at the subsystem and component levels. As the Site Defense System Fact Book explained, the project office instituted a new two-phased approach – Validation and Integration.[6] The validation phase focused on upgrading key technical elements, e.g. bulk filtering, discrimination, software development, operation in a nuclear environment and dormancy. Integration ensured that the Site Defense design is "abreast of newly emerging offensive and defensive capabilities."[7] Previously planned missile intercepts for the SPRINT II were cancelled. With these changes, the Site Defense Project became the Systems Technology Program.[8]

The Ballistic Missile Defense Organization

That same year, the Secretary of the Army announced that all ballistic missile defense efforts would be realigned under one organization and, on 20 May 1974, the SAFEGUARD System Organization was redesignated the Ballistic Missile Defense Organization (BMDO).[9] The same General Order established the Ballistic Missile Defense Advanced Technology Center (BMDATC). The BMDATC, a field operating agency under the BMD Program Manager replaced the ABMDA.[10] Despite this reorganization, the BMD Program Manger remained principal assistant and staff advisor to the Office of the Chief of Staff of the Army. The mission for the new organization was comparable to that of SAFEGUARD. The Secretary of the Army tasked the BMDO: (1) to deploy and operate the SAFEGUARD System; (2) to execute the Site Defense program; (3) to conduct research and development in advanced BMD technology; and (4) to manage the Kwajalein Missile Range (KMR) as a National Range.

On 1 March 1975, the BMDATC received its own mission, to "formulate and execute approved BMD programs of exploratory and advanced development in BMD technology within the guidance and direction of the BMD Program Manager."[11] In addition, it would "(a) provide the advanced technology foundation for improving ballistic missile defense capability; (b) provide a measure of the BMD technology art to avoid technological surprise by an adversary; and (c) assist in the development and assessment of future U.S. strategic offensive systems." Specifically the BMDATC focused on five technology areas: discrimination, data processing. optics, radar, and interceptors.

With SAFEGUARD's inactivation, the BMDO experienced many changes. The BMD Program Manager recommended that the PM position transfer from Washington to Huntsville. The Washington-based element would be streamlined and many functions transferred to Huntsville and BMDSCOM. Emphasis was placed on the continued operation of the BMDSCOM and the BMDATC. The reorganization of BMDSCOM, conducted in conjunction with a reduction-in-force, was completed on 10 December 1976.[12]

Ballistic Missile Defense in the 1970s

The period between 1974 and 1983 began with declining interest in BMD initiatives as demonstrated by the decision to cancel the SAFEGUARD program and to redirect the Site Defense program. The decision was also made to move the Homing Overlay Experiment (HOE) into "high gear" and accelerate development of a defense for U.S. ICBMs.[13] Although no longer in the forefront of military proposals, the BMD effort was not totally abandoned. In 1976, Secretary of Defense Schlesinger testified to the Senate that "we must continue a BMD effort of significant breadth and depth to ensure that we can keep pace with the continuing Soviet BMD efforts and improvements." He added, "Our continued effort is essential not only as a hedge against a sudden abrogation of the ABM Treaty, but also because our demonstrable competence in this field will continue to motivate the Soviet Union to negotiate additional limits on strategic arms."[14]

Fig. 3-3. Data collected during reentry measurement studies are important to a successful intercept. Reentry vehicles blaze through the skies over Kwajalein.

Two years later, amid growing concerns about Soviet missile capabilities, the Deputy Under Secretary of Defense Research & Engineering (Strategic and Space Systems) placed specific emphasis on "near-term defense concepts and technologies applicable to defense of our land-based missile forces in the 1980s." At the same time, Secretary of Defense Harold Brown, in his report to Congress, observed, "An aggressive BMD R&D program is vital to this nation's interest." Brown added that the technological base developed by the Systems Technology and Advanced Technology programs provided cost-effective alternatives for "maintaining survivability of our strategic retaliatory elements in the ICBM threat environment."[15]

The BMDO subsequently received orders to conduct a Minuteman Defense II study. While briefing the U.S. Congressional Budget Analysts, the BMD Program Manager explained, "The restrictions on deployment previously were thought to be such that a treaty-limited deployment would not be worthwhile. However, due to advancing technology, this is no longer true and a limited deployment can be useful." Meanwhile, BMDO summarized their program as an effort "to provide a hedge against the strategic uncertainties associated with the ballistic missile threat to the United States." They further explained that BMD research and development served "to keep the U.S. abreast of the potentialities of new component and system technologies to guard against Soviet technological surprise or a perception on their part of sufficient technological advantage to suggest the attractiveness of abrupt ABM Treaty abrogation."[16]

Although BMDO was limited by funding constraints and the Congressional ban on prototyping that remained in effect until 1981, it did achieve a number of breakthroughs in these years.[17] The two primary elements of the BMD program, the Advanced Technology Program (ATP) and the Systems Technology Program (STP), worked together to develop and evaluate innovative means to address BMD. As Major General Robert Creel, the BMD PM, explained, "From the ATP we want a futuristic, imaginative search for better ways to do the BMD job, while from the STP we require an objective evaluation of systems applications of emerging components and concepts."[18] In addition to traditional interceptors and sensors, BMDO scientists and engineers explored and validated new technologies to achieve its missions. Some of these instrumental initiatives are examined below.

Systems Technology Radar

Developed as part of the Site Defense Program, the Systems Technology Radar (STR) was a key element of the Systems Technology Test Facility (STTF) constructed on Meck Island in Kwajalein Atoll. Installation of the STTF began in May 1976 with data processing computers. The STR arrived on Kwajalein in September 1976. The full STTF achieved initial operation in November and full operating capability on 1 June 1977. Testing began immediately with planned Air Force ICBM tests.[19] The system demonstrated its tracking capability in June 1977. On 3 September 1977, the STTF successfully accomplished bulk filtering of low velocity tank fragments entering the search volume, and gathered discrimination data on reentry vehicles on 13 September.[20]

In 1978, officials reoriented the Systems Technology Program to emphasize the application of more mature technologies developed by the BMDATC. The STP discontinued system performance analysis of the terminal defense system to fund these new experiments/systems analyses. The exception was the STR program that demonstrated the STTF's ability to perform specific critical functions such as bulk filtering, track in reentry clutter and discrimination and those that established critical functions and performance levels for other system functions.[21] Verification testing, concluding in September, demonstrated that the lower-level and subsystem radar performance met and exceeded most baseline specifications. The STTF completed 50 tests of the Site Defense Radar and data processors in September 1980.

The STR, designed to provide data in terminal, low-altitude and midcourse operations, represented a major improvement over the SAFEGUARD Missile Site Radar.[22] The unmanned system was equipped with fully automatic electronic beam steering capable of transmitting thousands of beams per second. The STR also employed a "more versatile transmitted waveform in combinations with a more advanced signal processors [which permitted] better discrimination."[23] Given these advances, the STR could serve as a stand-alone radar system for defense of the Minuteman missiles. In addition, the radar was an important element in the underlay of the proposed layered defense concept of the late 1970s/early 1980s.

Designating Optical Tracker (DOT)

Recognizing the inherent limitations of ground-based radars, BMDO engineers explored the feasibility of airborne/spaceborne sensors to conduct target discrimination. One product of this investigation was the Designating Optical Tracker (DOT) program. The DOT, established in 1975, sought to determine the feasibility of a probe-launched long-wave infrared (LWIR) sensor to detect and track incoming ICBM warheads.

Fig. 3-4. The Designating Optical Tracker (DOT) enjoyed a perfect test record and demonstrated the viability of the onboard infrared optics technology. The DOT on its launch pad at the Kwajalein Missile Range.

Fig. 3-5. The DOT sensor package recovered from the Pacific to be prepared for the next test.

The DOT was an infrared telescope. The probe was launched by a Castor I rocket above the atmosphere in a series of tests conducted at Roi-Namur in the Kwajalein Atoll. Following each test, the telescope parachuted into the ocean to be recovered, refurbished and reused. In five consecutive tests between 1978 and 1982, the DOT demonstrated that a LWIR sensor could discriminate, designate, and track a reentry vehicle. The tests also collected signature data on targets and debris and provided research data on the impact of radar, celestial backgrounds, targets, optical chaff penetration aids, and atmospheric conditions on LWIR sensors.[24] The DOT set the standard for future LWIR technology.

Airborne Optical Sensors

As discrimination had always been a concern for researchers, the BMDO conducted several data collection and sensors projects in the 1970s and early 1980s. Concurrent with the DOT, researchers theorized that airborne sensors could provide an expanded tracking and discrimination capability. The Optical Aircraft Measurements Program (OAMP) was the first such experiment. Comparable in size to a compact car, the OAMP sensor was mounted into a modified Boeing 707 aircraft. The sensor recorded data in three infrared bands, with the first telescope equipped with simultaneous spectral and radiometric measurement capabilities. During the two-year period of 1982-3, the OAMP collected signature data on Soviet reentry vehicles and missile launches.[25]

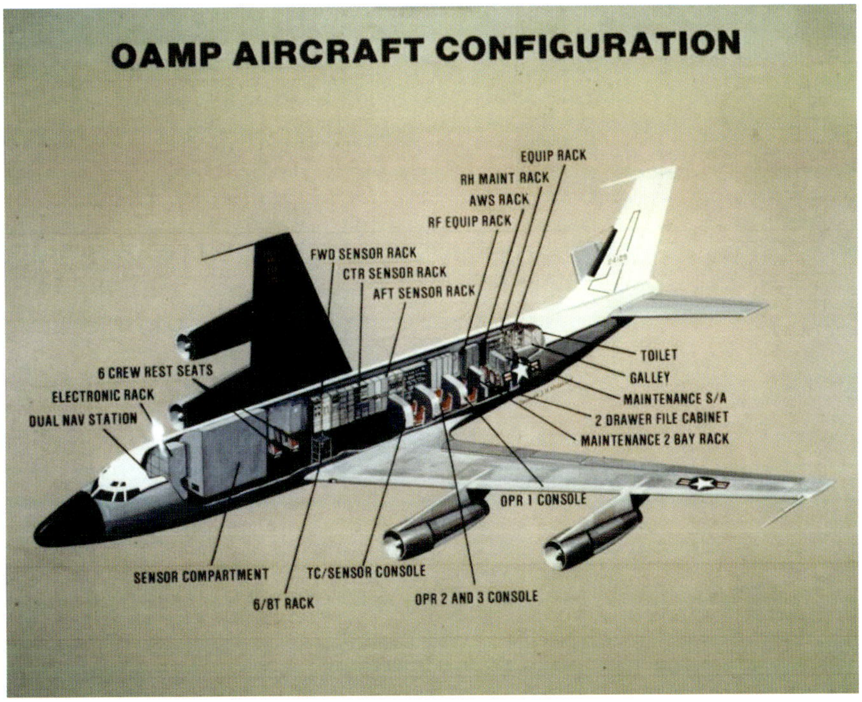

Fig. 3-6. The Optical Aircraft Measurements Program was an airborne sensor installed into a U.S. Air Force aircraft.

Building upon the DOT and OAMP programs, the Systems Technology Program received permission in October 1983 from Mr. James Ambrose, Under Secretary of the Army, to proceed with a new initiative: the Airborne Optical Adjunct (AOA). The BMDO created the AOA "to experimentally investigate the technical feasibility of using airborne optical sensors for detecting, tracking and discriminating ballistic missile reentry vehicles and handing over trajectory data to ground-based radars."[26]

To address the potential threat, the AOA program called for two OAMP sensors and a data processing unit to be installed in a C-135B aircraft. Funding restrictions later reduced the program to one sensor aboard a modified to an experimental Boeing 767.[27] Nevertheless, the Army awarded an initial five-year contract to Boeing Aerospace in July 1984. Subsequently renamed the Airborne Surveillance Testbed (AST), the optical sensor was the first BMD project to be incorporated into the next generation ABM initiative and it continues to play an important role in missile test programs and exercises.

Advanced Research Center

Beginning in the early 1970s, the Advanced Research Center (ARC) provided the BMD community with an integrated and centralized data processing capability specially designed to meet the software and hardware needs of ballistic missile defense. In FY75, the ARC had four missions:[28] developing methodologies for designing and implementing the massive real-time BMD software; testing large, advanced data-processing systems for applicability to BMD; testing validating and demonstrating software processes for specific BMD applications (simulations); and conducting systems analysis studies for new technical requirements. In many respects the mission remains unchanged as the ARC continues to provide a cost effective focal point for BMD data processing research and simulation.

Fig. 3-7. The Advanced Research Center's simulations capabilities have applications to all of the services. Soldiers from this command practice battle management techniques.

Work conducted at the ARC made great advances in data processing technology. Normal computer performance in the mid-1970s, for example, was measured at 20-30 millions of instructions per second (MIPS), an improvement over the SAFEGUARD systems.[29] During the mid-1970s, however, the ARC was testing the parallel element processing ensemble (PEPE) with an operation rate of 800 MIPS. Engineers designed the PEPE "to handle high correlation and high computation loads, as well as a high file-search load" to meet BMD requirements to include tracking and discrimination of warheads and decoys, controlling radar beams, etc. Other concepts under review during this time period included distributed data processing, micro-processing and missile borne data processors.

Directed Energy Research

Along with the various forms of radars and sensors, the Army was also experimenting with lasers. The concept of directed energy weapons has existed since ancient times.[30] By definition a directed energy weapon "generates radiant energy or energetic particles, focuses them into narrow beams and points and delivers them to targets." The source of this energy can be chemical fuel, electrical power, intense sources of heat, or high explosives. Meanwhile, "the beams consist of charged or neutral atomic particles or electromagnetic radiation and are capable of near-instantaneous delivery to targets."[31] During the 1970s, the Ballistic Missile Defense Advanced Technology Center explored two different directed energy technologies: neutral particle beams and high energy lasers.

The Advanced Research Projects Agency (ARPA) began initial research exploring military applications of directed energy weapons in the late 1950s. ARPA initiated the particle beam weapon program in 1958. The weapon would direct "a beam of atomic particles (electrons or protons) toward a target at or near the speed of light and could rapidly redirect its beam of particles among a multitude of targets." [32] Given the nature of the light beam, the Neutral Particle Beam (NPB) can penetrate clouds and is not adversely affected by poor weather conditions. In addition, the NPB can also penetrate the exterior body of the target and thus destroy the electronics and circuitry which control it.

The Army/BMDATC was the principal developer of particle beam technology throughout the 1970s. The two primary efforts were the exoatmospheric NPB accelerator program and the collective ion accelerator experiment. In 1980, the Defense Science Board recommended that the NPB remain a technology program. It transferred to the Defense Advanced Research Projects Agency (DARPA) in 1981, when they became the manager of all NPB programs. The BMDO, however, continued to oversee contracts and monitor the DARPA-funded programs.

Research in laser technology began with ARPA in 1962 as studies addressed the effects of high energy lasers in BMD. In the 1970s, the BMDATC addressed several critical technology issues related to chemical and high energy laser weapons. Included among these issues were producing high-intensity, high-quality ion sources, neutralizing particles in a high energy charged beam, developing high energy laser beams for ballistic missile defense, and developing an adequate data base for target-beam interactions.[33] By the end of the decade, researchers had demonstrated that lasers could work in conjunction with pointing and tracking devices to form an effective weapons system.

In the FY76 Defense Authorization Bill, the Department of Defense recommended White Sands Missile Range as a suitable location for a high energy laser range testing. In October 1978, it was reported that "Congressional officials are pressing the Army to begin space-based laser weapons development."[34] In response, the Army began to change the program from endoatmospheric tactical laser weapons application to conceptual designs for space-based laser weapons. Then, in 1980, following policy established by President Jimmy Carter, Secretary of Defense Harold Brown directed the services to explore all potential uses of lasers but to emphasize the use of lasers in space.[35] For the BMD organization, the focus became the potential use of lasers to destroy ballistic missiles in the boost or midcourse phase of their flight, before the deployment of the reentry vehicles.[36]

Low Altitude Defense (LoAD)/SENTRY

In the 1970s, as improved Soviet technology increasingly threatened existing intercontinental ballistic missiles, the Air Force developed a new ICBM, the MX or Peacekeeper missile. To improve its survivability, the Air Force explored a number of basing options, including mobile systems. It was the Army's role, in particular the BMD organization, to develop a suitable ABM system to protect the ICBMs. The response was the Low Altitude Defense (LoAD) system. In 1977, the BMDSCOM chartered a six month study entitled "Mobile ICBM Defense Concept Analysis" to review deployment issues.[37] The study team determined the circumstances under which LoAD could improve the survivability of the MX, assessed the feasibility of silo-based ICBMs, examined candidate MX defense concepts, and identified actions required by BMD to achieve a mid-1980s deployment.

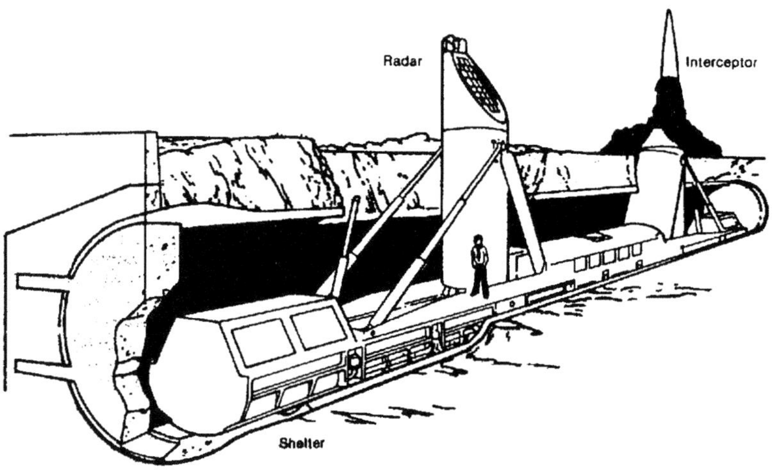

Artist's Concept of LoAD Unit for Defense of MX in Multiple Protective Shelters

Fig. 3-8. Designed to protect the Air Force's MX missiles, the LoAD/SENTRY was to be a mobile defensive interceptor.

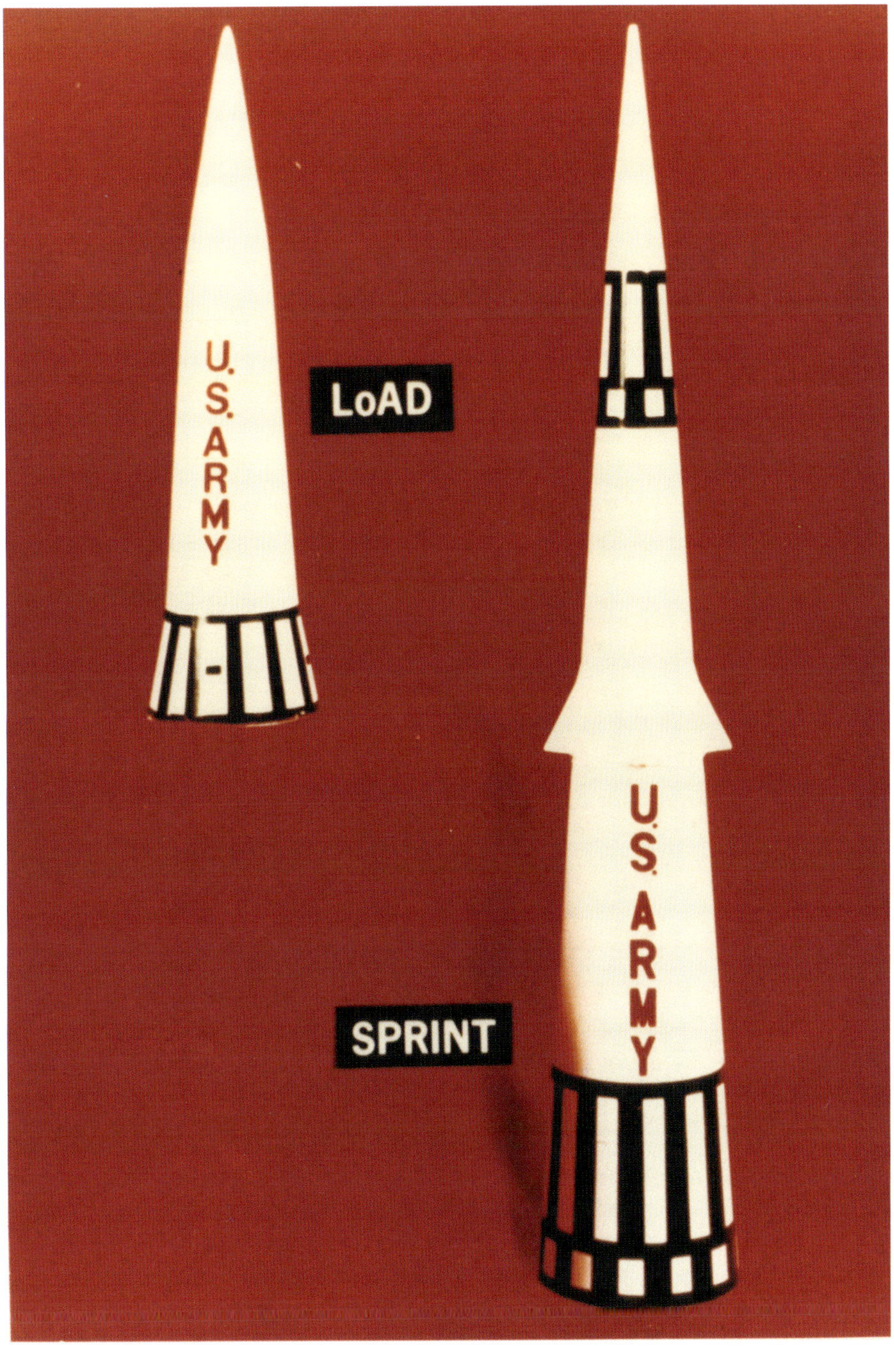

Fig. 3-9. This drawing illustrates the differences between the SPRINT, developed in the 1960s, and the smaller LoAD, a product of the 1970s.

At the same time, the Ballistic Missile Defense Organization began to review a new layered defense system described as being divided into an overlay and an underlay. The overlay focused on exoatmospheric interceptions employing a non-nuclear interceptor equipped with a number of small kill vehicles. This program was still in the early stages of development. The underlay, however, was an improved Site Defense system that engaged targets in the endoatmosphere that had escaped the initial defense layer.[38] While the overlay interceptor was not yet fully defined, the LoAD system was identified as a suitable underlay.

On 22 January 1977, the BMDSCOM chartered the Low Altitude Defense (LoAD) system. It was designed to operate at altitudes under 50,000 feet. The system would incorporate a series of radars, distributed data processors and nuclear-tipped interceptors.[39] Its size and design would complement any of the proposed deployments for the MX ICBM. In 1979, the Carter Administration selected a mobile basing mode for MX. The design called for 200 Peacekeeper missiles to be stationed in 4600 hardened shelters.[40] Periodically the ICBMs and decoys would be moved among the various shelters in the cluster. Similarly the LoAD battery, consisting of three missiles and a radar system, would be moved among the shelters in an underground system.

Congress lifted the prototyping ban in 1981. The new administration, however, did not concur with the mobile basing system. In 1982, Secretary of Defense Caspar Weinberger issued a BMD Program Directive to support all MX basing options, with particular concentration on a closely based system.[41] The directive also called for the development of a non-nuclear endoatmospheric weapon. Based on this guidance, the BMDO planned to convert the LoAD to a non-nuclear interceptor and renamed the program SENTRY.

The next year, the BMDSCOM terminated the SENTRY program. One factor was the ABM Treaty that would have placed restrictions on a LoAD battery and prohibited deployment of a mobile system.[42] In addition, funding constraints coupled with the decision to deploy the Peacekeeper in existing silos contributed to this decision.

Homing Overlay Experiment (HOE)

With the advances made in infrared sensors and computer technology, the Army was ready to address kinetic energy intercepts. The first such effort was the Homing Overlay Experiment (HOE) Task Force, charted by the Systems Technology Program in March 1977. There was a great deal of interest in this endeavor; one of the proposed elements to the overlay of the layered defense system was the HOE.

The two-phased HOE effort began with technology verification, followed by the flight demonstration program scheduled for 1982-1983. The BMD engineers designed the experiment to resolve specific development issues. These were Search, Acquisition, and Detection; Discrimination (including scan to scan correlation); Designation; Homing Guidance Accuracy; D^3 and Track in the Natural and Induced Environments; and, Sensor to Sensor Handover/Correlation).[43] The overall objective was to demonstrate the exoatmospheric intercept of a mock ICBM reentry vehicle using infrared homing sensors and non-nuclear kill vehicle and thereby reduce the lead-time required to deploy an exoatmospheric non-nuclear interceptor.[44]

Fig. 3-10. Noted for its distinctive web (insert) designed to capture an RV, the Homing Overlay Experiment achieved the first kinetic energy intercept colliding with its target at a speed of 20,000 mph.

Launched by two Minuteman motor stages, the HOE kill vehicle consisted of a computer, a long wavelength infrared optical sensor package for guidance, and a unique kill device.[45] When the missile reached a point above the atmosphere, a sensor and computer on-board the MINUTEMAN launch rocket would locate and track the reentry vehicle.[46] The computer would then relay tracking data to the intercept vehicle. As the target neared, the kill vehicle would be launched and using its own infrared sensors and computer would home in on the target. In the final stage before intercept, the kill vehicle would unfurl the spokes of a 13-foot radial net that would capture the reentry vehicle.

On 10 June 1984, in its fourth and final flight test, the HOE successfully completed the first kinetic kill intercept.[47] Launched from Meck Island, the HOE kill vehicle intercepted a mock ICBM reentry vehicle over the Pacific Ocean at an altitude greater than 100 miles. In this test, "the HOE and the warhead closed at more than 15,000 feet per second, and telemetry data shows that they smashed into each other nose to nose."[48] As Principal Deputy Assistant Secretary of the Army Amoretto Hoeber explained, "We tried to hit a bullet with a bullet and it worked."[49] Ultimately, the evolution from nuclear to kinetic energy intercepts, represented by the HOE system, was "the first major revolution in ballistic missile defenses since the United States began BMD research in the 1940s."[50]

Flexible Lightweight Agile Guided Experiment (FLAGE)

The next non-nuclear kill technology achievement came in the same year when the Small Radar Homing Intercept Technology (SRHIT) completed its first flight test.[51] The SRHIT program sought to assess guidance and control technology to develop a missile capable of intercepting small high-velocity targets (tactical ballistic missiles) at low altitudes. Subsequently renamed the Flexible Lightweight Agile Experiment (FLAGE),[52] the program's mission was to demonstrate an accurate endoatmospheric interceptor, quantify the achievable miss distance in low atmosphere, and validate a 6-degree of freedom system simulation for endoatmospheric nonnuclear kill.[53]

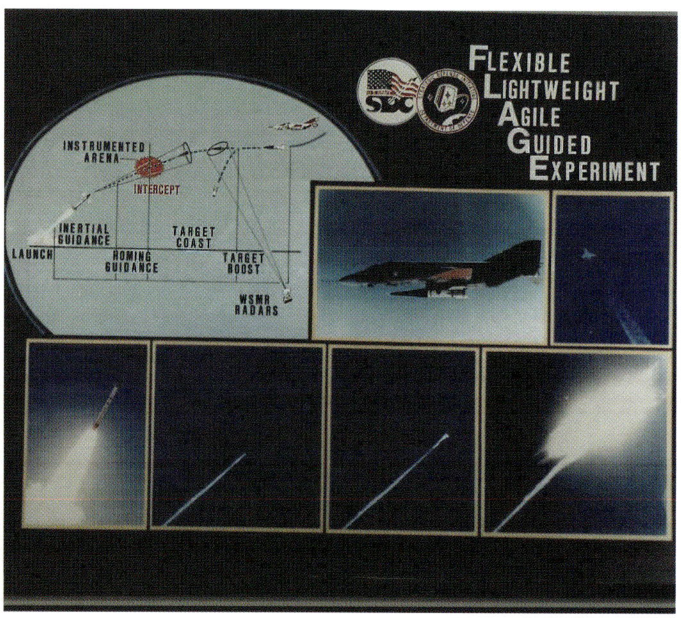

Fig. 3-11. This 27 June 1987 flight of the FLAGE shows the second successful intercept of the FLAGE program, a simulated RV launched from an aircraft.

Fig. 3-12. Guided by 216 attitude control motors, the Flexible Lightweight Agile Guided Experiment demonstrated the feasibility of kinetic energy intercepts at short ranges.

During flight, the FLAGE's on-board millimeter wave radar would lock onto a target.[54] To maneuver the interceptor toward the target, 216 shotgun shell-sized shell motors, located in a band behind the radar, were fired selectively. Having demonstrated successful intercepts against a stationary sphere and an air-launched target in 1986 the FLAGE was tested against a Lance short-range surface-to-surface missile in the next test. On 21 May 1987, in its seventh and final test, the FLAGE demonstrated the feasibility of a short-range nonnuclear intercept, destroying the Lance at an altitude of 16,000 feet within seconds of launch.[55]

The Continuing Threat

At this time, the primary threat remained the Soviet Union. As of January 1981, authorities estimated the Soviet arsenal included 1400 operational ICBM launchers and 950 sea-launched ballistic missile launchers.[56] Officials believed that this arsenal would easily give the Soviets a 3 to 1 advantage over the American ICBM arsenal. The increasing numbers of ICBMs led DoD to approve a pre-prototype demonstration of the LoAD to develop technology to protect American systems. The 1970s also saw the proliferation of short-range missiles. The Soviet Union exported large numbers of SS-1 Scud B missiles to Warsaw Pact nations, China and North Korea.[57] These nations in turn supplied information and materials to such nations as Egypt, Iran, Libya and Syria.

In addition, as the decade progressed, there existed in some quarters a sense of "urgency because of assertions by certain intelligence officials and scientists that the Soviet Union may have a dangerously significant lead in the development of directed energy weapons."[58] Retired Air Force Major General George J. Keegan repeatedly warned that Soviet laser technology could be deployed as early as 1981. However, Dr. Ruth Davis, Deputy Under Secretary of Defense for Research and Advanced Technology, testified that in her opinion both nations were at similar stages with regard to directed energy technology.

The Scowcroft Commission

In January 1983, in response to Congressional opposition to the proposed MX basing plan, President Ronald Reagan established the President's Commission on Strategic Forces, chaired by Lieutenant General Brent Scowcroft (USAF, Retired). Known as the Scowcroft Commission, the group would review modernization efforts and find an acceptable basing mode for the Peacekeeper ICBM.

The Commission issued its report in April 1983. Following their review of the Peacekeeper deployment issue, the Commission favored basing the ICBMs in existing MINUTEMAN silos. This deployment plan contributed to the demise of the SENTRY program, which had become firmly associated with the mobile basing option.[59] With regard to modernization, the report placed greatest emphasis on command control and communications and battle management planning.[60] Other specific recommendations were (1) continued Trident submarine construction and development of the Trident II missile; (2) bomber and cruise missile defense programs; and, (3) vigorous research in anti-submarine warfare and Ballistic Missile Defense. The commission viewed the BMDATC as an innovator – an institution that could freely initiate and nurture innovation, an "organization that could support greatness." Although valued as a deterrent to the Soviets, they concluded, however, that "No ABM technologies appear to combine practicality, survivability, low cost, and technical effectiveness sufficiently to justify proceeding beyond the stage of technology development."[61]

The Army's Revival

Near the end of the decade, NASA fulfilled a 1969 promise made to the Chief of Staff of the Army to consider Army officers as astronauts when it identified future manned space missions. In January 1978, NASA announced it had selected 35 new astronaut candidates for the Space Shuttle program, the first group selected since 1969. Major Robert Stewart, the first Army astronaut, was a mission specialist among this group of candidates. While these changes gave the Army a potential opening, it had to wait to exploit them.

After the Vietnam War, the United States faced a revived Soviet threat. In the 1970s, the Soviets changed from Khrushchev's emphasis on conflict escalation to Brezhnev's desire to field a force not overly reliant on nuclear weapons. This reversion to traditional Soviet doctrinal themes – a combined arms approach to warfare – emphasized balanced force development.[62] The new Soviet force was upgraded and expanded through the 1960s and 1970s while American attention was focused on the Vietnam War and possible active Chinese hostility.[63]

In the early 1980s the strategic and tactical situations changed. In Washington, Ronald Reagan's election brought the critics of détente to power. Nevertheless, the United States continued to follow the same defensive strategy President Truman enunciated in the late 1940s aimed at containing Russian military expansion in Europe.[64]

The Soviets reverted to their earlier Cold War strategy in Europe, picking the times and places for action. The U.S.S.R. upgraded its forces and began to build a fleet to cruise the Mediterranean. However, the American nuclear deterrent was still potent and the Soviet forces were locked into a single theater of operations in Europe, unable to aid geographically noncontiguous allies or clients.

After the Soviet invasion of Afghanistan in 1979, the Carter Administration increased the defense budget. This accompanied a renaissance of doctrinal thought in the United States Army begun in 1975. Nevertheless, in the early 1980s many believed the West faced an economic and military crisis. An aggressive Soviet Union could undermine the West's ability to use its nuclear arsenal with nuclear, biological and chemical weapons. Economically, Soviet domination of space could mean Russian domination of the commercialization of space. These factors helped shift American strategic thought from deterrence to defense.[65]

The 1980s also saw a growing disquietude in Soviet journals of military thought as various authors analyzed AirLand Battle Doctrine. The cozy world of Soviet military planning was disturbed by the ways new types of technology were assimilated into military theory, doctrine and equipment. Beginning in the 1970s, Russian and American military theorists began writing about changes new information technologies made possible in warfare, asserting that future armies would be very mobile, linked by communications devices giving commanders a common picture of the battlefield. These armies would mount attacks throughout large theaters of operation, not a linear front. Battles would simultaneously expand in space and be shortened in time. While the Soviets did not have the economic or the technological strength to pursue these ideas, the United States began to experiment with them. As early as 1983, Soviet planners expressed doubts about their ability to handle future competition with the American military threat based on doctrine refined using the new information technologies.[66]

The Soviet military theorists' misgivings were echoed in the social situation that Mikhail Gorbachev inherited when he became General Secretary of the Communist Party of the U.S.S.R. in 1985. He confronted a stagnant society beset by unexpressed internal doubts and problems. Economic stagnation meant that the Soviet leadership was preoccupied with the old Stalinist concern of industrial modernization, a key target of the Gorbachev reforms. The effort to jump start a command economy reduced the growth in the military budget, cut conscription levels, slowed conventional weapons production, and shifted key personnel in the defense sector of the economy to the civil sector.[67]

The National Space Policy and the Army

Much of this became clear only in retrospect. President Reagan did come to office intending to strengthen the military. He believed that although overall Army modernization was overdue, it was crucial to update the nation's space systems. On 4 July 1982, he announced a new National Space Policy. It included commitments to (1) explore and use space for peaceful purposes by all nations; (2) participate in international cooperative space-related activities to achieve scientific, political, economic, or national security benefits for the United States; (3) pursue activities in space that support the United States' right of self-defense; (4) develop Space Transportation System capabilities and capacities to meet appropriate national needs and to make the system available to commercial and government users; and (5) continue to study space arms control options and consider verifiable and equitable arms control measures that would limit testing and deployment of specific weapons if compatible with American national security.

In 1988, the policy was updated, reaffirming the national commitment to space exploration and addressing civil, military and commercial space use. It established six goals, to (1) strengthen American security; (2) obtain scientific, technological and economic benefits for the general population and to improve the quality of life on earth through space related activities; (3) encourage private sector investment; (4) promote international cooperative activities while protecting American interests; (5) cooperate with other nations to maintain the freedom of space for all activities that increase the security and welfare of mankind; and (6) expand human presence and activity beyond earth orbit into the solar system. These goals would be guided by six principles: (1) a commitment to the peaceful exploration and use of outer space for all mankind's benefit, including national security goals; (2) pursuit of activities that support the right of self defense and the defense of allies; (3) rejecting any claim of sovereignty over outer space or celestial bodies; (4) considering national space systems to be national property; (5) encouraging the commercial use and exploitation of space technologies; and (6) conducting international cooperative space related activities to achieve national scientific, political, economic, or national security benefits.

These events and issues gave the Army an impetus to explore the ways it could use space and space-based military assets. However, the direct stimulus to re-evaluating the role of space assets as well as ballistic missile defense was the Army-wide debate over doctrine that took place between 1975 and 1982.[68] It was only then that the Army determined the ground commander's needs required the Army to return to space. As AirLand Battle doctrine developed, the entire conception of the battlefield expanded. The Army now concerned itself with the Deep Battle (a need to see and strike deep) and with the Rear Battle (its own needs for expanded command and control). Space-related activities offered the ground commander unique platforms for observation, positioning and communications over a greatly expanded area of concern: the operational level battlefield.

As it had with missile defense, the Army proceeded in an orderly, deliberate way that involved developing concepts and long-range planning followed by investment in programs. It was prodded by its growing needs for the products that existing and planned space systems would provide ground forces. Although intelligence and surveillance capabilities of satellites garnered the most attention, the Army also used space assets to multiply its abilities to deter, detour and defeat an enemy. The other services formed space commands to centralize and coordinate their efforts to use space. In 1982, the Air Force, as the lead armed service in space, established U.S. Air Force Space Command (AFSPC) "to further consolidate Air Force operational space activities." As a major command, AFSPC "supports Air Force space operations, including satellite control and Department of Defense space shuttle flight planning, readiness, and command and control." In 1983, the Navy, dependent on a world-wide communications and intelligence network for its surface and submarine fleet operations, formed Naval Space Command at Dahlgren, Virginia. It was not until 1984 that an Army Staff Field Element was activated at AFSPC headquarters. This marked the beginning of the U.S. Army Space Command.

President Reagan and the Strategic Defense Initiative

President Reagan's announcement of the Strategic Defense Initiative (SDI) in March 1983 reemphasized space's role in national defense and gave added impetus to the Army's ballistic missile defense effort.[69] Between 1983 and 1989, the Army began to pay attention to its space role in both a conceptual and organizational sense as it reinvigorated its ballistic missile defense effort.

Before the Scowcroft Commission submitted their report, the Joint Chiefs of Staff (JCS) had begun to assess the vulnerability of the American ICBM arsenal. Following a series of 40 meetings, between June 1982 and February 1983, the JCS concluded that a missile defense effort was required. In February, Admiral James Watkins, Chief of Naval Operations, presented a briefing to the JCS recommending that "the United States should quit looking for a complex basing mode for the MX missile, deploy a small number of MXs in MINUTEMAN silos, and start developing a strategic defense that would provide the basis for a shift 'to a long-term strategy based on strategic defense.'"[70] During an 11 February meeting with President Ronald Reagan, the JCS unanimously recommended that the United States pursue a national security strategy which placed increased emphasis on strategic defenses. As General John Vessey, Chairman of the Joint Chiefs, observed "Wouldn't it be better to defend the American people rather than avenge them?" Their recommendation marked the end of a 37-year policy of offensive deterrence.

A long time opponent of the doctrine of mutual assured destruction, President Ronald Reagan introduced a new era in BMD on 23 March 1983. In a televised speech to the nation, Reagan announced his concept for the Strategic Defense Initiative (SDI), popularly known as "Star Wars".[71] Following a review of Soviet capabilities, Reagan suggested that security should rest upon more than the threat of "instant U.S. retaliation to deter a Soviet attack." Recognizing that he established "a formidable, technical task", the President proposed that the nation pursue a missile defense policy and called on "the scientific community in our country… to give us the means of rendering these nuclear weapons impotent and obsolete." Reagan concluded, "We seek neither military superiority nor political advantage. Our only purpose - one all people share - is to search for ways to reduce the danger of nuclear war."[72]

In National Security Directive 85, "Eliminating the Threat From Ballistic Missiles," Reagan ordered "the development of an intensive effort to define a long term research and development program aimed at an ultimate goal of eliminating the threat posed by nuclear ballistic missiles."[73] In addition, a study would be conducted to assess the role the role of BMD in the future security strategy for both the United States and its allies. This study would also provide guidance for research and development, funding the fiscal year 1985 budget.

The Fletcher and Hoffman Reports

Presidential guidance resulted in two studies, both published in October 1983. The Future Security Strategy Study, or Hoffman Report, sought to determine the strategic and policy implications of the Strategic Defense Initiative. The second, the Defense Technologies, or Fletcher Report, would assess the state of missile defense technology and recommend a technology program for the new missile defense program.

The Hoffman report was composed of a series of papers by two study groups. Mr. Franklin Miller, Assistant Secretary of Defense for Strategic Forces Policy, headed an interagency body and Mr. Fred Hoffman, of the Pan Heuristics Corporation, led a group of contractors. Two of the major findings were "the idea that missile defense could enhance deterrence (Miller group) and the view that an anti-tactical ballistic missile system could serve as [a] useful first step toward a national missile defense system (Hoffman group)."[74]

The Fletcher Committee composed of a group of fifty scientists and engineers led by Dr. James Fletcher, former NASA administrator, outlined two models for the new missile defense research program. Their report, completed in February 1984, recommended a "blueprint" for SDI. The recommended research areas were Systems Concepts; Surveillance, Acquisition, and Tracking; Directed Energy Weapons; Conventional Weapons; Battle Management and Command, Control, and Communications; Survivability; Lethality and Threat Vulnerability; and Selected Support Systems. Proposed funding levels for this version totaled $1.405 billion in 1984, $2.385 billion in 1985, $3.43 billion in 1986, $4.284 billion in 1987, $4.623 billion in 1988, and $4.766 in 1989. The alternative, funded at a lower level, was known as the fiscally constrained program. It was this program that became the guide for the Strategic Defense Initiative.

The Strategic Defense Initiative and the Organization for Missile Defense

National Security Directive 119 authorized the SDI program to explore the possibility of developing missile defenses as an alternative means of deterring nuclear war and assigned responsibility to the Secretary of Defense.[75] The Secretary issued an interim charter to establish the Strategic Defense Initiative Organization (SDIO) on 24 April 1984 and appointed Air Force Lieutenant General James Abrahamson as the first director.[76] Department of Defense Directive 5141.5, dated 21 February 1986 established the SDIO as a multi-service agency of the Department of Defense. The director reported to the Secretary of Defense.[77]

The SDI management focused their initial efforts on three tasks: ensuring continuity of relevant programs, tailoring programs to fit the needs of the SDI, and initiating new programs to expand and accelerate the pre-SDI effort in BMD. Emphasis was placed on treaty compliance and non-nuclear technologies. The overall goal, however, was to provide the technical knowledge necessary to support an informed decision, about the "feasibility of eliminating the threat posed by nuclear ballistic missiles of all ranges, and of increasing the contribution of defensive systems to U.S. and allied security." This decision was to be made in the early 1990s.

The SDIO was a multi-service organization. The Army's years of ABM experience, however, proved to be the foundation, as the Army repeatedly took the lead in project development. This experience, according to one report, allowed the SDIO to protect the technology base, increase the emphasis on proof-of-feasibility experiments with increased investment in high risk, high payoff approaches, and continue examining multi-layered defense.[78]

The U. S. Army Strategic Defense Command

Fig. 3-13. "They Shall Not Pass" - is the motto on the distinctive unit insignia created in 1987 for the USASDC. The illustration symbolizes the defensive shield protecting the world from an incoming threat.

As part of the Strategic Defense Initiative, the Army was responsible for directing and managing research associated with Surveillance, Acquisition, Tracking and Kill Assessment; Directed Energy and Kinetic Energy weapons technologies; and Survivability, Lethality, and Key Technologies. To facilitate development of this new proposal, the Army sought to align its effort with the SDIO structure. In July 1984, the BMDO became a part of the Strategic Defense Initiative Organization.[79] One year later, effective 1 July 1985, the BMDO became the U.S. Army Strategic Defense Command, a field operating agency of the Office of the Chief of Staff.[80] At this point the BMDATC and the BMDSCOM continued to exist as separate entities.

These two organizations dissolved into the framework of the U.S. Army Strategic Defense Command (USASDC) on 6 January 1986. To correspond to the series of program elements established by SDIO, they were replaced by a series of five Directorates (Weapons, Sensors, Systems Analysis/Battle Management, Survivability, Lethality and Key Technologies, and Advanced Technology) and five Project Offices (Airborne Optical Adjunct, Terminal Imaging Radar, High Endoatmospheric Defense Interceptor, Exoatmospheric Reentry-vehicle Interceptor Subsystem, and Ground-Based Laser).[81] Each of these was devoted to the development of a specific weapon system or radar. During this period, project offices were created and disestablished as directed by the budget and focus of the SDIO.

In October 1988, President George Bush recognized the significant role played by the USASDC. Under National Security Directive 219, Lieutenant General Robert Hammond, USASDC Commander, was named the Program Executive Officer for Strategic Defense. With this position, LTG Hammond reported directly to the Army Acquisition Executive.

Star Wars

From the beginning, opponents criticized the SDI concept as an unrealistic proposal, more akin to the movie "Star Wars" than actual, achievable capabilities. Both politicians and scientists argued that the Reagan administration was "ambiguous" in their goals[82] and relied heavily on "exotic" technologies.[83] Even as the program became better defined, critics questioned the feasibility of the SDI program. At the same time, they argued that it would lead to another arms race and the militarization of space.[84] These arguments would appear frequently during the history of the SDI program impacting budgets and systems development.

The SDI Concept for a Layered Defense

Researchers from the SDIO, the Army and the Air Force proceeded to apply the SDI concepts and created a tiered, or layered, defense against enemy missile systems.[85] This layered defense would facilitate the intercept of an incoming missile during the three phases of flight: boost, midcourse, and terminal. Each of the services was assigned elements designed to track or intercept during specific phases of the missile flight. The USASDC and the Army assumed the lead in the SDI effort.

The Three Phases of ICBM Flight

Boost Phase – The three to five minute period from the ignition of the enemy missile's propulsion rocket to burnout, propelling the missile payload through the atmosphere into space to the desired trajectory. The missiles exhaust plume enhances detection, but speeds of up to 15,000 mph make an intercept challenging. In the post-boost phase, the nose cone separates from the booster rockets and releases the reentry vehicle(s) (RVs) and penetration aids (PENAIDS) (decoys and chaff).

Midcourse Phase – This is the longest period lasting 20-25 minutes for ICBMs, less for SLBMs. During this phase, the RVs and PENAIDS are traveling in an arc toward their targets. In the weightlessness of space, PENAIDS travel at the same trajectory and speed as the heavier RVs.

Terminal Phase – The RV and decoys reenter the Earth's atmosphere. Friction and heat caused by reentry help to distinguish between the targets. Nevertheless there is only a short time-30 seconds or so-to react and intercept the RV.

The SDI defense concept for the boost phase incorporated the Boost Surveillance and Tracking System, the Space-Based Laser (SBL), and the Ground-Based Laser (GBL). The USASDC shared responsibility for the SBL with the Air Force, while it was assigned sole control over the GBL. In the midcourse phase, the SDI system architecture envisioned a Space-Based Surveillance and Tracking System (SSTS), a Space-Based Interceptor (SBI), a Neutral Particle Beam (NPB), and the Exoatmospheric Reentry-vehicle Interceptor Subsystem (ERIS). The Air Force directed the development of the SSTS and the SBI and shared responsibility with the Army, USASDC, for the NPB. The Army then directed the evolution of the ERIS. The final layer of defense, the terminal phase, employed the Airborne Optical Adjunct (AOA), the Ground-Based Radar (GBR), the Ground-Based Surveillance and Tracking System (GSTS), and the High Endoatmospheric Defense Interceptor (HEDI) The USASDC had the lead on all of these programs. All three primary elements, the Air Force, the Army and SDIO, shared in the development of the Battle Management/Command, Control and Communications systems (BM/C^3).

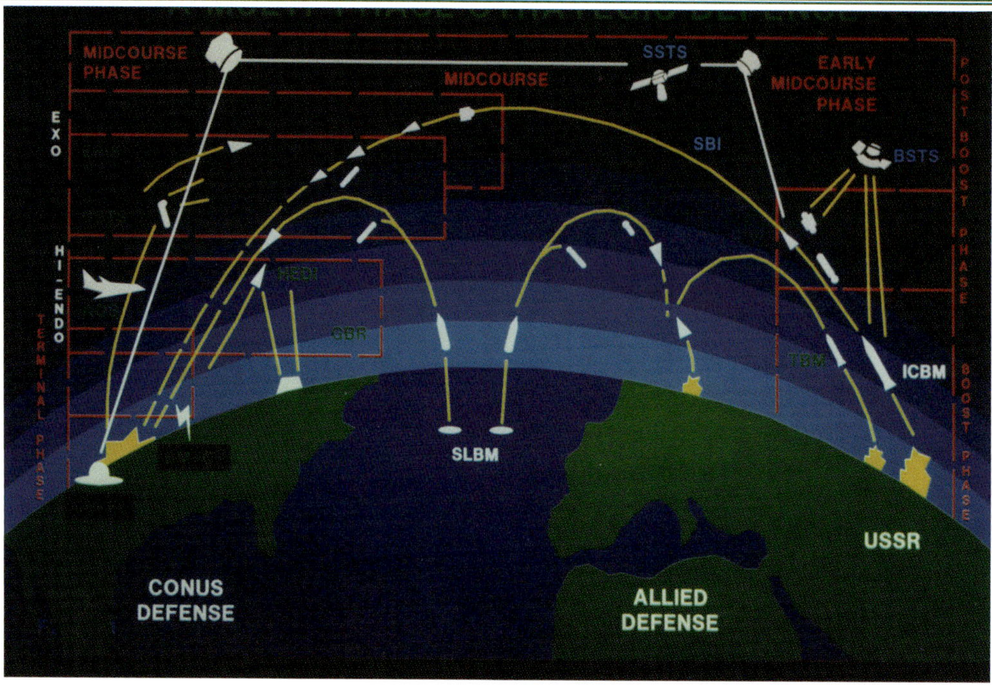

Fig. 3-14. The SDIO program called for a Multi-phase Strategic Defense. The layered architecture addressed the boost, mid-course and terminal phases of the target missile's flight.

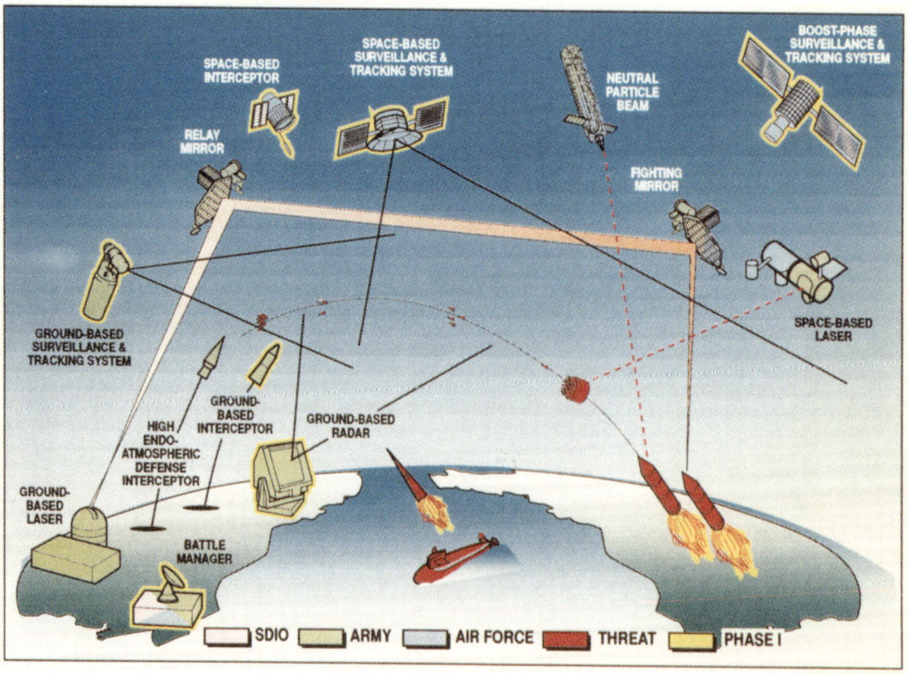

Fig. 3-15. This diagram of the Strategic Defense Initiative System Architecture Concepts attributes projects to the appropriate service or organization.

SDI-The Boost Phase

The Ground-Based Laser (GBL)

On 2 April 1984, the SDIO authorized the laser imaging technology program.[86] Two years later, on 26 March 1986, the USASDC created the GBL Project Office. Located at WSMR, New Mexico, the office oversaw the development of the ground-based free electron laser (FEL) technology integration experiment.[87] The goal was to develop a system that could intercept a target in the boost phase by bouncing the laser beam off relay mirrors based in space.[88] To this end, they explored the benefits of the radio frequency FEL and the more powerful induction FEL.[89] Initial tests showed that both approaches were feasible for full-scale development.[90] The Project Office subsequently elected to proceed with a dual laser concept. As the project continued to progress, the SDIO and USASDC began to explore the possibility of using the laser as an anti-satellite (ASAT) system.

Fig. 3-16. On 29April 1986, the ALTAIR radar on Kwajalein tracked its 100,000th deep-space satellite. In that same year, on 16 October, President Reagan signed Public Law 99-239, the Compact of Free Association between the United States and the Republic of the Marshall Islands.

Program redirections by SDI and repeated budget cuts, beginning in fiscal year 1988, however, forced frequent modifications and downscaling in the project. These events culminated in the eventual demise of the project in January 1991, six month after the official dedication ceremony for the new Ground-Based FEL facility.[91]

With the agreement of the SDIO, the Average Power Laser Experiment, a restructured version of the GBFEL, was transferred to the Directed Energy Weapons (DEW) Directorate. Research continued on laser programs under the auspices of the High Energy Laser Technology Division. In conjunction with this effort, the division also worked to evaluate the component design option of the FEL to use in a possible space-based FEL.

Fig. 3-17. A special facility for the Ground-Based Laser project was constructed at the White Sands Missile Range.

The Neutral Particle Beam (NPB)

In addition to this laser research, the DEW Directorate was involved in the development of the neutral particle beam technology. As defined by the SDI architecture, the NPB would be a space-based system with a variety of capabilities. An NPB would be used to penetrate the target to destroy electronics, ignite the explosives and highlight the target to aid identification. Given these anticipated capabilities the command also explored the effectiveness of the NPB as an ASAT system.

The NPB system itself is composed of a particle accelerator, beam focusing and pointing magnets, and a stripping device, to rid the beam of extra electrons. An NPB is created by accelerating negatively charged hydrogen or deuterium irons until they travel in a continuous wave or pulsed beam.[92] The resulting beam travels at a rate near the speed of light. Unlike a laser beam, the NPB does not interact with the magnetic fields in the atmosphere and thus travels in an unbendable beam. At the same time, however, an NPB is a line-of-sight system and cannot be retargeted with relay mirrors.

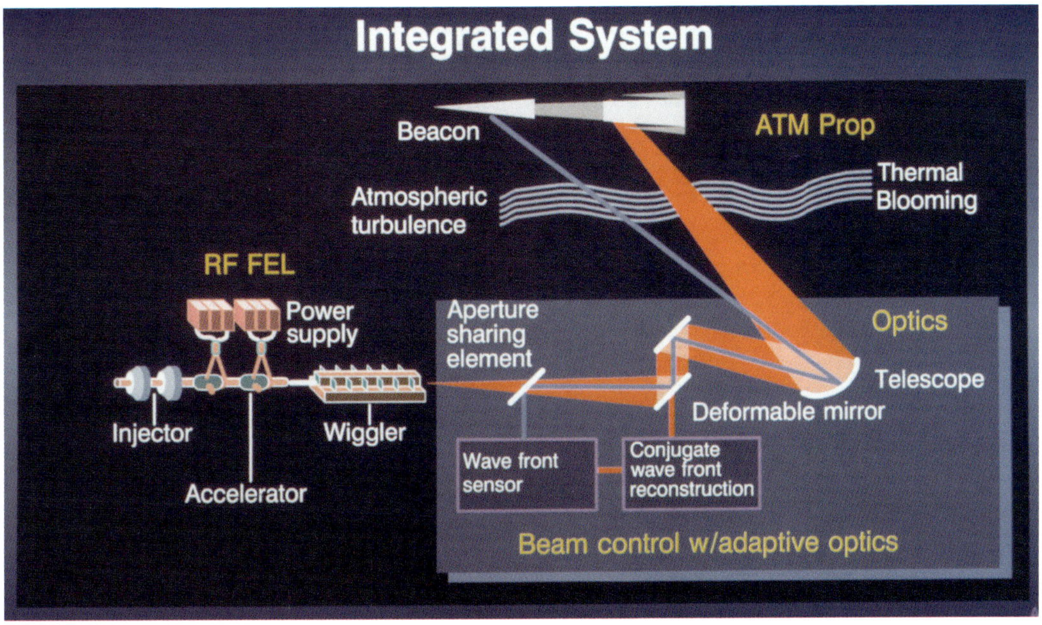

Fig. 3-18 and 3-19. These artists' concepts illustrate the proposed missions of the Ground-Based Laser. The first shows the Integrated System of the ground-based free electron laser. The second illustrates the system components for a theoretical ground- based laser anti-satellite system.

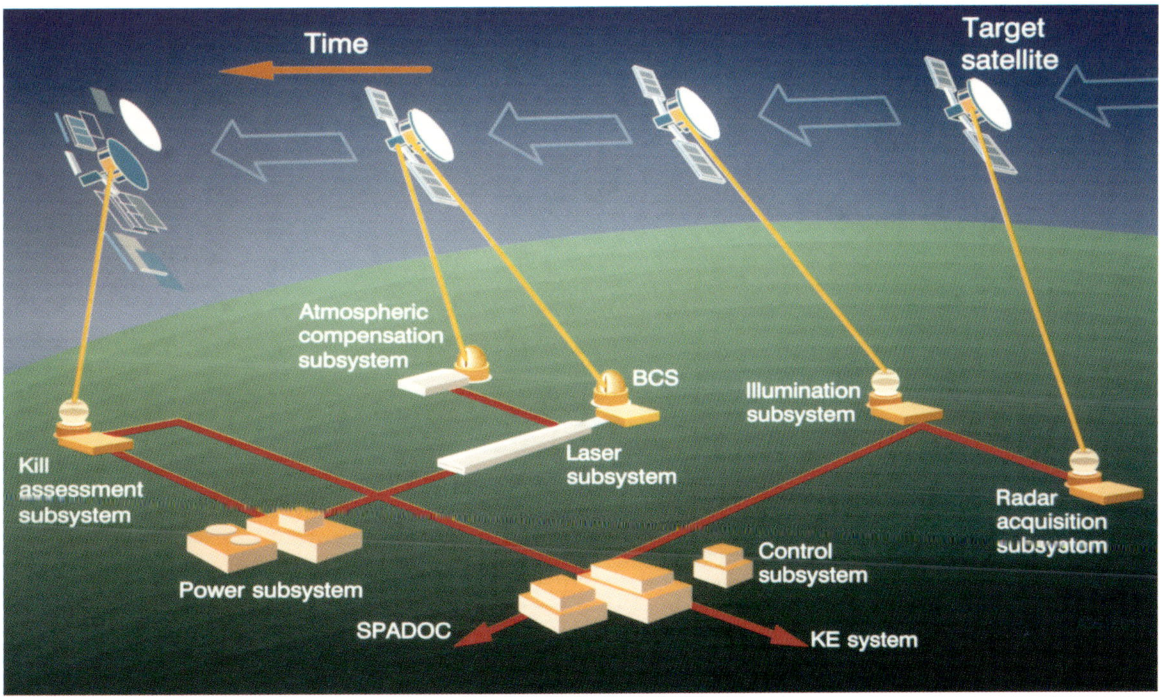

The Army was the principal developer of the NPB from 1974. As early as 1987, particle beam technology was described as the "closest to the required level of brightness of all directed energy options."[93] By 1992, the program had completed four of the eight objectives outlined in the 1984 directed energy plan. Specifically, these were the development of a beam neutralizer, lightweight magnetic optics, beam sensing and bore sighting methods and a sensor to measure the effect on the target.[94] In 1993, officials reported that "the program [had] made rapid progress and the last remaining technology demonstrations are being completed."[95] Budget cuts in the SDIO program ultimately resulted in the redirection of the directed energy efforts, with greater emphasis placed on laser technology.

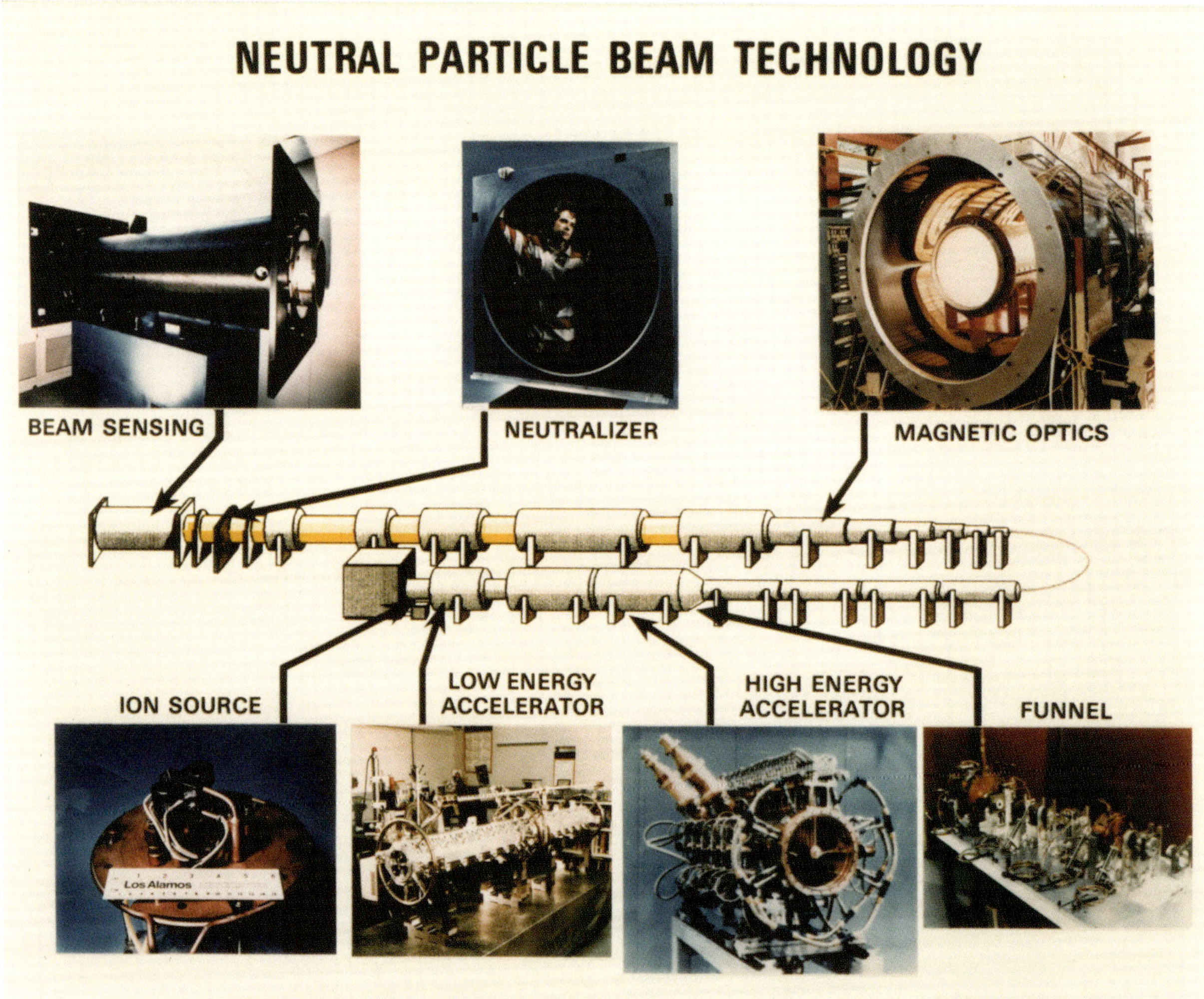

Fig. 3-20 and 3-21. By 1993, many of the Neutral Particle Beam technologies had reached maturity. The diagram above shows the elements of an NPB and their place in the finished product. A deployed NPB (depicted on the facing page) would be a space-based system which could shoot hydrogen molecules at about 60,000 kilometers per second.

TARGET
BEAM
• NEUTRAL HYDROGEN ATOMS
• NEAR SPEED OF LIGHT
BEAM OPTICS
RFQ
DTL
ACCELERATOR

SDI-Midcourse Phase

Exoatmospheric Reentry-vehicle Interceptor Subsystem (ERIS)

According to the initial SDIO system architecture, the interceptor for the midcourse phase was the ERIS, renamed the Ground-Based Interceptor (GBI) in 1990.[96] Based on the results of the High Altitude Defense Study, conducted in fiscal year 1983, the USASDC created the ERIS Project Office on 1 July 1984.[97] The SDIO subsequently identified the program as a high priority effort in 1986. Its mission was to resolve technical issues associated with the development of lightweight, low-cost, non-nuclear interceptors for midcourse defense. In addition to these concept definitions, the ERIS project was tasked to develop "key components, including miniaturized seeker/optics, advanced propulsion and controls and innovative low-cost avionics and terminal maneuver propulsion and controls."[98]

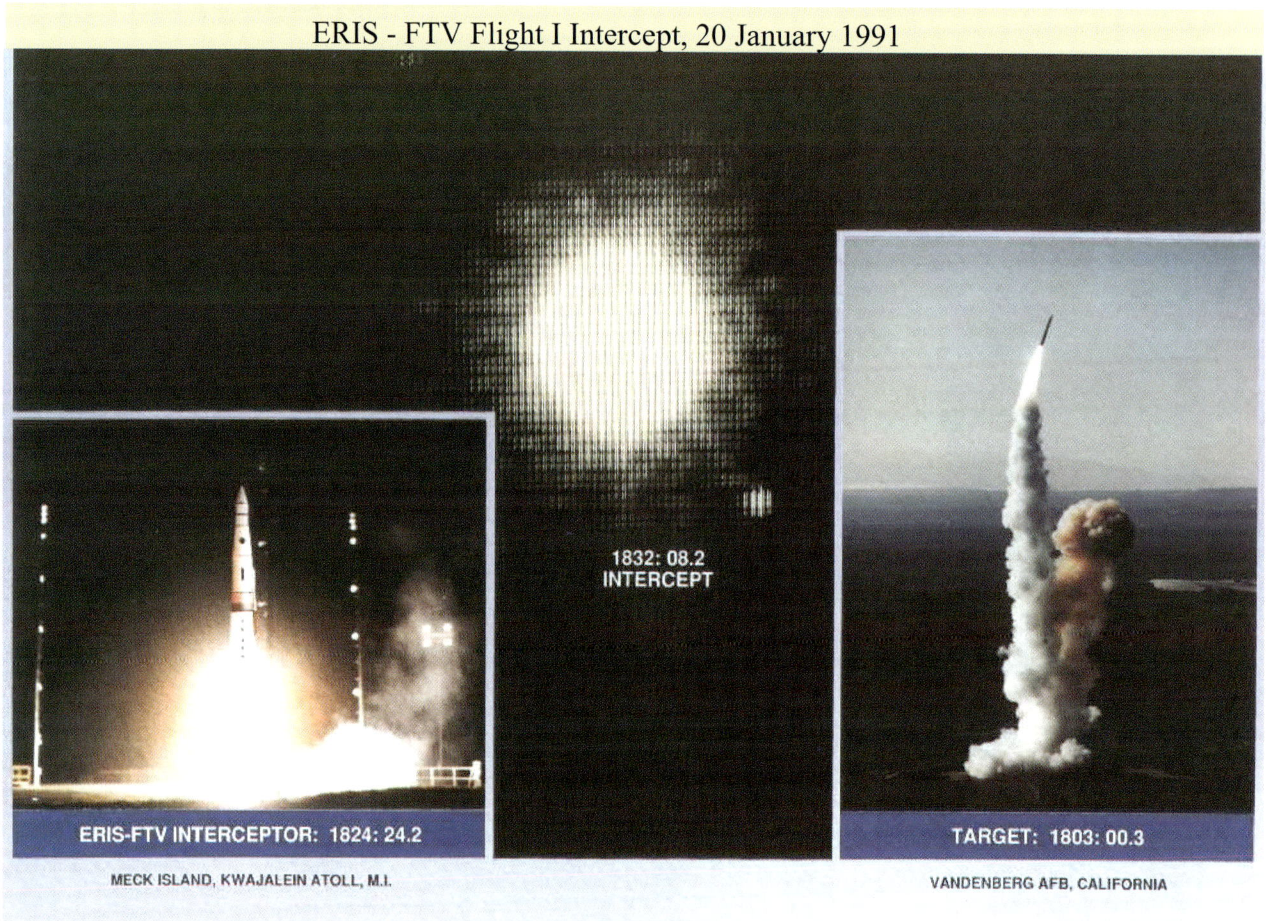

Fig. 3-22. The Exoatmospheric Reentry-vehicle Intercept System was the first SDI project to achieve an intercept as seen in this collection of photographs from January 1991.

Employing some of the technology from the Homing Overlay Experiment and existing materials, development of the ERIS system began in 1985 with the contract award to Lockheed Missiles and Space Company. Constructed of surplus Minuteman ICBM second and third stages, the experimental ERIS missile would incorporate a kill vehicle with an LWIR scanning seeker, a data processor and flight divert attitude control propulsion motors in a two stage rocket booster.[99] The 160-kg ERIS interceptor would receive information from external sensors and, based on this data, select the appropriate target by comparing flight signatures.[100]

Fig. 3-23. ERIS at sunset, before a test flight.

The first major milestone of the ERIS functional technology verification program was met in April 1989, when the integrated system test vehicle left the manufacturer's facility to begin the test phase. There was another two years of testing before the first flight test. Nevertheless less than a decade after the HOE intercept, on 28 January 1991, launched from an underground facility on Meck Island in the Kwajalein Atoll, the ERIS test vehicle successfully detected the target amidst decoys, and intercepted the mock ICBM warhead launched from Vandenberg AFB. The test, "the first time an SDI experiment attempted an interception in a counter-measures environment," exceeded expectations for this initial mission.[101]

The second and, due to budget cuts, final test was conducted on 13 May 1992 against a Minuteman I ICBM. The primary focus of this effort was in data collection on the guidance, acquisition, track and divert functions. Although a direct intercept was not achieved, the mission met its objective of demonstrating target handover, acquisition and resolution of threat and the collection of radiometric data on the target and decoys.[102]

SDI-Terminal Phase

Airborne Surveillance Testbed (AST)

The Ballistic Missile Defense Organization selected Boeing as the prime contractor for the Airborne Optical Adjunct, later renamed the Airborne Surveillance Testbed, in July 1984.[103] The project was chartered later that year. The purpose of the AST was to prove that "an infrared sensor, data processor, and associated communication links, can be integrated on an aircraft."[104] Perhaps more importantly, the effort was to show how this system could be used "to acquire, track, discriminate, designate, and hand over track data on ballistic missile threats in real time to a ground-based radar."

Fig. 3-24. The 5,000 pound sensor with its 38,400 detectors which flies aboard the AST aircraft.

In 1987, the AST became the first element of the terminal phase and the SDI program itself to enter the test phase. In August 1987, the modified Boeing 767, with its 86-foot long cupola, passed its first airworthiness tests. In July 1988, Hughes Aircraft Company delivered its sensor, the most complex, long-wavelength infrared sensor built to date. The integration and installation process began in preparation for the 1990 mission fight tests.[105] These tests successfully demonstrated the feasibility of the airborne seeker.

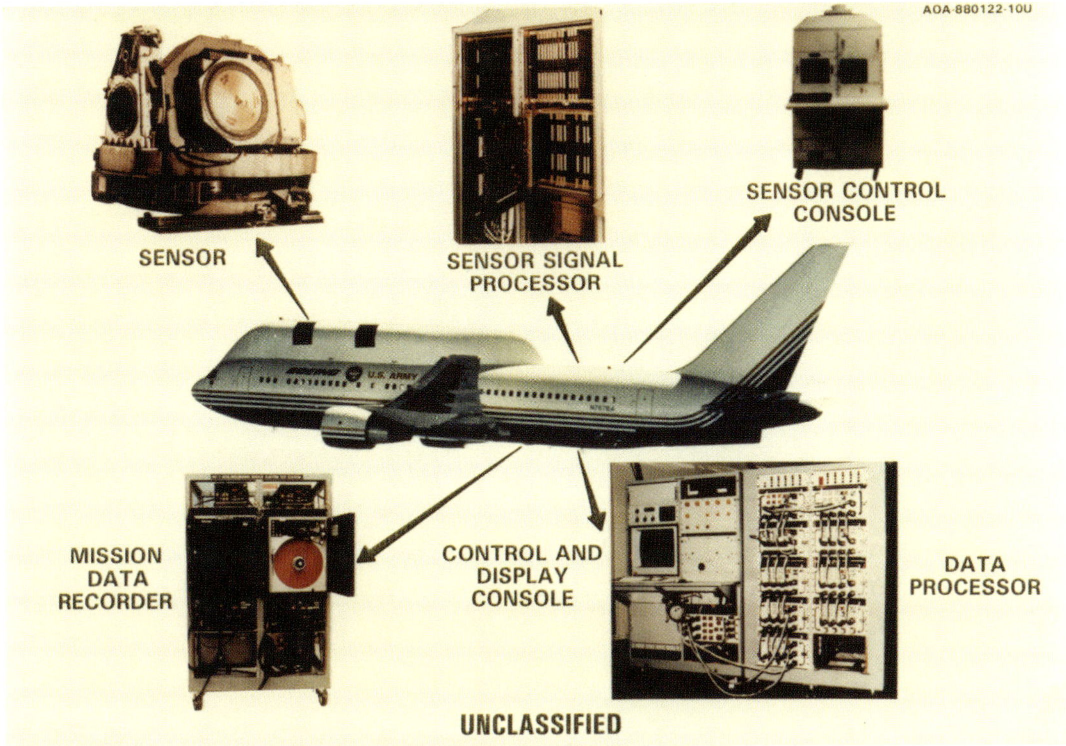

Fig. 3-25. The Airborne Surveillance Testbed, an airborne system with its heat detecting telescope, remains an important asset to Army data collection efforts.
This diagram illustrates the various components of the AST system.

The AST program further demonstrated its capabilities on 29 June 1991 in a seven-hour mission. During this flight, the AST performed its first real-time discrimination of multiple reentry-objects.[106] Despite the frequent threat of termination due to cost growth, the AST moved from the developmental to the experimental phase. The AST continues to provide optical and tracking support to the command and other services.

Ground-Based Radar (GBR)

As mentioned above, the AST would hand over data to a GBR facility.[107] This project began in 1984, as the BMD Radar Project Office. In 1986, it was renamed the Terminal Imaging Radar (TIR) Project Office and assigned the mission "to develop and validate an ABM treaty compliant defense radar technology testbed that [can] perform high altitude discrimination in real-time."[108] This phased array radar would have the ability to relay data to the various interceptor subsystems. In addition, by operating in the X-band, the system will be able to "propagate thru rain… [and] nuclear effects," ensure the measurement precision need for discrimination, and "defeat jammers and chaff."[109]

Fiscal year 1988, saw further developments in the program with the addition of a GBR-Experiment (GBR-X),[110] to be constructed at USAKA, and a GBR-Midcourse, still in the conceptual stage. At the same time SDIO ordered that the project office be redesignated the GBR Project Office. On 15 June 1990, the Defense Acquisition Board granted the SDIO and GBR approval to move into the demonstration and validation phase, beginning a series of experiments and testing on the radars.[111] It remained to be decided, however, whether to have mobile or fixed-based facilities.

Fig. 3-26. Model of the turret facility of the Ground-Based Radar-Experimental planned for Kwajalein.

Ground-Based Surveillance and Tracking System (GSTS)

Another element in the terminal defense stage of the Strategic Defense System is the GSTS.[112] At the urging of the Defense Acquisition Board, the GSTS Project Office evolved out of a research effort initiated by the Systems Analysis/Battle Management Directorate. Established on 14 November 1988, the aim of the Project Office was "to design and fabricate an LWIR sensor housed in a ground-launched rocket that could locate, track and discriminate real targets from decoys in the event of a ballistic missile attack."[113]

In October 1988, McDonnell Douglas Space Systems Company won the contract to manufacture and test its design for "a reusable, fully flight qualified sensor payload and a ground-based data processor."[114] Funding limitations put some constraints on the options being explored, but production continued steadily towards the series of flight tests planned for fiscal year 1996. Nevertheless, the 1992 decision by Congress to defer deployment on the proposed National Missile Defense site and limited funding to the SDIO resulted in the termination of the GSTS.[115] Ambassador Henry Cooper, SDIO Director, signed the termination letter on 8 October 1992.

Fig. 3-27. Terminated before flight tests could be conducted, this drawing shows the sensors in the payload section of the Ground-Based Surveillance and Tracking System.

High Endoatmospheric Defense Interceptor (HEDI)

The interceptor designed for this terminal phase[116] was the High Endoatmospheric Defense Interceptor (HEDI).[117] Originating from a study on high altitude defense, this Project Office was created on 20 February 1985.[118] Its goal was to develop a nonnuclear interceptor capable of destroying an ICBM reentry vehicle within the Earth's atmosphere, operating at altitudes between 50,000 and 200,000 feet.

The HEDI Project Office and its contractor, McDonnell Douglas Corporation, made steady progress in the program until 1989, when budget cuts forced the redirection of the contract. Several tests were, at that time, either altered or deleted from the schedule.[119] In many respects, the HEDI project became a technology demonstration program.

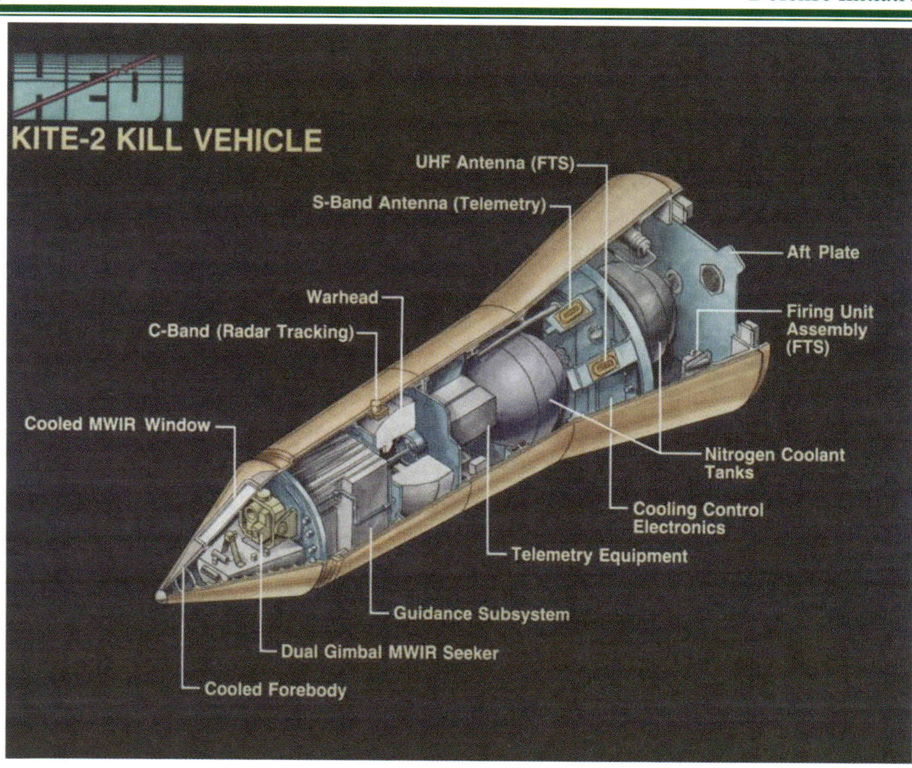

Fig. 3-28. The High Endoatmospheric Defense Interceptor incorporated a number of innovations as seen in this cut-away drawing.

Despite these cuts, on 26 January 1990 the HEDI Project Office conducted its first flight test at White Sands Missile Range. The Kinetic Kill Vehicle Integrated Technology Experiment (KITE), which self-destructed prematurely, still succeeded in demonstrating the viability of "the nose cone shroud and on-board seeker window."[120] This and other tests ultimately proved the feasibility of the shrouded sapphire window technology, cooled optics, two color seekers, advanced propellants, and other innovations.

During the summer of 1990, SDIO Director Ambassador Cooper approved the Endoatmospheric/Exoatmospheric Interceptor (E_I) program as a "logical follow-on to the HEDI KITE program."[121] At the same time, HEDI was identified as a "viable candidate for the lead ground-based interceptor for the SDS [Strategic Defense System] architecture." Using both a medium wavelength and an LWIR seeker, the E_I would expand the range of SDI's terminal defense interceptor from "tens of kilometers to hundreds of kilometers."[122] In September 1991, the KITE-2 test again prematurely detonated, the KITE-2A flight, however proved successful. On 25 August 1992, the KITE-2A gathered data on all the required objectives, proving that "the necessary technology is in hand to perform an intercept of reentry vehicles within the earth's atmosphere using an infrared homing seeker and a non-nuclear warhead."[123] Despite these successes, officials favored interceptors above the earth's atmosphere, and the subsequent budget constraints led to the termination of the HEDI project office at the end of fiscal year 1992.[124]

Fig. 3-29. The rail-launched High Endoatmospheric Defense Interceptor missile achieved a successful intercept in this Kinetic Intercept Technology Experiment 2A conducted on 25 August 1992.

ABM Treaty Interpretations

As the United States began to move forward in the development of a new missile defense system, opponents questioned the compliance of the proposed SDI system with the 1972 ABM Treaty. Initially, the Reagan Administration held that the proposed research programs involved only subcomponent testing and was therefore allowed under the treaty. Soviet President Mikhail Gorbachev, however, disagreed, calling the proposed program illegal.[125]

In July 1985, President Reagan presented an address to the nation on the Strategic Defense Initiative. Quoting Soviet Marshal Grechko's 1972 testimony to the Supreme Soviet, Reagan argued that "the treaty on limiting ABM systems imposes no limitations on the performance of research and experimental work aimed at resolving the problem of defending the country against nuclear missile attack."[126] In 1985, following a lengthy review of the treaty, the Reagan administration concluded that a "broad" interpretation was valid. As introduced by U.S. National Security Adviser Robert McFarlane, on 6 October 1985, space-based and mobile ABM systems and components that are based on "other physical principles" (i.e., lasers, particle beams) may be developed and tested but not deployed. According to the administration, these technologies are not covered by the treaty, as they did not exist when the treaty was written. They are thus addressed in Agreed Statement D, which stated that "specific limitations on such systems and their components would be subject to discussion."[127]

Strategic Defense System Phase I

At the end of 1986, officials decided to enter a missile defense system, to be deployed in the early 1990s, into the defense acquisition process. The Strategic Defense System (SDS) Phase I was the product of this decision. In 1987, the Defense Acquisition Board conducted two reviews of the SDI program, which concluded in part that "'there is presently no way of confidently assessing' the system's price or its effectiveness."[128] Nevertheless based on the overall DAB assessment, in September 1987, Secretary of Defense Caspar Weinberger approved the SDS Phase I baseline architecture and authorized six components of SDI to enter Demonstration/ Validation phase.

The six Phase I components included a space-based interceptor, a ground-based interceptor (the ERIS), a ground-based sensor (the GSTS), two space-based sensors (the boost surveillance and tracking system and the space-based surveillance and tracking system), and a battle management system. With this layered deployment, the architecture concept would provide a defense against Soviet missiles in all stages of their flight. There were however two drawbacks to the proposal: it was costly, and the space-based elements were vulnerable to Soviet anti-satellite systems. To enhance survivability, the SBI was replaced in1990 with the Brilliant Pebbles concept of 300 orbiting interceptors.[129] With the adoption of Brilliant Pebbles, the requirement for a boost surveillance and tracking system was also eliminated.

Other USASDC Initiatives

As these programs evolved from the theoretical to the demonstration stage, the command continued to explore new areas in interceptor, sensor and related technology.[130] Advances have been made in the realm of optics, sensors and data processing, which have subsequently been applied to existing and planned systems. The DEW Directorate continued work on a variety of lasers and neutral particle beams. At the same time, they sought to develop rapid retargeting technology, laser radar, and phased array mirror capabilities.

Fig. 3-30. State of the art technology is proven as the Lightweight Exoatmospheric Projectile tracks a simulated target in this hover test.

In addition to the electrothermal gun and hypervelocity launcher, the KEW Directorate worked on a D2 projectile. With regard to miniaturization, they produced the Lightweight Exoatmospheric Projectile (LEAP), a miniaturized infrared sensor system and kill vehicle for ground or space-based rockets.[131] The LEAP successfully performed a required hover test on 18 June 1991. They subsequently conducted several productive flight tests, but had not completed an intercept.[132] Transferred to the Navy in 1993, the LEAP continued to be tested as part of the Navy's Terrier/LEAP program. Four flight tests conducted between 1992 and 1995, demonstrated that the LEAP could be integrated into a sea-based tactical missile for exoatmospheric BMD.[133] As a result, the LEAP technology formed the basis of the Navy Theater Wide program.

Anti-Satellite (ASAT)

On January 6, 1989, the Defense Acquisition Board authorized the development of an Anti-Satellite (ASAT) program for deployment in the mid-1990s.[134] In March, the Army, "based largely on the Army track record with ground-based interceptors," was given the lead in this joint service effort which included both the Navy and the Air Force.[135] The program was initiated to counteract an already deployed Soviet ASAT system that proponents argued, "held many of our critical intelligence and communications satellites at risk"[136] To address this threat, the DAB requirements included both kinetic energy (KE) and directed energy (DE) approaches.

As a Department of the Army-funded program, the ASAT was distinct from the SDI effort. Nevertheless it did draw upon the KE and DE research conducted by the USASDC and its contractors. Thus SDI funding, as in the case of laser research, directly impacted the ASAT development.[137] Despite the delays in DE-ASAT progress, the KE-ASAT continued with only a few setbacks. The proposal for two versions of a KE-ASAT, one ground-launched, the other sea-launched, was however scaled back to one. In August 1990, the USASDC awarded a demonstration/validation contract to Rockwell International Corporation, to develop a ground-launched KE-ASAT.[138] The first tests for a component of this single site system, a visual light sensor, were planned for January 1992. Following significant budget reductions in 1991, and program restructuring, the Army recommended cancellation of both ASAT programs.[139] Funding for the KE-ASAT was restored after several prominent senators wrote to President George Bush in support of the effort. In fiscal year 1992, the Congress directed that the ASAT program be "updated to reflect the lack of a Soviet threat and the proliferation of militarily significant space capabilities to a growing number of countries throughout the world."[140]

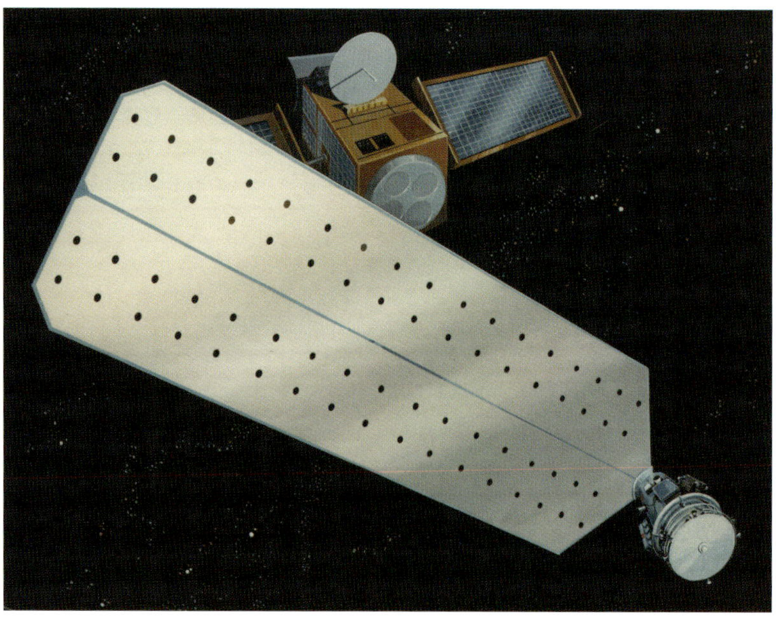

Fig. 3-31. This drawing illustrated the concept of a kinetic energy intercept of a satellite.

By June 1993, continued budget cuts had forced the termination of the KE ASAT Joint Program Office. The Defense Authorization Act for fiscal year 1994, however, directed that the program should be converted into a technology program, managed by the command.[141] The ASAT program continued, although at a slower rate. The work culminated in a hotfire strapdown test, conducted in September 1995. This test demonstrated the kill vehicle's ability "to 'fly' a pre-determined simulated flight path by firing its divert/attitude control system thrusters." The system also successfully acquired and tracked the target with its on-board computers. Two years later, the prototype concluded a successful hover test, in which the sensor acquired and locked onto a simulated moving target.[142]

The KE-ASAT program experienced repeated funding problems throughout its history resulting in program rescheduling and other setbacks. In 1998, the U.S. Space Command's Mission Needs Statement for Space Control included a requirement for an ASAT capability. In the same year, however, President Bill Clinton used a line item veto to eliminate funding for the ASAT and 42 other programs. This action was subsequently deemed unconstitutional by the U.S. Supreme Court and funding was restored. Surviving on Congressional plus-ups, the KE-ASAT program transferred to the U.S. Army Aviation and Missile Command effective October 2001.

High Energy Laser Systems Test Facility (HELSTF)

In 1974, the United States Congress directed the Department of Defense to create a "national" tri-service high energy laser test facility, to address the "proliferation of site development work at various government and contractor facilities."[143] DoD awarded a contract in 1981 for the construction of the site at White Sands Missile Range, New Mexico, and by 1984 it was nearly complete. The mission of the High Energy Laser Systems Test Facility (HELSTF) was to support Army and DoD laser research and development, test and evaluation. It is also to integrate and operate lasers and related instrumentation, facilities, and support systems and conduct and evaluate laser effects tests on materials, components, subsystems, weapons and systems.[144] The HELSTF became operational on 6 September 1985 with an Air Force Lethality and Target Hardening program experiment for the SDIO. In this test, the Mid-Infrared Advanced Chemical Laser destroyed a Titan booster rigged to simulate the conditions of a thrusting rocket booster.

In October 1989, Secretary of the Army Michael P.W. Stone directed the transfer for the HELSTF from the Army Materiel Command to the USASDC, in order to centralize high energy laser research within one command. The actual transfer came one year later on 1 October 1990.[145] Under this new leadership, the mission of HELSTF expanded to include a full range of research, development, test and engineering functions to include test and evaluation, laser damage and vulnerability support, intelligence evaluation resources, advanced system integration center, range instrumentation, space surveillance, and anti-satellite contingency capability.[146] The HELSTF site has been instrumental in the command's subsequent directed energy programs.

The Army Returns to Space: New Organizations

As the Army continued to make progress in missile defense, experimenting with anti-satellite weapons systems as well as laser and particle beam weapons, its long dormant interest in space-based systems began to revive. Because there was not an already existing critical mass of interest for space as there was for missile defense, the way forward was more difficult. First, the Army had to reinvigorate its interest and learn to recognize space-based systems as force multipliers.

In 1983, the Army Science Board's study *Army Utilization of Space Assets* concluded the Army was not using space systems to their full potential. The study concluded that to achieve better exploitation of space systems there must be a high-level commitment backed by sufficient resources. Operation Urgent Fury, the 1983 invasion of Grenada, highlighted the scramble for limited space assets between different services and government levels. The Army had relied on the systems fielded by the other services too long, and frequently received the "leftovers" in a crisis situation. The Combined Arms Grenada Work Group recommended the Army develop, own, and control its own satellites to assure critical communications in such operations.[147] Later in 1983, an Army Space General Officer Working Group was founded to provide direction for Army space efforts.[148] In 1984, the Army Science Board studied the Army's use of space to support its missions, concluding the Army made limited use of space assets and was neither active nor influential in designing and operating most of the space systems then in use. In August 1984, an Army Space Council was created as a coordinating body to approve proposals and provide direction for the Army's involvement in and use of space. The Council met in Washington and coordinated programs that were divided among various staff offices organized by function.

In September 1984 the Vice Chief of Staff of the Army (VCSA), General Maxwell Thurman, activated an Army Staff Field Element at AFSPC headquarters, the nascent form of the U. S. Army Space Command (USARSPACE). The Field Element acted as liaison to AFSPC and initiated planning for Army participation in the unified U. S. Space Command. The Staff Field Element was also responsible for exchanging information about space policy, strategy and plans, monitoring Army space-related education and training developments, representing the Army Space Office at HQ Space Command and providing technical information to Space Command regarding Army space efforts. In October 1984, the Army Space Council met to discuss the Army's emerging role in space and produced guidance for future Army efforts. The Army assembled a staff organization to manage its space activities after the other services. For many years, as the role of space in military operations expanded, the Army's interest and influence decreased, but this would change.

By the end of 1984, the Army Management Structure for Space had four components: (1) an Army Space Council chaired by the VCSA; (2) the Army Space Working Group, chartered to support the Space Council with recommendations and act as its coordinating body; (3) the Army Space Office, part of the ODSCOPS, serving as a focal point for space-related matters serving as a liaison to the Joint Staff and the Office of the Secretary of Defense; and (4) the Army Staff

Field Element of AFSPC. The Army Space Office identified five high priority tasks: (1) developing an Army space policy, (2) creating an inventory of existing Army space-related requirements and programs, (3) crafting "near-term enhancements" to Army space involvement in "key areas," (4) developing "Army space-related requirements based on an operational concept for space support to warfighting," and (5) developing "Army options to support a potential unified command for space."[149]

The Army Space Institute and the Army Space Agency

Army space activity increased and reached a critical mass in 1985. In January of that year, the Training and Doctrine Command (TRADOC) directed that the Combined Arms Combat Development Activity (CACDA) form a Space Directorate. Rearranging resources, the directorate was duly formed and given responsibility for developing concepts, doctrine and operational requirements to make the best use of space to support operations. In May, the VCSA, General Maxwell Thurman, directed an Army Space Initiatives Study (ASIS) Group be formed to analyze the Army's role in space and the ways it should use space. In August, CACDA's Space and Concepts Directorates published "Army Space Operations." In September 1985, the Staff Element at AFSPC was renamed the Army Space Planning Group and became the Army element of the newly formed U. S. Space Command. The Army Space Planning Group was under the operational control of the new unified command, but remained subordinate to the Army ODCSOPS.

In 1986, the Army Space Planning Group became the Army Space Agency.[150] The name change did not affect the organization's mission. It would still "assist USCINCSPACE in planning enhancement of space support to ground force components in AirLand Battle doctrine and mission requirements" and "provide Army input to the strategic defense planning process," while providing "support to TRADOC's requirements, concepts and doctrine work." It would also be an "operationally oriented point of contact at USSPACECOM for the U.S. Army Strategic Defense Command [USASDC], the U.S. Army Space Programs Office [ASPO] and the military satellite communications [MILSATCOM] communities" and "assist the ODCSOPS in determining Army space roles, missions, requirements and master plan development."[151]

Between July and December 1985, the ASIS group, directed by Brigadier General William J. Fiorentino, prepared the Army Space Initiatives Study.[152] The Fiorentino group provided (1) an extensive analysis of space and space-related activities in order to develop an operational concept for Army space activities, (2) a plan to acquire and manage qualified space personnel, (3) an Army investment strategy for space, (4) a management strategy and (5) an implementation plan.[153]

In December, the ASIS group presented the results of its study to the Army Space Council. The published four volume study concluded that if used properly, space systems would increase the Army's mission capabilities along the entire spectrum of conflict. However, the study group found that responsibility for developing, coordinating and using these space capabilities was fragmented among the Army's many commands. The group made more than two hundred recommendations to improve the Army's use of space systems and products.

The Army Space Initiatives Study report contained an investment strategy, educational, training and personnel management recommendations, a suggested Army organization for space, an implementation plan, a technological assessment and projections and a discussion of threats. Specifically, the report advocated making the Office of the Deputy Chief of Staff for Operations and Plans the senior Army staff proponent for space, recommended that the Combined Arms Center at Fort Leavenworth become the Army proponent for space and the Command and General Staff College become the lead Army school for space education. The study urged the formation of an Army Space Command as the Army component of USSPACECOM and advocated the Army integrate the use of space and space products into its doctrine. Concretely, the report called for the creation of a Space and Special Weapons Directorate within the Office of the Deputy Chief of Operations and Plans, establishing an Army Space Institute (ASI), the Army Space Technology Research Office (ASTRO) and the Army Space Agency. The Fiorentino study also counseled making the Army Materiel Command responsible for managing space research and development. In addition, the report advocated conducting Mission Area Analyses to discover the potential uses of space systems and capabilities, training soldiers about space systems and creating an additional specialty indicator to trace personnel with experience, education and training in space systems.[154] The four-volume report did not discuss space-related aspects of ballistic missile defense, anti-satellite weapons, or theater missile defense space issues. In the two years following the report's release many of its recommendations were implemented.

The Army Space Institute was established in June 1986 to serve as a clearinghouse for matters relating to the Army's use of space.[155] Functioning this way, as the TRADOC proponent for space and space systems, it would be responsible for developing Army space concepts, doctrine, training, force structure, materiel requirements, techniques and procedures that would apply space systems and technology to "enhance the execution of AirLand Battle Doctrine and support the Strategic Defense Initiative."[156] The ASI maintained a tactical focus. It consistently concentrated on reaching the small unit commander in order to familiarize him with space systems and their use and provided training and support to tactical units. This approach was markedly different from the ways space systems had been treated before ASI was established. Before 1986, the focus of military space systems was on the strategic level and the systems were dedicated to supporting the missions of the Strategic Command and the North American Air Defense Command. The ASI approached its mission aggressively and predicted that space systems would be available at the battalion and company levels. In 1987, the ASI Commandant predicted a future in which advanced positioning systems would allow commanders to know the locations of their subordinate units continuously, space-based communications systems would make line of sight limitations on ground-based radios meaningless which would allow smaller units to act as a whole even though separated by great distance or rough terrain, and that a battalion intelligence staff section would have instant access to real-time satellite imagery and weather information.[157]

Demonstrating the Utility of Space-Based Systems

Over the next year, working at the direction of the VCSA, General Maxwell Thurman, the Institute prepared for the Army Space Demonstration Program (ASDP). The program would serve as ASI's primary experimental vehicle, to show the ways current space-related products could support battlefield commanders and their units, down to the squad level.[158] General Thurman wanted the program to inform the Army of the ways space-based systems would support AirLand Battle Doctrine and not test the technology.[159] The first four proposed demonstrations included the Global Positioning System (GPS) Receiver Position/Navigation, GPS Azimuth Determination, weather and terrain analysis and lightweight small satellite (LIGHTSAT).[160] The Global Positioning System Receiver Position/Navigation demonstration showed the system's capabilities. The Azimuth Determination demonstration showed how useful it would be to mount GPS receivers on combat vehicles in order to orient them and their associated weapons systems. The weather and terrain analysis demonstration provided corps and division commanders with weather support using WRASSE commercial weather receiver systems. LIGHTSAT was intended to demonstrate and evaluate the operational value of lightweight, relatively inexpensive, limited purpose satellites and associated expendable booster vehicles as a cost effective method of providing space-based support to operational and tactical commanders throughout the world. Among the uses envisioned for LIGHTSAT were reconnaissance, intelligence collection, surveillance and target acquisition (RISTA). The lessons learned garnered from the demonstrations would be used to help design future systems.

Fig. 3-32. Artist's drawing of a Global Positioning System satellite.

Fig. 3-33. Drawing of a GPS satellite web.

Fig. 3-34. Drawing of a Military Strategic Tactical and Relay 3 satellite.

By 1989 the new equipment's capabilities had been demonstrated to some Army commanders and units. The first equipment items shown were the Small Lightweight GPS Receiver (SLGR), the WRAASE weather receiver and AN/PSC3 TACSAT radios. The SLGR was a handheld receiver that gave accurate position and navigation data to tactical users. The weather receivers, deployed to Air Force weather teams supporting divisions, separate brigades and other units, used the network of weather satellites to provide them with accurate weather forecasts. The tactical radios could relay and transmit voice and data messages directly between users in the same theater of operations or store and forward messages anyplace in the world using the network of geosynchronous communications satellites. In addition, the research and development undertaken to use GPS to determine accurate azimuth information led to the creation of prototype receiver/processors with special antennae.[161] By August 1990, the objective that General Thurman established for the Army Space Demonstration Program was being realized. After Iraq invaded Kuwait, threatened Saudi Arabia as well as other Persian Gulf states and the stability of a substantial portion of the world's energy supply, a coalition led by the United States deployed troops first in Operation Desert Shield and then in Operation Desert Storm. Many of the tactical units deployed to the Gulf participated in the Army Space Demonstration Program and now wanted this equipment.

The Army was also coming to grips with the issue of developing space expertise. As it re-entered space and participated with the other services in USSPACECOM, personnel managers realized that trained officers would have to fill space-related positions in the new Army Space Agency and on Army staffs. Personnel managers needed to develop the expertise while they were creating the positions to justify the appropriate training programs. The ASI had to develop the training at the same time its combat development actions began to define what training was necessary. In 1986, shortly after the space activities skill code was established, ASI proposed to redefine it, while realizing this did not address the basic need to build expertise.[162] In 1987, a new Space Activities skill code definition was sent to the VCSA with more specific qualifications in duty assignment, military training and civilian schooling.[163]

The Army Astronaut Program

The Army had long had an interest in manned space flight. In January 1959, NASA dealt a blow to the Army's hopes for continued involvement in space exploration when it published the selection criteria for astronauts from the military services. One requirement, stipulating that an astronaut be an experienced jet aircraft pilot, eliminated Army personnel from consideration as astronaut candidates.[164] In 1964, NASA dropped the requirement for pilot experience for crew members, but only in an effort to recruit "scientist-astronauts" to conduct research on space flights. Most of these candidates had superior academic qualifications, usually a doctoral degree in the natural sciences, medicine or engineering, or equivalent experience.[165] Because few of its officers had advanced training in these fields, the Army once again found itself excluded from the manned space program.[166]

Undaunted by these developments, Army commentators and officials continued to press NASA to assign Army officers as astronauts. In a 1968 article in *Military Review*, Major Thomas C. Winter, Jr. argued that the Army should be part of a Manned Orbiting Laboratory, which the space program thought it would deploy in the early 1970s. Using equipment originally designed for the Apollo flights, the program would place a manned laboratory in earth orbit for as long as six weeks at a time. Proclaiming control of space crucial to the national interest, Major Winter contended the Army should enter this program to sponsor scientific research to support its missions. He advocated that selected Army officers pursue graduate schooling for doctoral degrees in space-related disciplines at leading universities to acquire the necessary knowledge and experience to become astronauts. He also recommended that the officers spend time working in the NASA Apollo applications program conducting research and acquiring proficiency in crucial skills.[167]

Senior Army leaders echoed Major Winter's sentiments. In February 1969, General William C. Westmoreland, the Chief of Staff of the Army, took up a similar line of reasoning in a letter to Dr. Robert R. Gilruth, Director of NASA's Manned Spacecraft Center in Houston, Texas. After congratulating Dr. Gilruth on his many accomplishments, Westmoreland voiced concern that the Army still lacked representation in the astronaut program. Emphasizing that the Army had more than 18,000 qualified aviators, the general expressed the conviction that "these men are capable of absorbing the training in the pilot-astronaut program and of contributing to the expanding projects in space exploration." He encouraged the NASA director to review his space projects and the criteria for selecting astronauts to ascertain how the Army might increase its participation in the program.[168]

Gilruth's response held out slim hope for General Westmoreland. The NASA director pointed out that NASA already had enough astronauts for the Apollo flights and until it identified future manned space missions it did not intend to select any more astronauts.

However, he noted that future space crews would incorporate a variety of disciplines, including pilots, engineers, scientists and physicians, for which the Army could easily supply talented candidates. Despite the director's reassuring tone, the Army would wait ten years before one of its officers entered the astronaut program.[169]

In January 1978, NASA announced the selection of 35 new astronaut candidates for the Space Shuttle Program, the first chosen since 1969. This group included the first women and racial minorities chosen; additionally, two new astronaut job titles were created, pilot and mission specialist. Both civilians and military officers were among the candidates; one of the latter was Major Robert L. Stewart, who would become the Army's first astronaut.

Events leading to the formation of this group of astronauts began in the late 1960s as NASA officials began to develop plans for a reusable launch vehicle and orbiter to put people in space. This concept evolved into the shuttle, a space plane that would carry astronauts into orbit and return them safely to earth. NASA viewed the shuttle as an inexpensive way to launch people, satellites, probes, an orbiting station and military hardware into space.[170]

Fig. 3-35. Robert Stewart, the first Army Astronaut, a few meters from the Space Shuttle Challenger, floating untethered.

Major Stewart, along with the other 34 candidates, began a rigorous training and evaluation period at the Johnson Space Center in Houston for assignment to future space shuttle flight crews. After clearing this initial hurdle, Stewart and his colleagues became astronauts in August 1979. Stewart, who held a Master of Science degree in Aerospace Engineering, emerged from the training as a mission specialist, responsible for shuttle operations in areas affecting shuttle experiment procedures. Mission specialists conducted space walks, handled payload and maintenance activities and other operations as needed. Mission specialist qualifications included an advanced degree in engineering, life, physical sciences or mathematics, along with specific age, physical and medical requirements.[171]

In December 1976, NASA and the Department of Defense drew up rules governing the assignment of military personnel to the Shuttle program in a Memorandum of Understanding (MOU). The agreement set the tour of duty at five years with the possibility of a one-year extension. At the end of their tours, personnel either retired or resumed duty with their respective services. Any military officer detailed to the shuttle service reported directly to NASA with respect to his astronaut responsibilities. Individual officers remained subject to the Uniform Code of Military Justice and NASA prepared and maintained fitness and effectiveness reports in accordance with the regulations of each member's service. NASA also reimbursed the services for all pay and allowances made to personnel detailed to the agency.[172]

On his initial mission in 1984, Lieutenant Colonel Stewart and another astronaut were the first to perform an untethered space walk using the manned maneuvering unit, or jet pack, on Space Shuttle *Challenger*. He also took part in a classified military mission in 1985. Altogether Stewart logged 289 hours in space. After he left the astronaut corps, he became a brigadier general and deputy commander of the U.S. Army Strategic Defense Command in Huntsville, Alabama. Colonel Sherwood Spring, later head of the Army Space Program Office, became the Army's second astronaut in 1980. As a mission specialist aboard a 1985 shuttle voyage, he launched three communications satellites and performed two space walks to assess construction techniques in space.[173]

Fig. 3-36. Launch of a Space Shuttle flight.

In 1986, the Pentagon established the Military Man in Space program as part of Shuttle operations. The Air Force was the over-all Executive Agent and the Office of the Deputy Chief of Operations and Plans, Department of the Army (ODCSOPS, DA) became the Executive Agent for the Army program. The object of the Military Man in Space Program was to evaluate, through experiments proposed by each uniformed service and approved by DoD, ways in which military operations on earth could be improved using space-related facilities and technologies. In 1987, the Army proposed three experiments that it thought would improve its war fighting capabilities, Terra View, Terra Scout and Terra Geode. These three experiments played significant roles in the future of manned space flight.[174]

Terra View is a four-phase experiment to make observations of ground sites. The first three phases were designed to be conducted on shuttle flights while the fourth phase would be conducted on the space station. Terra View's first phase determined what Army astronauts could detect from space of military value using cameras and binoculars while observing training areas both inside and outside the continental United States. In Terra View's second phase, the Army augmented the astronauts' visual equipment with communications equipment to allow them to pass information directly to ground commanders in real time. Army Colonel Jim Adamson participated in this portion of Terra View. Phase Three used Army experts instead of astronauts to observe ground activity and communicate tactical information to the ground commander. This phase encompasses two other Army Military Man in Space experiments, Terra Scout and Terra Geode. Lessons learned from the site observations and direct communications between the Shuttle and ground sites were used to determine the Army's communications and observation requirements.

The Army Intelligence Center and School developed and sponsored Terra Scout. Its intent was to determine what an experienced imagery interpreter can observe of military value from the Space Shuttle. The Shuttle crewmembers used the Spaceborne Direct View Optical System, an optical device that uses a manual pointing and tracking system with manually controlled zoom lens. Army Astronaut Lieutenant Colonel Jim Voss and Payload Specialist Chief Warrant Officer Tom Hennen performed the first phase of Terra Scout during Space Shuttle Mission STS-44 in November 1991.

In January 1987, the Army Chief of Engineers proposed using a military geologist's observations from earth orbit to evaluate terrain conditions for tactical movement. Terra Geode itself is a four-phase experiment. The results of the first two phases, based on NASA astronauts' observations, helped refine the experiment's design and strengthen the justification for an expert observer to explore potential Military Man In Space applications fully. Military astronauts using standard equipment available to NASA under the Earth Observation Program conducted the experiment's first phase. Dr. Kathy Sullivan, a NASA astronaut with a geology background, conducted the second phase observations during a five day space shuttle mission launched 24 April 1990. She demonstrated the feasibility of terrain analysis from earth orbit and was able to make basic observations of ground targets, determine soil color, type, ground cover, and other terrain data. She also provided guidance for improving the conduct of the next phase of the experiment. Dr. Sullivan completed Phase II of Terra Geode during another shuttle flight into space in 1992. The third phase will be carried out by an Army geologist on the Shuttle and will be the demonstration and validation phase to prove the value of employing the capabilities of a trained expert military observer. The experiment's final phase would integrate lessons learned into possible Army requirements for a space station and for permanently stationing military geologist/terrain analysts there. The Army has selected three officers and one warrant officer as primary, backup and alternate Payload Specialists.

In 1987, as its participation in NASA burgeoned, the Army established an Army Astronaut Detachment at the Johnson Space Center. That same year, the Army formalized its relationship with NASA in a new MOU that governed the assignment of personnel at the astronaut detachment.[175] In 1988, the unit fell under the control of the new Army Space Command (ARSPACE), the Army's central organization providing operational space support.[176]

Fig. 3-37. Army Astronaut Lieutenant Colonel Nancy J. Currie aboard the space shuttle, maneuvering the remote arm.

The Army's renewed interest in space and space-related assets began with its participation in the TENCAP program and blossomed as it underwent a doctrinal renaissance and training revolution that resulted in AirLand Battle Doctrine. The demands of the new doctrine forced the Army's leadership to look toward the ultimate high ground to satisfy a commander's new critical information requirements. By mid-1985 the ASIS group was developing a report that would give the Army a vision for the potential of space. Mixed in with the vision were a series of practical recommendations to realize the vision. The study advocated a division of labor between the Army Staff, ARSPACE, ASI, AMC and UASSDC. The object was to give the Army the tools it would need to satisfy its current and future needs.

As the Army began debating the ways it should use space, it began developing doctrine and operational concepts and created a space command headquarters. It also grappled with the issue of creating a cadre of space-trained soldiers and began promoting Army space exploitation. However, this was only a beginning, as the Army still had to create a doctrine that would exploit space assets. That the ultimate end users of space-related information did not participate in forming their own requirements led to an imperfect acquisition strategy. Most important was the difficulty of getting the majority of the Army's senior leadership to wholeheartedly support operational space exploitation roles and missions.

As the Cold War abruptly ended, the Army was faced with a new strategic environment. The world grew smaller as the United States had fewer overseas bases. As the Army began to change from a forward deployed force to one that could project power, it would depend more on space capabilities for surveillance, warning, communications, navigation, meteorology and geodesy.

End Notes

[1]SAFSCOM General Orders 4, dated 12 February 1971 and SAFSCOM General Orders 7, dated 19 March 1971
[2]Quoted in Clarence A. Robinson, Jr., "Missile Engineering, Prototype Site Defense Construction Set," *Aviation Week & Space Technology*, 29 April 1974.
[3]System Charters SAFEGUARD Ballistic Missile Defense System dated 24 March 1971 and 7 May 1971.
[4]Each module consisted of three radars and 100 interceptors. By the early 1970s commercial software and data processing systems had developed to the extent that off-the-shelf technology could be incorporated into advanced systems at lower cost.
[5]Hardsite Defense was shortened to Site Defense in April 1972.
[6]Site Defense System Fact Book dated 1 December 1974.
[7]"Testimony of Secretary of Defense James R. Schlesinger to the Senate Committee on Armed Services Conducting Hearings on Military Procurement for FY76 and the 3-Month Transition Period" Ballistic Missile Defense Systems Command, Annual Historical Review FY76 and 7T, Volume II.
[8]Effective date was 28 October 1975.
[9]General Orders 12, dated 22 May 1974. Similarly, the SAFEGUARD System Manager, System Organization and the Systems Command were renamed the Ballistic Missile Defense (BMD) Program Manager, BMD Program Office and BMD Systems Command (BMDSCOM) respectively. The SAFEGUARD Systems Site Activation Command was assigned to BMDSCOM under General Orders 3 issued on 17 June 1974.
[10]The role of the ABMDA had been to develop advanced technologies to minimize deployment times. The BMDATC was a Class II activity with its own competitive area and line in the DOD budget.
[11]See BMDATC General Orders 4, dated 8 July 1975 and BMDATC General Orders 5, issued 15 October 1975
[12]In the Huntsville competitive area, which included small field offices in New York, New Jersey, North Carolina, Florida, California, and Montana, the authorized personnel strength reduced from 577 civilians and 46 military personnel to 345 civilians and 31 military personnel. The personnel strength had already been reduced in July 1973 to 58% of that allowed prior to the ABM treaty.
[13]BMD Overview Briefing Presented to Drs. Fletcher and Agnew on 21 July 1983.
[14]Secretary of Defense Schlesinger Testimony to the Senate Committee on Armed Services Conducting Hearings on Military Procurement for FY76 and the 3-Month Transition Period. Schlesinger continued stating that this "R&D … assists in the design and evaluation of our strategic offensive systems … [and] assists our intelligence agencies in the assessment of Soviet BMD capabilities."
[15]Secretary of Defense Harold Brown Annual Report to Congress dated 2 February 1978, quoted in Ballistic Missile Defense Organization, *Ballistic Missile Defense Organization Annual Historical Review Fiscal Year 1978, 1 October 1977 – 30 September 1978.*
[16]*Ibid.* p. vii.
[17]In recognition of these advances, in October 1977, Secretary of the Army Clifford Alexander authorized the first Army Award for Project Management outside the Department of Army Readiness Command, be awarded to Brigadier General John G. Jones for his outstanding accomplishments on the BMD program.
[18]United States Senate, Senate *Hearings on FY 77 Authorizations, Part 12, Research and Development*, pp. 6679, 6682-84, and 6686-87. Quoted in Baucom, *Origins*, pp. 98-99.
[19]Summary of BMDSCOM Activities FY76/7T, Unpublished typescript.
[20]Bulk filtering refers to the ability to quickly eliminate lightweight objects such as tank debris and traffic decoys from further consideration. Discrimination is the more precise ability to differentiate between precision decoys and RV's. New Technology, Inc., *History of Ballistic Missile Defense Developments: A Synopsis* (Huntsville: BMDATC, 1983), p. 31.
[21]See, for example, Historical Office, *Ballistic Missile Defense Systems Command Annual Historical Review, 1 October 1976—30 September 1977*, Volume I, pp. IV—13-14.
[22]The SAFEGUARD MSR was 50 times larger that the STR. Similarly the battle management data processor was 10 times larger than its proposed replacement which was also 50% more capable. Briefing to Drs. Fletcher and Agnew, 21 July 1983.

[23]Clarence A. Robinson, Jr., "Strategic Shifts – 3: Missile Defense Radar System Tests Set," *Aviation Week and Space Technology* 20 September 1976. The STR "featured separate waveforms for search, bulk filtering, tracking, discrimination, and interceptor tracking and guidance." McDonnell Douglas Astronautics Company, *Ballistic Missile Defense: A History of Achievement* (Huntington Beach, California: 1982).
[24]Frances Martin, et al., *Four Decades of Progress*, p. 53
[25]Bell Aerospace, "Optical Aircraft Measurement Program," http://www.ball.com/aerospace/oamp.html.
[26]Public Affairs Office, U.S. Army Strategic Defense Command, "Airborne Optical Adjunct (AOA) Project," 1985.
[27]As part of the modification to the Boeing 767, the aircraft was equipped with an 86-foot long and 10-foot high cupola to house the optical sensor.
[28]Ballistic Missile Defense Advanced Technical Center, Annual Historical Review FY75.
[29]Kenneth J. Stein, "New Missile Defense Systems Studied," *Aviation Week and Space Technology* 11 October 1976: 46-47.
[30]For a brief overview of the evolution of DE weapons see Baucom, *Origins*.
[31]Major General Donald L. Lamberson (USAF), "DoD's Directed Energy Program: Its Relevance To Strategic Defense," *Defense* June 1983: 16-20.
[32]Matthew Nichols, "Early Concepts for Space-Based Ballistic Missile Defense," unpublished paper presented at the Conference of Army Historians, August 2002.
[33]Clarence A. Robinson, Jr., "Amy Pushes New Weapons Effort," *Aviation Week and Space Technology* 16 October 1978.
[34]*Ibid.*
[35]Baucom, *Origins*, p. 109. Space was viewed as the optimum medium as particles in the atmosphere caused thermal blooming and scattering of the beam.
[36]Ballistic Missile Defense Organization, *Ballistic Missile Defense Organization Annual Historical Summary, FY81,* (Huntsville: BMDO, n.d.) p. 1.
[37]Historical Office, *Ballistic Missile Defense Systems Command Annual Historical Review, 1 October 1976—30 September 1977*, Volume I, pp. IV—118-119.
[38]The endoatmosphere is defined as the distance between the Earth's surface and 300,000 feet altitude.
[39]The small phased array radars were 1/40th the size of the site defense. The interceptor measured 15 feet, one half the size of a SPRINT missile. The missiles would be capable of accelerating to 8,500 feet in only 1.5 seconds and would reach an altitude of 4,000 to 5,000 feet in about one second. Baucom, *Origins*, p. 117 and Office of Technology Assessment, *MX Missile Basing* (Washington, D.C.: 1981). As defined in the Ballistic Missile Defense Organization FY84 Annual Historical Review, n.d. "a distributed data processing system is one in which there exists a multiplicity of interconnected processing resources able to cooperate under system-wide control on a single problem with minimal reliance on centralized procedures, data, or hardware."
[40]*MX Missile Basing*, p. 8. Each of the 200 missiles would be based in clusters of 23 shelters. Each cluster would contain one MX, 22 decoys, 23 shelters, one large transporter truck and one maintenance truck.
[41]BMD Program Office, "FY1982 Historical Input for BMP," undated, p. 2 cited in Walker, Martin, *Four Decades*, p. 45.
[42]The treaty for example would also have limited the number of LoAD batteries to 18, as each battery included a radar. See *MX Missile Basing*, p. 142.
[43]Ballistic Missile Defense Organization, *Ballistic Missile Defense Organization Annual Historical Review Fiscal Year 1979* (Huntsville: BMDO, n.d.).
[44]This type of kill vehicle contains no warhead and destroys its target by direct impact - kinetic energy.
[45]The HOE measured 70.6 feet and weighed a total of 68,081 pounds, 1200 kilograms attributed to the kill vehicle. The kill device, a radial net, has been likened to the folded skeleton of an umbrella with weights attached to its ribs.
[46]Baucom, *Origins*, p. 103.
[47]The first HOE flight occurred on 7 February 1983. The test achieved 85% of its goals. Some opponents accused the Army of employing homing beacons on the target and charges on the interceptor. The GAO refuted accusations that the HOE tests were rigged in their report "Ballistic Missile Defense: Records Indicate Deception Program Did Not Affect 1984 Results" published in July 1994.
[48]Lockheed Missiles and Space, "Homing Overlay Experiment," http://lmms.external.lmco.com/newsbureau/photos/hoe1_caption.html.

[49]Quoted in *U.S. Army First in Space and Strategic Defense*, USASDC Public Affairs Office pamphlet.
[50]Baucom, *Origins*, p. 103.
[51]The SRHIT successfully completed its first flight test to assess missile performance and stability on 20 January 1984. This was the first in a series of nine tests.
[52]At the request of the Army Staff the name change took place in January 1986.
[53]U.S. Army Strategic Defense Command, *U.S. Army Strategic Defense Command Annual Historical Review Fiscal Year 1987* (Huntsville, Alabama: USASDC, n.d.).
[54]The FLAGE measured 148.94 inches in length and 9 inches in diameter. The missile weighed 504.9 a total of pounds.
[55]Traveling at a rate of 3200 feet per seconds, the entire flight lasted 7 seconds. The tests took place at White Sands Missile Range.
[56]"Ballistic Missile Defense, The Army's Strategic Weapons Program," .n.d, p. 5. Clarence A. Robinson, "Soviets Grasping Strategic Lead," *Aviation Week & Space Technology* 30 Aug 1976: 14-18
[57]Trolone, "Ballistic Missile Threat," p. 6.
[58]Malcolm W. Browne, "Weapon that Fights Missiles Could Alter World Defense Focus" *New York Times* 4 December 1978: 1, 3.
[59]Report of the President's Commission on Strategic Forces, 6 April 1983, Brent Scowcroft Chairman, p. 18, quoted in Ruth Currie-McDaniel and Claus Martel, *The U.S. Army Strategic Defense Command, Its History and Role in the Strategic Defense Initiative*, 3rd edition (Huntsville, Alabama: USASDC, 1989), p. 25. The report stated "vulnerability of such silos in the near term ... is not sufficiently dominant part of the overall problem of ICBM modernization to warrant ... ABM defense of these silos."
[60]*Ibid.*
[61]*Ibid.*, p. 18.
[62]Dennis M. Gormley, *Double Zero and Soviet Military Strategy: Implications for Western Security* (London: Jane's Publishing Company, 1988), pp.13-14.
[63]Until 1964, when China exploded its first nuclear weapon, there was only a Soviet strategic threat. This was a complication but not a threat because the Chinese had no missiles and few bombs. However, there was no way to gauge the speed of Chinese ICBM development and no way to determine how much saber rattling was genuine and how much was bombast. After the Sino-American rapprochement in 1972 this was a moot point. The post-Cold War period brought renewed concerns about Chinese military development.
[64]This point was made by John M. Collins, *U.S.-Soviet Military Balance, 1980-1985* (Washington, D.C.: Pergamon-Brassey's, 1985), p. 4.
[65]See the statement of President Reagan and the Pentagon related to the MX missile in "Text of Reagan and Pentagon Statements on MX Missile Proposal," *New York Times*, 23 November, 1982, A14.
[66]See Gormley, pp. 36-43, especially pp. 37-39.
[67]See Gormley, p. 40 and Stephen Kotkin, *Armageddon Averted: The Soviet Collapse, 1997-2000* (New York: Oxford University Press, 2001).
[68]The standard works include Robert A. Doughty, *The Evolution of U.S. Army Tactical Doctrine, 1946-1976* (Fort Leavenworth: Combat Studies Institute U.S. Army Command and General Staff College, 1979), Paul H. Herbert, *Deciding What Has To Be Done: General William E. DePuy and the 1976 Edition of FM 100-5 Operations* (Fort Leavenworth: Combat Studies Institute, U.S. Army Command and General Staff College, 1988), Jonathan M. House, *Combined Arms Warfare in the 20th Century* (Lawrence: University Press of Kansas, 2001) and John L. Romjue, *From Active Defense to AirLand Battle: The Development of Army Doctrine, 1973-1982* (Fort Monroe: Historical Office, U.S. Army Training and Doctrine Command, 1984). DePuy's own views may be found in Romie L. Brownlee and William J. Mullen III, *Changing An Army: An Oral History of General William E. DePuy, USA Retired* (Washington, D.C.: Government Printing Office, n.d.) and Richard M. Swain (comp.), *Selected Papers of General William E. DePuy: First Commander, United States Army Training and Doctrine Command,* edited by Donald L. Gilmore and Carolyn Conway (Fort Leavenworth: Combat Studies Institute, U.S. Army Command and General Staff College, 1994). The then Commandant of the U.S. Army Command and General Staff College expressed his point of view in John H. Cushman, *U.S. Army Operational Doctrine as Expressed in FM 100-5 and the Defense of Central Europe* (McLean: MITRE Corporation, 1978). Retrospective accounts include William E. DePuy, "FM 100-5 Revisited." *Army* 30 (November 1980):12-17, L. D. Holder, "Doctrinal Development, 1975-

1985," *Military Review* 65.5 (May 1985):50-52, Donn A. Starry, "A Tactical Evolution-FM 100-5," *Military Review* 58.8 (August 1978):2-11, Donn A. Starry, " A perspective on American Military Thought," *Military Review* 69.7 (1989): 2-11 and Huba Wass de Czege and L. D. Holder, "The New FM 100-5," *Military Review* 62.7 (July 1982):53-70. A recent account of the early stage is Richard Lock-Pullan, "'An Inward Looking Time': the United States Army, 1973-1976," *The Journal of Military History* 67 (April 2003): 483-512. For more general information on the subject of military innovation, see Colin Gray, *Weapons Don't Make War: Policy, Strategy and Military Technology* (Lawrence: University Press of Kansas, 1993), Stephen Peter Rosen, *Winning the Next War: Innovation and the Modern Military* (Ithaca: Cornell University Press, 1981), McGregor Knox and Williamson Murray (eds.), *The Dynamics of Military Revolution, 1300-2050* (New York: Cambridge University Press, 2001), Williamson Murray and Allan R. Millett (eds.), *Military Innovation in the Interwar Period* (New York: Cambridge University Press, 1996).

[69]This new thinking concentrated on developing new antiballistic missile technologies, including reduced-sized high speed integrated circuit computer processors, High Endoatmospheric Defense Interceptors (HEDI), the Ground-Based Laser and the Airborne Optical Adjunct. For greater detail see the section on missile defense.

[70][Linton Brooks], "CNO and the Strategic Defense Initiative," n.d., quoted in Baucom, *Origins*, p. 190.

[71]The name was coined by opponents who questioned a reliance on "exotic technology."

[72]Speech by President Ronald Reagan, "Peace and National Security, A New Defense," 23 March 1983.

[73]President Reagan issued the directive on 25 March 1985.

[74]Donald Baucom, "Missile Defense Milestones 1944-2000,"
http://www.acq.osd.mil/bmdo/bmdolink/html/milstone.html.

[75]Issued 6 January 1984.

[76]Effective 15 April 1984.

[77]William H. Taft IV, Deputy Secretary of Defense, issued a revision to Directive No. 5141.5 on 4 June 1987 which transferred supervision of the SDIO program to the Deputy Secretary of Defense.

[78]In addition to the Army, Air Force and Navy, other participants in the SDIO were the Defense Nuclear Agency and the Defense Advanced Research Project Agency. Teledyne Brown Engineering, "The Relevance of Previous Anti-Ballistic Missile Programs to the Strategic Defense Initiative," Special Report SS89-USASDC-3221, Contract # DASG60-87-C-0042, 28 March 1989, 9-5.

[79]General Orders 26, dated 24 October 1985. The effective date for this transition was 1 July 1985.

[80]Once again headed by lieutenant general, the USASDC was headquartered in Washington, D.C.

[81]The eight research areas of the Fletcher Report were consolidated into five program elements: surveillance, acquisition, tracking and kill assessment; DEW; KEW; survivability, lethality and key technologies; and systems concepts/battle management.

[82]Paul Warnke, a director of the Arms Control and Disarmament Agency in the Carter Administration observed that "S.D.I. is all things to all people. To the President, it is saving peoples' lives. To Defense Secretary Weinberger, it is a technological stepping stone from missile defense ... To others, it was simply a means of defending missiles. To some, it was a bargaining chip in arms-control negotiations, while to others ... it was untouchable." Quoted in Philip Boffey, et. al. *Claiming the Heavens The New York Times Complete Guide to the Star Wars Debate* (New York: Times Books, 1988), p. 65.

[83]Keith Payne, *Strategic Defense: "Star Wars" in Perspective*, Foreword by Zbigniew Brzezinski (Langham, MD: Hamilton Press, 1986).

[84]Lou Cannon, "Reagan's Big Idea. (Ronald Reagan's Strategic Defense Initiative)" *National Review* 22 February 1999, http://www.findarticles.com/cf_natrvw/m!282/199_Feb_22/53703734/print.jhtml and Boffey, et. al. *Claiming the Heavens*.

[85]Reprinted from Chapter 12, Walker, Martin and Watkins, *Four Decades of Progress*.

[86]Reprinted with additions from Chapter 13, Walker, Martin and Watkins, *Four Decades of Progress*.

[87]As designed the laser would be "several football fields long and situated on a 20-square mile site." Heike Hasenauer, "Army Takes the Lead in ASAT," *Soldiers,* August 1989. A FEL uses electrons that have been "freed" from atomic nuclei and are accelerated to near the speed of light in a particle accelerator and then "wiggled magnetically" to produce a beam. Unlike a regular laser beam, a FEL "can be 'tuned' to any wavelength, from microwave to the ultraviolet" with research continuing to explore means to extend the range up to other spectrums. Boffey, et. al., *Claiming the Heavens*, pp. 35.

[88]Besides boost phase intercept capabilities, the projected benefits of a GBL system were target imaging and kill assessment, an unlimited magazine and target discrimination. Briefing, "GBFEL Site Dedication, White Sands Missile Range, New Mexico, 2 July 1990.
[89]Colonel Nicholas Barr_n, et. al., "U.S. Army Strategic Defense Command Ground-Based Laser Project Office A History and Lessons Learned, 1986 to 1991," 19 April 1991.
[90]By 1986, FEL systems had already demonstrated the most powerful laser to date operating at only 42% efficiency (converting electrical power into laser light). Boffey, et. al., *Claiming the Heavens*, pp. 35-36.
[91]Briefing, "GBFEL Site Dedication." The dedication ceremony took place in July 1990. The USASDC completed termination of the Ground-Based Laser Project Office on 1 August 1991.
[92]Directed Energy Weapons Fact Sheet, prepared by the USASDC Public Affairs Office, dated March 1992.
[93]Office of Technology Assessment, *Strategic Defenses*, p. 154.
[94]GAO Report "Ballistic Missile Defense Information on Directed Energy Programs for Fiscal Years 1985 Through 1993." GAO/NSIAD-93-182, dated June 1993. The objectives not yet achieved included: generation of a scalable high-power beam, development of electrical power source, test of integrated ground test accelerator on the ground and Test of NPB operation in space.
[95]Directed Energy Weapons Directorate, "Information Paper," dated March 1993.
[96]Reprinted with additions from Chapter 14 of Walker, Martin and Watkins, *Four Decades of Progress*.
[97]On 9 November 1985, Secretary of the Army John O. Marsh, Jr., signed the charter for the ERIS Project Office and appointed the Army's first civilian project manger, Mr. James Katechis.
[98]John Bosma and Richard Whelan, *Guide to the Strategic Defense Initiative* (Arlington, Virginia: Military Space, 1985), 257.
[99]Historical Office, *U.S. Army Strategic Defense Command – Annual Historical Review Fiscal year 1986* (Huntsville: USASDC, n.d.), II-151.
[100]The integrated system test vehicle and ERIS flight version measured 13 feet 2 inches from the nose tip of the kill vehicle to the end of the adapter section. During flight the ERIS would reach speeds of about 4 miles per second.
[101]"Strategic Defense test flight 'an unqualified success,'" *The Redstone Rocket*, 6 February 1991: 3.
[102]Ground-Based Interceptor Project Office Historical Report for Fiscal Year 1992.
[103]Reprinted with additions from Chapter 15 of Walker, Martin and Watkins, *Four Decades of Progress*.
[104]Historical Office, *U.S. Army Strategic Defense Command Annual Historical Review Fiscal Year 1987* (Huntsville: USASDC, 1990), 239.
[105]Boeing determined "a sensor with a 20-degree by 100-degree field of view with a detection range of about 100 miles could provide coverage of a 100,000 square mile defended area." Bosma and Whelan, *Guide*, 118. The aircraft carried 25 linked computers. The sensor system included 38,000 detectors.
[106]"Airborne surveillance testbed successfully tracks targets," *The Redstone Rocket*, 26 June 1992: 4.
[107]Reprinted with additions from Chapter 15 of Walker, Martin and Watkins, *Four Decades of Progress*.
[108]Historical Office, *AHR FY86*, II-144.
[109]Media Briefing presented by GBR Project Manager, Colonel Arthur C. Meier II, on 5 June 1989.
[110]Developed by Raytheon, the GBR-X would be a dual-field-of-view radar, able to see at both short and long distances, by incorporating both limited-field-of-view and full-field-of-view techniques. The radome measured 80 feet in diameter. The radar enclosure would be 43' W X 50'D X53'H and weigh 750,000 pounds.
[111]Vincent Kiernan, "Defense Panel Gives Green Light to SDI Ground-Based Radar," *Space News* (30 July –5 August 1990), 17.
[112]Reprinted with additions from Chapter 15 of Walker, Martin and Watkins, *Four Decades of Progress*.
[113]Currie-McDaniel and Martel, *Strategic Defense Command*, 70.
[114]"SDC awards contract to McDonnell Douglas," *The Redstone Rocket*, 5 October 1988: 2.
[115]Ground-Based Surveillance and Tracking System Project Office Historical Report for Fiscal Year 1993, dated 22 October 1993. A delayed deployment precluded the development of the GSTS as a precursor to the Brilliant Eyes sensor system.
[116]Reprinted with additions from Chapter 15 of Walker, Martin and Watkins, *Four Decades of Progress*.
[117]The rail launched HEDI would be the fastest interceptor traveling at Mach 15, about 10,000 mph. The kill vehicle weighed 800 pounds.... Hasenauer, "Army Takes the Lead."

[118]Mr. Alan Sherer, the second civilian to selected as Program Manager was named the HEDI PM later that year. On 29 April 1991, Mr. Sherer was the first civilian to be named Project Manager of the Year.
[119]Historical Office, *U.S. Army Strategic Defense Command Annual Historical Review, Fiscal Year 1989*, unpublished. The USASDC awarded the HEDI contract to McDonnell Douglas in January 1986.
[120]The shroud protects the sensor and its sapphire crystal window during the early part of flight. The window is cooled with liquid nitrogen after the shroud is removed. Marsha Taylor, "SDC conducts successful kinetic kill vehicle test," *The Redstone Rocket*, 31 January 1990: 5.
[121]HEDI Historical Report for Fiscal Year 1990, dated 19 November 1990.
[122]"U.S. Army's HEDI Concept Evolves Into More Capable Two-Stage Interceptor," *Aviation Week & Space Technology* 7 January 1991: 62.
[123]No intercepts were attempted during these HEDI tests. HEDI Historical Report for Fiscal Year 1992, dated 23 November 1992.
[124]Press Release, "Termination of Contract for High Endoatmospheric Defense Interceptor," 16 September 1992.
[125]Boffey, et.al., *Claiming the Heavens*, pp. 198-200.
[126]Marshal Grechko quoted in Reagan's 13 July 1985 "Radio Address to the Nation on the Strategic Defense Initiative," http://www.reagan.utexas.edu/resource/speeches/1985/71385a.htm. The speech adds that the Soviets have continued to pursue missile defense research for the last 20 years, to deploy an ABM system around Moscow, and have begun to construct new facilities, such as the Krasnoyarsk radar, which violate the treaty.
[127]Boffey, et.al., *Claiming the Heavens*, pp. 198-200.
[128]Joseph Cirincione, "A Brief History of Ballistic Missile Defense," Speech presented at a conference in Como, Italy, 2-4 July 1998, adapted from *The Persistence of the Missile Defense Illusion*, http://www.ceip.org/Programs/npp/bmdhistoyr.htm. Mr. Cirincione was a Senior Associate with the Carnegie Endowment for International Peace.
[129]Donald Baucom, "The U.S. Missile Defense Program, 1944-1994: Fifty Years of Progress," dated 14 November 1994. Brilliant Pebbles were described as "stand-along, 'un-garaged' interceptors" weighing 10-25 kilograms equipped with on-board sensors and computers.
[130]Reprinted from Chapter 16 of Walker, Martin and Watkins, *Four Decades of Progress*.
[131]Equipped with a state of the art imaging infrared seeker, high-density lightweight electronics, low-drift fiber optic IMU, and a compact high energy liquid or solid-divert propulsion system the LEAP kill vehicle weighed 15 pounds.
[132]Public Affairs Office Fact Sheet, "Lightweight Exoatmospheric Projectile (LEAP)," April 1992; Vincent Kiernan, "SDI: Setbacks Expected – Critics Blast Program After Last Month's LEAP Failure," *Space News*, 13 July 1992, 3 and 29; and "Third LEAP test misses intercept," *BMD Monitor*, 2 July 1993; Robert Holzer, "Navy Panel May Speed Missile Tests," *Defense News*, 31 July – 6 August 1995, 38.
[133]Missile Defense Agency Fact Sheet, "Sea Based Midcourse Defense," dated 2002.
[134]Reprinted with additions from Chapter 16, Walker, Martin and Watkins, *Four Decades of Progress*.
[135]Headed by Brigadier General J. Morgan Jellett, the Kinetic Energy Anti-Satellite Joint Program Office was established on 27 February 1989. "DAB Selects Ground-Based Mode for Initial KEW ASAT," *Defense Daily*, 15 December 1989, 405. The Army also has previous experience with ASAT technology. During the NIKE-ZEUS era, the command was directed to develop an ASAT capability using the ZEUS missile system. This ability was validated in 1962.
[136]Historical Office, *U.S. Army Strategic Defense Annual Historical Review Fiscal Year 1989*, unpublished.
[137]Vincent Kiernan, "Lengthy Delay Hits Laser ASAT Work," *Space News*, 20 August 1990: 1.
[138]James Asker, "Rockwell Selected as Sole Contractor for $100-Million ASAT Design Effort," *Aviation Week and Space Technology*, 23 July 1990.
[139]Anti-Satellite Joint Program Office Historical Report for Fiscal Year 1991, dated 5 December 1991.
[140]Anti-Satellite Joint Program Office Historical Report for Fiscal Year 1992, dated 27 January 1993.
[141]Gerda Sherrill, "SSDC strapdown test a success," *The Redstone Rocket*, 21 September 1994, 6.
[142]USASSDC News Release, "KE ASAT hover test is highly successful," 12 August 1997. Equipped with an on-board seeker, processor, and guidance equipment, the 94-pound device unfolds "a sail-like device … shortly before impact to strike and disable the target" and minimize space debris.
[143]Reprinted with additions from Chapter 18 of Walker, Martin, and Watkins, *Four Decades of Progress*.
[144]Briefing, "U.S. Army Strategic Defense Command High energy Laser Systems Test Facility," dated

5 December 1991.
[145]General Orders 12, dated 8 May 1991. This transfer "is consistent with OSD desire to consolidate strategic test facilities and weapons development under one command."
[146]HELSTF Historical Report for Fiscal Year 1991, dated 5 December 1991.
[147]Combined Arms Center History Office, *U.S. Army Combined Arms Center Annual Historical Review 1988* (Fort Leavenworth, 1989), pp. 270-275.
[148]The following, unless otherwise noted, is based on Chapter Two (Army Space History) of the Army Space Reference Text.
[149]"Message: Army Management Structure for Space, 19 November 1984."
[150]Provisionally, the new agency would "represent Army interests in developing space-related strategic defense planning." It would also "investigate and report to appropriate Army organizations on space-related technology research and development programs of other services and DoD activities which may apply to the mission requirements of Army forces." See "Message regarding the provisional Activation of the U.S. Army Space Agency, 28 July 1986."
[151]"Message: Army Element of USSPACECOM, 18 October 1985."
[152]It was aided by the Intelligence and Security Command, the RAND Corporation Arroyo Center and the TRADOC Assistant Chief of Staff for Intelligence.
[153]TRADOC Military History Office, *Annual Historical Review 1986* (Fort Monroe: TRADOC Military Office, 1987), p. 80 (secret—information used is unclassified).
[154]TRADOC Military History Office, *Annual Historical Review 1986* (Fort Monroe: TRADOC Military Office, 1987), p. 81 (secret—information used is unclassified). See also *Mitchell, Apogee, Perigee and Recovery*, pp. 45-48.
[155]Combined Arms Center History Office, *U.S. Army Combined Arms Center Annual Historical Review 1991* (Fort Leavenworth, 1992), p. 331 and Combined Arms Center History Office, *U.S. Army Combined Arms Center Annual Historical Review 1994* (Fort Leavenworth, 1996), "Chapter 7: Combat Developments."
[156]"Briefing, U.S. Army Space Institute, January 1989, Mission" (chart 2.).
[157]Steven Siegel, "Army Space Institute," *Army Trainer* Summer 1987:20-21.
[158]Marquis Shepherd, "Army Unit to Bring Technology of Space Down to Earth for Troops," *Kansas City Times*, 13 January 1988:B-3.
[159]Combined Arms Center History Office, *U.S. Army Combined Arms Center Annual Historical Review 1987* (Fort Leavenworth, 1988), pp. 145, 160, 161, 221-226
[160]Between December 1986 and March 1987, the Deputy Chief of Staff for Research, Development and Acquisition (DCSRDA) received demonstration submissions from organizations across the Army. From these, the DCSRDA selected five and presented them to the Army Space Council in April 1987, at the same meeting the ASI presented the proposed Army Space Concept. The VCSA ordered ASI to review the space demonstration program to ensure the proposed demonstrations melded to the proposed concept. In June, the Institute presented a final list for the proposed Army Space Demonstration Program, which was approved in August 1987.
[161]For a very brief treatment, see Donald W. Evans, *U.S. Army Use of Space-Based Systems*, www.globalsecurity.org/space/library/report/1990/EDW.htm, accessed on 4 February 2003.
[162]Combined Arms Center History Office, *U.S. Army Combined Arms Center Annual Historical Review 1986* (Fort Leavenworth, 1987), pp. 88-93
[163]Combined Arms Center History Office, *U.S. Army Combined Arms Center Annual Historical Review 1987* (Fort Leavenworth, 1988), p. 226. At this time, the Army was not seeking the same kind of expertise it had developed earlier. Now, the emphasis was on applied knowledge, ways in which the new tools serve as a force enabler and enhancer.
[164]In 1958, at NASA's request, the four armed services submitted the names of 508 pilots, including 35 Army officers who had graduated from test pilot school. In January 1959, NASA concluded that only 110 of them met the minimum standards of age, height, education, physical condition and jet pilot flying time. None were Army personnel. Eventually, NASA pared the 110 down to the seven Mercury astronauts after a series of written tests, interviews and physical and psychological examinations. Army pilots generally failed to qualify because they flew only helicopters and rotary engine aircraft rather than jets. See William E. Burrows, *This New Ocean: The Story of the Space Age* (New York: Random House, 1998), pp. 288-289.

[165]While many of these "scientist-astronauts" successfully completed the NASA training program, only one, Harrison H. Schmidt, ever flew on an Apollo mission. Edward G. Gibson, Owen K. Garriot and Joseph P. Kerwin flew on various Skylab crews. Not until the shuttle program of the early 1980s did most of the scientists, now called mission specialists, fly on actual missions.

[166]See Chapter 5 of David Compton, *Where No Man Has Gone Before: A History of Apollo Lunar Exploration Missions* (Washington, D.C.: National Aeronautics and Space Administration, Office of Management, Scientific and Technical Information Division, 1989) (NASA Special Publication-4214, NASA History Series, 1989).

[167]Major Thomas C. Winter, Jr., "The Army's Role in Space," *Military Review*, 48.7 (July 1968):82-86.

[168] General William C. Westmoreland to Dr. Robert C. Gilruth, 10 February 1969, Johnson Space Center Collection, 071-11.

[169]Dr. Robert C. Gilruth to General William C. Westmoreland, 25 February 1969, Johnson Space Center Collection, 071-11.

[170]Burrows, *This New Ocean*, pp. 518-523.

[171]Joseph D. Atkinson and Jay M. Shafritz, *The Real Stuff: A History of NASA's Astronaut Recruitment Policy* (New York: Praeger, 1985), pp. 131-171.

[172]Memorandum of Understanding between the Department of Defense, the Army, the Navy and the Air Force and the National Aeronautics and Space Administration Concerning the Detailing of Military Personnel for Service as Shuttle Crew members, 17 December 1976.

[173]Heike Hasenauer, "The Army Astronaut Program," *Soldiers* 45.2 (February 1990):18.

[174]See the ninth chapter, "Military Man in Space," in U.S. Army Space Reference Text, http://www-tradoc.army.mil/dcscd/spaceweb/internet2.htm, accessed 3 March 2003. Material in the following paragraphs is based on this source.

[175]Memorandum of Understanding between the National Aeronautics and Space Administration and the Department of the Army, 17 May 1987.

[176]ARSPACE traces its roots to the Army Staff Field Element established to act as liaison to the U.S. Air Force Space Command and initiate planning for Army participation in the unified U.S. Space Command (USSPACECOM) in 1984. In 1985, this staff field element became the Army Space Planning Group, the Army component of USSPACECOM. In 1986, the planning group was designated the Army Space Agency (ASA), a Field Operating Agency of the DCSOPS. In 1988, the ASA became ARSPACE and in August 1992, ARSPACE became a subordinate command of the U.S. Army Space and Strategic Defense Command, a predecessor of the U.S. Army Space and Missile Defense Command. An overview of the events leading up to creating ARSPACE may be found in Lieutenant Colonel Patrick Gagan, "The Army Space Command," *Military Review* 58.3 (March 1988):44-51.

Chapter 4
Renewed Interest in Space and The War in the Persian Gulf, 1985-1991

The Army Returns to Space

In the years after 1958, the Army's starring role in space was diminished until it became a mere glimmer. The service became a passive consumer, dependent upon others to decide its needs. This loss was described by an Army War College Strategic Studies Institute fellow in 1985: "Although the Army now heavily depends on space systems for communications, command and control, reconnaissance and weather information, its role has declined from being the lead service in space operations in the late 1950s to that of the customer of the services provided by space systems."[1] The spark that reignited Army interest in space came from President Reagan's Strategic Defense Initiative speech of March 1983. The basic antiballistic missile technology research that provided the SDI's underpinnings was done by the ABMA and by Nike-Zeus. Its successor organizations would start paying the Army dividends.

Work on SDI galvanized other parts of the Army. The chairman of the Army Space Council, the Vice Chief of Staff of the Army General Maxwell Thurman, started several initiatives. A formal space policy was drafted, the military personnel system identified officers who had space-related education, skills or background and a space activities skill code was created to keep track of them. At the same time, officers were sent to civilian university graduate schools in space-related disciplines to meet an anticipated demand for their services. While these initiatives were proceeding, the Army Space Council realized there was no clearly defined role for the Army in space. To remedy this oversight and develop an Army Space Master Plan, an Army Space Initiatives Study (ASIS) Group was established at Fort Leavenworth in 1985.[2]

The Army had numerous organizations with responsibilities involving space. The result was a hodge-podge grouping of offices and staff organizations competing with each other for resources and attention. An earlier report concluded

> Individuals and groups with interest in space can be found in the BMD Program Office, ODCSOPS (Office of the Deputy Chief of Staff for Operations and Plans), ODCSR-DA (Office of the Deputy Chief of Staff for Research, Development and Acquisition), OACSI (Office of the Assistant Chief of Staff for Intelligence), Long-Range Planning, the Army Space Program Office, and elsewhere. There appears to be little coordination of effort and a distinct need exists for better integration of the space program.[3]

The plethora of organizations led to competition for personnel and proponency and resulted in great confusion. The chaotic rush to participate in the "next new thing" led to creating new offices with space-related responsibilities that competed with already-established organizations. This absence of command unity led to anarchy. The many competing organizations resulted in too many diverse organizations being managed by too many high-ranking officers, all of whom declaring space as their "rice bowl." Unity of command required that the Army streamline its efforts and eliminate duplication and confusion.[4]

In the mid-1980s, two organizations rapidly developed and focused the Army's interest in space: the Army Space Institute at Fort Leavenworth, and the Army Space Agency (ASA) in Colorado Springs.[5] At this time, the ASI was the more dynamic organization of the two as it approached its mission, to show the Army how to use space, with a missionary zeal.

The Institute's focus was tactical and its mission was to make space products available to provide support to the Army at the small unit level. Before 1986, most military space systems supported the strategic missions of STRATCOM and NORAD. Now ASI wanted these systems to support tactical units as small as an infantry squad. The vehicle used to disseminate the wonders of space-based products to tactical units was the Army Space Demonstration Program. By June 1987, a series of space demonstration concepts had been created. They included experiments with LIGHTSATs, commercial weather receiver systems (WRAASE), Global Positioning System (GPS) receivers and satellite early warning systems. The initiatives were formally approved in August 1987. Over the next three years, ASI provided briefings about these systems to the Army's Major Commands and was working on demonstration projects.

By the time Iraq invaded Kuwait in 1990, Army units were aware of the various space-related products that were available and were demanding they be issued space-related devices like GPS receivers. The ASI was deactivated in 1990 and replaced by the TRADOC Program Integration Office for Space (TPIO-SPACE) as the Army demobilized after the Cold War. The Combined Arms Combat Development Agency and later Combined Arms Command-Combat Developments leadership did not believe there was enough support for space applications in the Army to warrant the Institute's relatively large investment in manpower and resources. As part of this reorganization, responsibility for the ASDP was given to ARSPACE and the ASDP was renamed the Army Space Exploitation Demonstration Program (ASEDP). Under its new name, ASEDP has continued to make inroads into getting space-based products into the hands of the people who need them, helping to operationalize and normalize the use of space by the warfighter. Its philosophy, goals and objectives remained unchanged.[6] The ASEDP stayed in ARSPACE until 1997, when a command reorganization placed it in the new SMDC Battle Lab.

The ASI's aggressive efforts to bring space products to the Army provided several lessons to the senior leadership. First, the use of space systems should not be confined to strategic-level missions because tactical units could also use the information they provide. The demonstration program showed these systems could provide commanders with better unit location information, weapon targeting data, communications, weather information and intelligence information. At the same time, ASI discovered that many space systems were unsuitable for tactical use. This led to their experiments in the LIGHTSAT program (to demonstrate and evaluate the operational capabilities of lightweight, relatively inexpensive, limited purpose satellites to provide space-based support to operational and tactical commanders for reconnaissance, intelligence collection, surveillance and target acquisition). The ASDP also convinced the Army of the utility of modifying off-the-shelf electronic products for its own use. By showing flexibility, ASI was able to use existing technology in the most effective manner. The ASDP also showed the Army's space community that it must be willing to train soldiers in their units on the space systems so they might better understand their capabilities. This willingness to train soldiers in the field if necessary stood the Army in great stead during the Gulf War.

U. S. Army Space Command Activated

Fig. 4-1. The unit insignia of the U.S. Army Space Command, authorized in December 1988, symbolizes the Army's responsibilities for missile defense and strategic defense planning and the significance of satellites in navigation, communications and surveillance.

As ASI was pursuing its vision, the Army activated an operational command to manage its space functions, U.S. Army Space Command (ARSPACE). The first Army space organization at Colorado Springs was an Army Staff Field Element, founded in 1984 as a liaison office to AFSPC. In 1985, it was renamed the Army Space Planning Group as a planning function was added to its liaison mission. In 1986, when USSPACECOM was created, the planning group was renamed the Army Space Agency and was designated as "the foundation of the Army's operational capability in space."[7] In 1988, ASA was reorganized and replaced by U.S. Army Space Command. The new command retained its predecessor's planning and coordination functions and received added responsibility for the Consolidated Space Operations Center Detachment, the U.S. Army NASA-Johnson Space Center Detachment and three Regional Space Support Centers. As ASI was deactivated, ARSPACE received responsibility for the space demonstration program, reassigning the Army Signal Command's Defense Satellite Communication System (DSCS) platform and payload control mission to its purview extended its operational role.[8]

The ARSPACE was the Army component command of USSPACECOM and was a Field Operating Agency of the ODCSOPS.[9] Directly tied to the Army Staff in the Pentagon, ARSPACE had five command roles. It would provide "USSPACECOM an Army perspective in planning for DoD space systems support to land forces and strategic defense operations" to ensure "integration of Army requirements." It would respond to "USCINICSPACE-directed taskings" and command "assigned forces" as well as plan "DoD space operations in support of Army strategic, operational and tactical missions."

Initially the command was also given five missions. Aside from supporting USCINCSPACE as its Army component,[10] it would command the Defense Satellite Communication System Operation Centers (DSCSOCs) and manage joint tactical use,[11] plan for the possible fielding and operation of "Strategic Defense System (SDS) elements and anti-satellite (ASAT) weapons, should the United States choose to deploy them."[12] The command was also charged with assuring the Army's access and use of space-based capabilities to accomplish the goals of AirLand Battle Doctrine[13] and preparing for personnel and facility growth.[14]

The Future Security Environment Working Group Report and ARSPACE

The Report of the Future Security Environment Working Group validated the Army's new concentration on space-based assets and the creation of ARSPACE.[15] The working group concluded that the "rapid pace of technological innovation will probably continue over the next twenty years.... New technologies will revolutionize war in the same way that the Industrial Revolution changed warfare." These changes will lead to the "possible alteration of tactics, operational possibilities and possible strategic choices." The group also posited that only the superpowers would have the wherewithal to "sustain full spectrum change," although the possibilities remained open for niche changes dominated by regional powers. "We will see new areas of strategic concern and renewed possibilities for 'discarded options.'" The group's report explored emerging technologies and tried to ascertain "the implications of the new technologies for warfare."[16]

The working group identified nine types of emerging technologies that would influence warfare in the future. While not prescient, the technologies on the list were not generally known to the public or to the defense establishment at large. They included stealth technologies, unmanned vehicles, stand-off very high accuracy weapons and advanced strategic defense systems. The group also called for examining new cheaper space-based systems including newer GPS, anti-satellite weapons and satellite defenses, ballistic missile defense as well as advances in communications, reconnaissance, surveillance and weather technologies. The report then identified potential newcomers to space: India, China and Japan; space would no longer be the preserve of the Western powers and the Soviet Union. The group report then mentioned new sensors and processing technology, the ways greater use of computer-aided design (CAD) would ease and improve the "man-machine interface," and the importance of biotechnology weapons as well as directed energy and radio frequency weapons.[17]

The group members believed that these new technologies would change the face of warfare considerably, possibly ushering in a revolution in military affairs. Using weapons based on these technologies would "extend the battlefield to unprecedented depths" and at the same time, expose both sides to "increased infrastructure vulnerability." They believed future military operations would increase in speed and become more dependent upon information. This, in turn, would require "theater-wide integration of C^3I to support and a very rapid operational tempo (OPTEMPO)." Additionally, the weapons' increased destructiveness made the opening stages of a war more crucial than before. All this would lead to increased changes in military organization, doctrine and philosophy of command.[18]

The creators of AirLand Battle Doctrine anticipated many of these changes. They posited that future warfare would involve very mobile forces, linked by communications devices giving army and company commanders a common picture of the battlefield. Future armies would mount attacks throughout large theaters of operation, not along linear front. Battles would simultaneously expand in space and be shortened in time. Terms and concepts that first appeared in World War II, such as "deep attack," "flexible defense" and "follow-on forces attack," were refined using the new information technologies. This new approach may be observed in Army Field Manual (FM) *100-18 Space Capstone Doctrine*, which began circulating in draft in 1988. The draft noted that AirLand Battle doctrine "focused on a battlefield that was expanding in depth, duration and technology. Maturing technologies were found to be applicable to military missions."

When it spoke of future doctrine the manual emphasized the Army would capitalize on "emerging space capabilities," exploit those capabilities that "contribute to the successful execution of Army missions" and assure "access to space" in order to use space-based capabilities to accomplish "strategic, operational and tactical missions." These areas of responsibility included "ballistic missile defense, anti-satellite capabilities," the national test range, "national communications," the Military Man-in-Space Program and fulfilling "Army joint service taskings." The draft manual defined the operational and tactical missions as communications, reconnaissance and target acquisition, weather and environment monitoring, position location and navigation, fires support and support of the military man-in-space program.[19]

The Gulf War: The First Space War

Although not explicitly stated, the draft manual was explaining the role of space as a force enhancer. This was the focus that ASI and ARSPACE were publicizing and proselytizing. The demonstration of space-related technology as a force enhancer took place during the Gulf War. Following the Iraqi invasion of Kuwait in August 1990, the United States launched the largest military operation it had undertaken since the withdrawal of the last troops from Vietnam in 1973. More than 500,000 troops were sent to Saudi Arabia to protect the interests of the United States and its allies in the Persian Gulf region. As the United Nations imposed economic sanctions on Iraq and the U.N. Security Council condemned the invasion, the U.S., using bases in Saudi Arabia, began a logistics build-up, Operation Desert Shield, under the command of General H. Norman Schwarzkopf, commander-in-chief, Central Command. The efforts of the president and secretary of state resulted in

Fig. 4-2. ARSPACE personnel in Saudi Arabia during Operation Desert Shield/Desert Storm.

assembling a coalition of more than thirty nations to oppose the Iraqi dictator Saddam Hussein's invasion and pillage of Kuwait.[20] Operations Desert Shield and Desert Storm tested the Army's space-based technologies. Desert Storm has been called the "first space war" by some commentators because every aspect of military operations depended, to some extent, on support from space-based systems. The Army used these systems for position/navigation, weather, communications, imagery and tactical early missile attack warning. The assistance rendered was invaluable and the new technology, combined with AirLand Battle Doctrine, changed the way the Army fought. The conflict represented a watershed in the development of these systems.

Position/Navigation in the Desert

Navigation in the desert has always been problematic. Maps, if they exist, are not current and one area may be indistinguishable from another. Maps may also be next to useless because there are few terrain features on which to orient one's position. Navigation by the sun and stars may be hampered by clouds and sandstorms. While it is possible to navigate with map and compass, a better method of finding one's way was crucial to military success. Although other parts of space-based force enhancement can seem quite arcane, the value of one tool that emerged from the Gulf War was easily and quickly understood: the Global Positioning System.

Fig. 4-3. Global Positioning System satellite.

The origins of the GPS may be traced to the 1960s and is part of the larger human quest to locate itself in featureless terrain. Predictably, the first customer for this system was the Navy. Using maritime chronometers, sextants and tables to determine local noon and one's position at sea or in featureless terrain on land depends upon clear weather. A space position/navigation system that would work in all kinds of weather was on many wish lists. Work began in earnest in the mid-1970s, but the first satellites were not launched until the late 1980s. At the time of the Gulf War, only a partial system was in place.

The GPS is a position/navigation tool that uses a network of satellites that function as spaceborne beacons continuously transmitting a signal that can be used by a receiver to determine the operator's location. It is used for military, commercial, scientific and recreational purposes today, from mapping to surveying to air traffic control to search and rescue operations. The system itself has three segments, space, user equipment and control. The first segment, space, consists of a constellation of satellites placed in orbit allowing a receiver to pick up signals from several of them-one can determine one's location in two dimensions if the receiver picks up signals from three satellites; three dimension location information may be obtained if the receiver gets signals from four satellites. There were "16 usable (experimental and operational) satellites" in service at the time of Desert Storm "providing approximately 24 hours of two-dimensional coverage and 19 hours of three-dimensional coverage."[21]

The user segment consists of different types of receivers as well as test equipment, antennae and software. The two types of receivers used during the Gulf War were the "manpack/vehicular (M/V) models" and the commercial small, lightweight, GPS receivers (SLGRs). The M/V models weighed between 10 and 20 pounds and could "receive the precision-coded signals" resulting "in close to 10-meter positioning accuracy." The SLGRs were hand held and could receive signals with "15- to 30-meter accuracy."[22] The SLGR "fits in the side pocket of BDU trousers, weighs a little over four pounds and operates on two lithium batteries."[23] The control segment consisted of several tracking stations in Hawaii, Diego Garcia, Ascension Island, the Marshall Islands and Colorado Springs. The stations track each satellite, compute orbital and clock corrections, and transmit that information to the Master Control Facility, which sends the corrections back to the satellites.

Fig. 4-4. Soldier using a Small Lightweight Global Positioning System Receiver in Operation Desert Shield/Desert Storm.

The GPS may be the ideal system for the soldier. It continuously provides accurate position and velocity data from any location in the world while weather and other environmental conditions have no effect on its performance. It fits the 1986 CACDA definition of a perfect position/navigation system. This definition demands that such a system must provide coverage throughout the world, the user can be passive, an adversary can be locked out of the system, it must be capable of handling a large number of users without becoming saturated, and it must be able to resist electronic interference measures employed by a foe. It also must be unaffected by natural disturbances, provide real-time responses to its users, and be available for combined operations. There must be no difficulty allotting frequencies and it must provide a common grid reference for all users. The data it provides cannot be changed by differences in altitude (for land and air forces) nor by changes in time of day or year. It must provide accurate data to a moving vehicle and be portable enough to mount on a vehicle. Finally, the equipment must be relatively simple to maintain by the unit's soldiers.[24]

The Army began GPS field demonstrations in 1989. Many of the units deployed during Desert Storm the following year clamored for the equipment. The ASI and ARSPACE organized "train-the-trainer" programs at Fort Bragg, Fort Stewart and Fort Campbell as the SLGR receivers were distributed. However, "as more units deployed to the Gulf, this train-the-trainer effort could not be sustained." A training support package was prepared and delivered to units receiving the SLGRs, but distributing the packages was "limited by competing demands for other critical supplies, reducing their effectiveness as a training tool."[25]

The GPS was a success in Desert Shield and Desert Storm; most users were pleased with the system and the Center for Army Lessons Learned (CALL) reported that "comments…did not generally relate to system problems but to the fact that there were not enough receivers to go to all of the users who wanted them."[26] Lack of training led to troops' misunderstanding the system's capabilities and limitations. For example, some users thought they were more accurate than they really were and others believed GPS only worked in specific parts of the world. Nevertheless, the system was a great success. In a letter to the ARSPACE commander, Major General J. H. Binford Peay, III, commander of the 101st Airborne Division (Air Assault), touted its wonders.

> The SLGR is working wonders and is the most popular piece of equipment in the desert. We use it for everything and it is used by everybody…cooks, log resupply, navigation by aviation, fire support officers and commanders. Navigation is the singularly most difficult thing in the desert. Maps are inaccurate and the terrain features do not facilitate orientation. The entire area operations is one big enemy avenue of approach and without the SLGR, firepower would be hampered and under-utilized.[27]

The system allowed combat units to navigate quickly to their objectives, helped guide convoy movements and supported resupply operations. Iraqi minefields were discovered and marked using GPS data. Forward artillery observers employed GPS when using artillery or close air support, and batteries exploited the system to conduct field artillery surveys on the fly. Signal units used GPS to help position communications units. The SLGRs and the M/V units were used in a variety of combat roles in the desert. However, the rush to deploy units resulted in a series of problems. Most of them had their roots in the lack of formal training on the system. The CALL reported, "There were not enough GPS receivers available to cover all the applications for which they could have been used….The only receivers available to some infantry brigades were with Air Force or fire support elements." Sometimes these elements accompanied reconnaissance sorties solely "to provide GPS support." With only one receiver allotted to each field artillery battery, the commander had to decide whether to use GPS as either a navigation tool or a survey control tool.[28]

The problems enumerated were symptomatic of a more general difficulty that was only partly attributable to the lack of training. There was a fundamental lack of familiarity with the way GPS functioned and its designed function. This was due in part to soldiers' general ignorance of the ways in which space-based products could aid them to carry out their missions. Because they had never been exposed to it, they had not developed the intuitive sense of its strengths and limitations that come from using it regularly and considering it a normal part of their equipment.[29]

Weather Forecasting and Space-Based Systems

Unlike their general lack of knowledge about GPS, senior commanders understood they needed responsive weather reporting and forecasting before Desert Shield started. Earlier in 1990, TRADOC presented a concept for a Division Standardized Command Post. The new concept would allow the division staff to shed excess vehicles and equipment, making it easier to maneuver and deploy. Instead of an Air Force weather team attached to division headquarters, along with their communications and weather equipment, the new division weather team would be sharply reduced in size and would only disseminate weather information, not produce it. Several divisions and the Intelligence School relayed caustic remarks back to TRADOC about their new concept.[30] The objections illustrated that senior commanders understood the role weather plays in operations, the value they placed on having weather reports and forecasts tailored to their individual needs, and the importance they placed on being able to collect and disseminate weather information to their subordinate units themselves.

During the Gulf War, the primary weather imagery receiver the Army used was the WRAASE commercial weather receiver.[31] It was selected because it could get information directly from civilian weather satellites as they flew over the Middle East, including imagery, television and infrared observations.[32] The military system, the Defense Meteorological Satellite Program (DMSP), comprised polar-orbiting satellites that provided indirect support to Army at echelons below corps and direct support to the Army Service Component Command of Central Command.

The only differences between these satellite types were the spatial resolution of the imagery and the amount of time between consecutive imagery. Geostationary imagery resolution was on the order of 10 kilometers, providing very large-scale views of the weather and taking a new picture of the same portion of the earth every half hour. Polar-orbiting satellite imagery resolution was on the order of 2-4 kilometers, providing a smaller scale look at the weather. The DMSP imagery had a resolution on the order of 0.4 kilometers, allowing meteorologists to identify smaller scale weather phenomena. Polar-orbiting satellites pass over every part of the earth about once every twelve hours.[33]

The units deployed with the WRAASE receivers. As Desert Shield began, the intelligence section of the XVIII Airborne Corps and the 30th Engineer Battalion (Topographic) requested ASI provide them with additional weather support. ASI responded by integrating weather imagery and terrain analysis systems. Two FORSCOM Automated Intelligence Support System computers were outfitted with the Weathertrac commercial software package and networked with the WRAASE receivers. The ASI noted, "this combination allowed the staff weather officer to enhance the visible and infrared imagery available from the weather satellites as they pass over Saudi Arabia 8-10 times a day. With the limited knowledge of Saudi weather…this satellite weather information provides the one means of seeing the battlefield." The relationship between satellite weather and satellite terrain imagery data was formalized when the 30th Engineer Battalion established a Topographic Technology Exploitation Cell (TTEC) to analyze satellite imagery, combining weather and terrain data and producing updated maps.[34]

Weather satellites and the data they delivered were used in novel and unexpected ways in Desert Storm. When combined with multi-spectral imaging, the data aided in target planning, as well as planning, executing and redirecting ground movement. Despite its recognized utility, tactical units did not have access to all the available weather information.

After the war, CALL identified three trends in satellite weather support, including integrating weather and terrain analysis through the TTEC and distributing weather support receivers throughout the operational theater. The CALL reported, "U.S. Central Command took steps to procure more receiver terminals to enable the use of weather data at all levels of command. New, lightweight prototype desktop receivers were distributed to ensure the Army had access to real-time weather data from a variety of weather satellites." The third trend was the demand for raw weather data by analysts outside the staff weather office. The Center recommended this demand be satisfied by collocating satellite weather receivers with unit intelligence and terrain analysis staffs.[35]

Multispectral Satellite Imagery

The Army also used multispectral satellite imagery to update its maps of Saudi Arabia, Kuwait and Iraq. The Defense and Army mapping communities gave the forces on the ground up-to-date maps. These maps relied on information obtained from two types of satellites and two types of ground systems.

The satellites were LANDSAT and SPOT. LANDSAT is a U.S. Department of Commerce earth resources satellite system that provides coverage of the entire earth every 16 days and takes multispectral pictures at 30-meter spatial resolution. The width of one pass is 185 km. Imagery can be used to create maps to about 1:80,000 scale. Imagery must be purchased and cannot be shared indiscriminately because of copyright restrictions. When the Gulf War took place, two LANDSAT satellites were operating. SPOT is a French satellite that performs the same functions as LANDSAT and can view every part of the earth every 26 days. It has three different bands at 10- and 20-meter resolution. Imagery can be used to produce maps to a scale of approximately 1:25,000. The width of one pass is approximately 60 km. Images are available commercially and cannot be shared.[36]

Ground systems consisted of Multispectral Imagery (MSI) Workstations and FORSCOM Automated Intelligence Support System (FAISS). The MSI workstations were part of the ASDP to show potential users "the value of multispectral imagery for producing image maps, conducting image analysis and providing up-to-date broad area views of the battlefield. The workstations consisted of high-speed desktop computers" running a commercial program, Earth

Resources Data Analysis System, that performed a wide variety of tasks relating to "image analysis, image enhancement, data merging and terrain visualization." The FAISS was used as an "intelligence analysis workstation." Division terrain analysis teams could use the system to automate terrain analysis.[37]

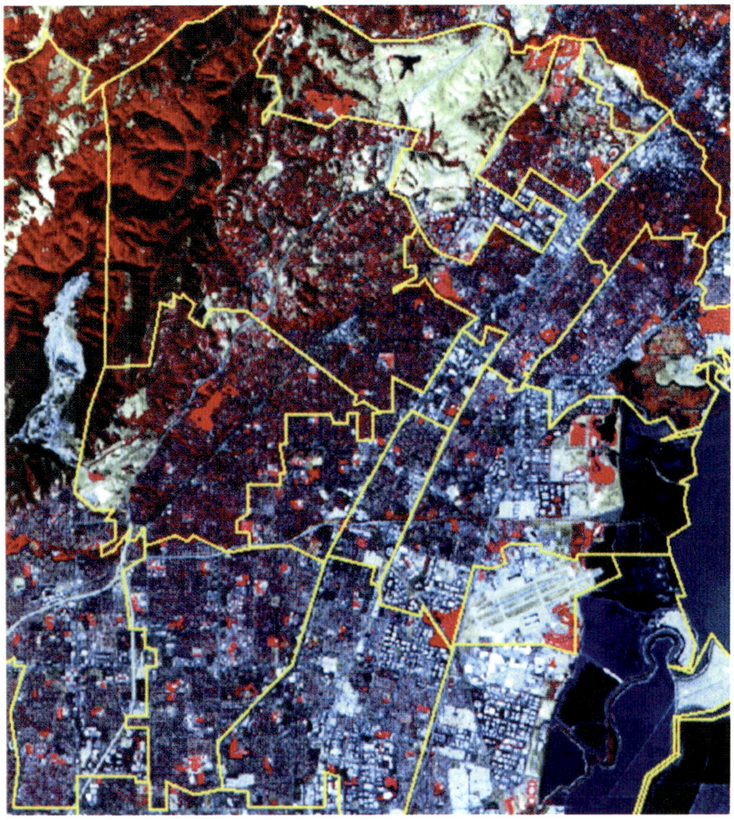

Fig. 4-5. An example of a multispectral satellite image.

The impact of multispectral imaging technology through the TTEC was felt on corps-level operations. According to an ASI report,

> Two thirds of the intelligence preparation of the battlefield [IPB] can now be combined using as current information as the last satellite pass allows. One month old LANDSAT imagery combined with weather satellite passes is providing a quantum leap in the ability of the commander to see his battlefield. IPB can be accomplished on the fly and not remain a pre-deployment or pre-exercise pursuit.[38]

The slow procurement process for LANDSAT imagery left the "topographic units without up-to-date imagery until November." The Army was also unable to get the money to pay the royalty rights for the large amount of SPOT images already in the possession of the Air Force. These delays hindered the topographic analysts' work and left Divisions with very little time to exploit available capabilities.[39] Nevertheless, "MSI was excellent for tactical planning. It provided accurate, updated maps, broader coverage and allowed planners the best available product before deployment to Saudi Arabia. The MSI terrain analysis supported the development of obstacle updates, proper routes, water locations, soil type, trafficability, etc."[40]

Space-based Communications Systems

The Army has been interested in using space-based systems for communications purposes since the first satellite systems were placed in orbit. Civilian and military satellite communications systems were of paramount importance to the command and control network the Army built during Operations Desert Shield and Desert Storm. An extensive voice and data communications network was needed to support the units in Saudi Arabia.[41]

The network used during the Gulf War consisted of military and civilian satellite communications systems. The Military Satellite Communications (MILSATCOM) system had three parts, (1) the Defense Satellite Communications System (DSCS), (2) the Fleet Satellite Communications (FLTSAT) System and (3) the Air Force Satellite Communications (AFSAT) System. The DSCS provided the greatest anti-jam transmission capacity while the other two had smaller transmission capacities, with no anti-jam capabilities.[42]

The Army had approximately 200 DSCS ground mobile force terminals that were normally placed in corps, division and echelons above corps headquarters. The FLTSAT and AFSATCOM systems had portable terminals and were used by command networks. All three systems were shared by government users. However, before the Gulf War, tactical units had made minimal use of these systems in exercises or contingency operations. In Desert Storm, the tactical users had priority and MILSATCOM services were provided from all resources. The Army deployed more than 1,500 terminals to Saudi Arabia (more than 75 percent were single channel portable military and commercial sets). The satellite networks were used for inter- and intra-theater communications, the latter was especially important given the lack of a communications infrastructure in the theater of operations. Approximately 50 percent of the communications traffic was carried by the DSCS terminals; the commercial INTELSAT system carried another 25 percent, while the remaining quarter was carried by FLTSAT, AFSATCOM and commercial terminals.

Operations Desert Shield and Desert Storm used much of the existing capacity of military and commercial communications satellite systems. Satellites were moved to better serve the operation and experimental satellites were used because of the high demand. The rapid movement and dispersion of units on the battlefield meant that maneuver units at levels below those usually issued with satellite communications receivers required them. MILSATCOM was used through the division level, but the rapid movement of the units meant that units frequently moved beyond line of sight and FM transmission and relays could not be established.

During the Gulf War, satellite communications was the backbone of long haul and intra-theater connectivity. The operations in the Persian Gulf War saw the beginning of three trends in the Army's use of satellite communications. Once satellite communications systems were the purview of higher headquarters. Since 1991, however, tactical units have made greater use of satellite communications systems, especially when deployed to places with rudimentary or nonexistent communications infrastructure. In the Gulf War, the DSCS was used by brigade-sized units. Second, the Army used commercial satellite systems to supplement its own communications network. Finally, the demand for communications support outstripped the capabilities of the available military systems. Part of the problem in the Gulf War stemmed from user inexperience that resulted in poor site selection, self jamming, and inadequate frequency planning that overloaded the satellite systems. The Army used this resource inefficiently because it had a limited amount of equipment, minimal control over satellites and complicated coordination procedures.

Theater Missile Defense

Space-based systems also played an important part in tactical early missile attack warning by supplying critical information on missile launches.[43] The early warning system was based on the Defense Support Program (DSP) satellite system developed in the 1970s. This system used a constellation of satellites equipped with infrared sensors to detect missile launches and determine trajectories and impact areas. During the Gulf War, after Patriot Air Defense units deployed to Saudi Arabia, USSPACECOM developed the Tactical Event Reporting System (TERS). The TERS modified a strategic system for tactical use and was designed to make tactical missile warning data available to the tactical commander in near real-time.

Fig. 4-6. Photo of a Scud fragment.

Fig. 4-7. Photo of damage caused by a Scud strike.

Operating the TERS was fraught with problems. Soldiers were not trained to use the equipment but, in retrospect, this proved to be a minor problem because the system itself "left much to be desired."[44] The original DSP system was designed to track Soviet strategic missiles that flew longer, further and had brighter infrared signatures than tactical Scud rockets. Therefore, TERS could not predict specific impact areas nor could it provide vectoring data to Patriot air defense batteries. The system was used to warn allied forces of impending missile impact.[45] However, the warnings were not timely because it generally took about two minutes to transmit them, leaving very little response time.[46] Finally, "Brigades operating away from the corps air defense artillery umbrella experienced difficulty receiving missile warning alerts."[47] Despite these shortcomings, TERS represented a breakthrough in early missile warning systems, a breakthrough that was exploited after the war.

Lessons Learned from the Gulf War

In this brief period the Army began to explore the possibilities inherent in using space-based systems. The activities of the ASI and ARSPACE brought these systems down to the tactical level. However, institutionalizing these changes has proven difficult because of institutional inertia and the short life of combat lessons learned.

The Gulf War demonstrated that space-related systems and products can successfully support the Army's operations. Units used GPS to navigate, control convoys and resupply operations, mark and breach minefields and for artillery surveying and fire direction. Tactical units can use weather receivers to obtain crucial weather information quickly. When weather information was combined with multispectral satellite imagery, maps using the latest intelligence can be created and distributed in a timely manner. Tactical missile detection has used space-based systems to warn units of incoming rocket attacks. As will be related, each of these capabilities has been improved since the end of the Gulf War.

It was also obvious that few commanders fully grasped the potential of the space-based systems to which they had access. Few understood how military space-related systems and their products can help them improve their tactical practices and their grasp of the operational art. This is a failure of imagination that can be remedied by fully integrating the uses of space into the Army educational system's curricula. As related above, both ASI and ARSPACE exposed tactical units to space-related systems and products. However, before the Gulf War, most units had not become acquainted with them. When the deployment began, both the Army Space Command and the Army Space Institute organized ad hoc training on the GPS and WRAASE weather receiving systems, allowing large numbers of soldiers to become acquainted with, use, and understand the idea of space support in position/navigation and weather intelligence. If schooling includes lessons on the use and deployment of space assets, then unit exercises will also use them.

Commercial space systems played a large role in the Gulf War and had a large impact on the military. Although the military DSCS carried about half of the communications traffic in the war, the INTELSAT system carried another quarter-the commercial system supplemented the military system. The WRAASE weather receiver was a commercial product and the topographical units' services expanded because of the commercial equipment and software bought during the war. Even the much heralded GPS could not be distributed to the majority of units until the Army bought and sent commercial receivers to the Persian Gulf.[48] Using commercial systems presented unique situations for the wartime commander. For example, although the Iraqis continued to receive weather forecasting information from three National Oceanic and Atmospheric Administration satellites and while the U. S. government feared this information could be used to launch Scud attacks, the satellites remained in service because they also supplied weather forecasting data to American allies in the region.[49] In a second instance, the Air Force could not share SPOT imagery with the Army because the latter could not pay the image royalties to the SPOT Corporation.[50]

A final enduring lesson from the Gulf War is the relatively short shelf life of combat experience. If the Army is to retain its interest in space and space-based systems and products, the Army's space community must make a greater effort to capture and disseminate the lessons it learns from observation and historical study of training, exercises and combat operations.

The Post-Gulf War Operation in Somalia

As the armistice took hold along the Iraq border, the United States found itself involved in Somalia. Beset by a lingering civil war that had destroyed all central authority, Somalia suffered from starving refugees, factional fighting and the proliferation of weapons. All of these troubles produced an anarchic situation. The problems of Somalia led the U.N. to commit peacekeeping forces to the area. In 1992-3, the United States mounted Operations Restore Hope and Continue Hope. The collapse of all central authority in Somalia, the inability to distinguish friend from foe, and the extremes of the Somali climate, presented new and unfamiliar challenges to the United Nations whose previous experience had been limited to peacekeeping operations in states that had not yet descended into chaos.[51] American troops committed to Somalia faced many of the same physical conditions they had faced in the Saudi Arabian desert.[52] Years of civil war had left very little in terms of dependable infrastructure. In these conditions space-based systems provided direct support to the deployed soldiers.

Standard map coverage for the region was either unreliable or nonexistent. At the beginning of the deployment, the division's standard was based on an old Russian map series. The TENCAP systems were used to produce the initial maps for the 10th Mountain Division's deployment. In fact, this imagery provided the commanders with their first reconnaissance of the area and was the initial source for terrain mapping. LANDSAT imagery was eventually purchased[53] to make maps of the uncharted areas of the Somalia-Ethiopia border. A problem highlighted in the after action report and lessons learned process was integrating signals intelligence into tactical planning and rapidly producing tactical maps to support ground operations.

Communications was a problem as the division acting as the Army forces command used INMARSAT as its primary communications medium in the initial phases of deployment. Single-channel tactical satellite radios were the primary vehicle for communicating over long distances until a long-haul communications system could be installed. The ARSPACE supported the division's deployment with SLGRs, multispectral imagery processing equipment and INMARSAT terminals. Initially, the 10th Mountain Division did not have any SLGR sets, INMARSAT terminals, trained WRAASE operators or any good maps. Within thirty-six hours of its alert by FORSCOM, ARSPACE sent equipment and trainers to the division's home station at Fort Drum. Using assets at Fort Drum, Fort Bragg and in Somalia, the division was able to provide for its communications and imagery needs. In addition, the division used SPOT imagery to update its maps and GPS to provide the troops with accurate position and navigation data.[54] The Seascape weather satellite receiver supported the Joint Task Force headquarters in Somalia with timely Defense Meteorological Satellite Program weather forecasts.[55]

In the late 1980s and early 1990s, as a result of the Strategic Defense Initiative and Operation Desert Storm, the armed forces became increasingly dependent upon space to wage war successfully. Space resources played a critical part in intelligence, communications, mapping, missile warnings, navigation, targeting and weather reporting and forecasting. At the same time, these assets were vulnerable to attack from potential adversaries. A determined enemy might easily destroy or nullify reconnaissance, communications and navigation satellites, paralyzing American forces.

The Army found itself increasingly dependent upon space to conduct its operations. The typical soldier relied on space-based systems to determine his position, locate the enemy, communicate with friendly forces, and fire "smart weapons." For the Army, space was becoming the new "high ground," an important part of firepower and information dominance on the battlefield of the future. It became crucial for the Army and the other armed services to take steps to improve their space technology and astronaut programs.[56]

End Notes

[1]Colonel Jan V. Harvey and Colonel (ret) Alwyn H. King, "Space: The Army's New High Ground," *Military Review* 65.7 (July 1985):38-51. The quote appears on p. 39; Colonel Harvey was the SSI fellow.
[2]Major Linas A. Roe and Major Douglas H. Wise, "Space Power is Land Power: The Army's Role in Space," *Military Review* 66.1 (January 1986):16.
[3]Cited by Harvey and King, 47.
[4]For a sketch, see Lieutenant Colonel Patrick Gagan, "The Army Space Command," *Military Review* 58.3 (March 1988):44-51, especially 47-49. He includes a chart ("Selected Army Organizations with Space Responsibilities: Offices (commands, agencies, activities, centers, directorates, elements, councils and groups) and their Senior Officers") on p. 47 that lists 23 Army organizations with an interest in space in 1985.
[5]The ARSPACE traces its roots to the Army Staff Field Element established to act as liaison to the U.S. Air Force Space Command (AFSPC) and initiate planning for Army participation in the unified U.S. Space Command (USSPACECOM) in 1984. In 1985, this staff field element became the Army Space Planning Group, the Army component of USSPACECOM. In 1986, the planning group was designated the Army Space Agency, a Field Operating Agency of the DCSOPS. In 1988, the ASA became ARSPACE and in August 1992, ARSPACE became a subordinate command of the U.S. Army Space and Strategic Defense Command, a predecessor of the U.S. Army Space and Missile Defense Command.
[6]The program's goal was to demonstrate to field commanders the latest relevant space technology from the academic, commercial and government research and development communities. Its governing philosophy maintained that (1) products from space-based systems are critical to rapid force projection operations, (2) using space-based products allows the force to dominate the contemporary battlefield and (3) space-based capabilities significantly enhance combat effectiveness. It maintained its three-fold set of objectives, to (1) educate tactical commanders on ways to use space-based assets, (2) assist in defining requirements for Army development, and (3) demonstrate new technology for possible development by the Army. A partial list of its post-Gulf War successes includes Satellite Multispectral Imagery Mapping, GPS Tracking, Command, Control and Communications (TRAC^3), the Mission Planning and Rehearsal System (MPRS), Space Enhanced Command and Control, the Joint Tactical Ground Station (JTAGS), the Gun Laying and Positioning System, Advanced Communications Satellite Technology (ACTS), the Digital Reconnaissance Tool, Commercial Space Package, the Army Space Support Team, the Army Theater Missile Defense Element (ATMDE), the Laser Boresight and the Global Broadcast Joint In-Theater Injection Terminal.
[7]U.S. Army Space and Missile Defense Command Historical Office and Science Applications International Corporation, Space Warriors: The Army Space Support Team (unpublished ms held at USASMDC Historical Office), 1-5.
[8]Space Warriors, 1-5. In 1990 ARSPACE received the mission to command the DSCS Operations Centers (DSCSOCs).
[9]Unless otherwise noted the next two paragraphs concerning command missions and roles are based on "31 July Information Paper."
[10]There were three facets to the first mission. The ARSPACE would run the Consolidated Space Operations Center (CSOC) Detachment and would perform duties in the GPS Mission Control Complex and the GPS Master Control Station. U.S. Army Kwajalein Atoll Complex contributed sensor information for the ARSPACE's Space Surveillance Network and supported USCINCSPACE's space control mission. It controlled the Army Astronaut Detachment at the Johnson Space Center. The detachment's purpose was to enhance the Army's ability to execute AirLand Battle Doctrine using manned space capabilities.
[11]The Defense Satellite Communications System was a super high frequency satellite subsystem of the Defense Communications System designed to provide secure voice and high data rate communications worldwide. ARSPACE was responsible for operations and maintenance of all the DSCSOCs.
[12]The ARSPACE was responsible for ground-based SDS elements (radars, surveillance tracking systems and interceptors). It would be the focal point for their combat and materiel developments weapons and be responsible for developing a concept of operations and force structure for their use as the Army was given the lead to develop system architecture for ASAT weapons.

[13]In terms of supporting AirLand Battle, ARSPACE would work with Corps and Division Tactical Operations Centers to support planning and training and contingency mission execution using the full spectrum of space support, position and navigation, communications, reconnaissance, weather and terrain sensing satellites. In terms of Theater Missile Defense, it would demonstrate the ability to provide theater warning.
[14]Growth in personnel strength reflected its increased responsibilities and facility growth was based on the idea of building a 50,000 square foot building to accommodate the command.
[15]Report of the Future Security Environment Working Group, October 1988.
[16]Report of the Future Security Environment Working Group, October 1988, pp. 26-27.
[17]Report of the Future Security Environment Working Group, October 1988, pp. 27-34.
[18]Report of the Future Security Environment Working Group, October 1988, pp. 33-34. The working group concentrated on the Soviet Union as the great potential enemy and believed future warfare would be a fight between peer enemies. The end of the Cold War saw the involvement of the United States Army in many smaller operations against forces that were not its peers in equipment or space-based assets. The most recent example was Operation Anaconda executed against al-Qaida holdouts in the Shah-i-Kot valley of Afghanistan. Even though American forces had unrestricted access to space-based intelligence systems as well as spy planes and unmanned aircraft, all these measure failed to discover that there were no civilians in the valley nor were they able to portray accurately the enemy's size, location, principal weapons and course of action. The enemy found simple low-tech ways to hide from the overhead systems (in caves, in rock crevasses, under trees or by pulling earth colored blankets over themselves). The lessons to be learned are that a technologically backward enemy can still surprise his high tech foe and that, while useful, technological superiority is not a panacea and ought not to be viewed in that way. See Sean Naylor, "The Lessons of Anaconda," *New York Times on the Web*, 2 March 2003, http://www.nytimes.com/2003/03/02/opinion/02NAYL.html?pagewanted=all&position=top, accessed on 2 March 2003.
[19]Space Capstone Doctrine [Draft] (FM 100-18) FY 88, pp. 1-2, 2-4, 4-2.
[20]See Joseph P. Engelhardt, comp., *Desert Shield and Desert Storm: A Chronology and Troop List for the 1990-1991 Persian Gulf Crisis* (Carlisle Barracks: Strategic Studies Institute, U.S. Army War College, 25 March 1991) for a list of the nations that made up the coalition and a list of forces and/or equipment each contributed to the effort.
[21]Center for Army Lessons Learned (CALL) Newsletter No. 91-3, *The Ultimate High Ground!: Space Support to the Army; Lessons from Operations DESERT SHIELD and DESERT STORM*, "Chapter 1, Position/Navigation (POS/NAV)," http://call.army.mil/products/newsltrs/91-3/chap1.htm accessed on 10 January 2003.
[22]CALL Newsletter No. 91-3, "Chapter 1, Position/Navigation (POS/NAV)," http://call.army.mil/products/newsltrs/91-3/chap1.htm accessed on 10 January 2003.
[23]U.S. Army Space and Missile Defense Command Archives, Army Space Command Box 4, *STAR*NET ARSPACE Newsletter*, 1.1 (15 February 1991).
[24]*The Army Position and Navigation Master Plan* (May 1986), p. II-1.
[25]CALL Newsletter No. 91-3, "Chapter 1, Position/Navigation (POS/NAV)," http://call.army.mil/products/newsltrs/91-3/chap1.htm accessed on 10 January 2003.
[26] CALL Newsletter No. 91-3, "Chapter 1, Position/Navigation (POS/NAV)," http://call.army.mil/products/newsltrs/91-3/chap1.htm accessed on 10 January 2003
[27] "Letter from Major General J.H. Binford Peay III to Colonel Ronan Ellis, Commander, ARSPACE, 16 October 1990." The SLGR continued to work wonders and after the war, the ARSPACE commander noted, "The SLGR was our greatest success in terms of volume and probably our single most significant contribution to Desert Storm. See "Memorandum from Colonel Michael Keaveney on Assuming Command, 1 April 1991."
[28]CALL Newsletter No. 91-3, "Chapter 1, Position/Navigation (POS/NAV)," http://call.army.mil/products/newsltrs/91-3/chap1.htm accessed on 10 January 2003.
[29]In 1991, the new ARSPACE Commander wrote, "The SLGR was our greatest success in terms of volume and probably our single most significant contribution to Desert Storm." "Memorandum from Colonel Michael Kearney to ARSPACE, Subject: Assuming Command, 1 April 1991."
[30]See Space Warriors, p. 1-17 reports that one of the less temperate replies was, "stupid, absolutely absurd!"
[31]The information in the following paragraphs, unless otherwise noted is from CALL Newsletter No. 91-3, Chapter 2, "The Battlefield Environment, Section A - Weather," http://call.army.mil/products/newsltrs/91-3/chap2.htm accessed on 10 January 2003 and "Briefing, WRAASE."

[32]The Army received its weather information from Geostationary and Polar weather satellites. Geostationary satellites placed over the United States (GOES), and Europe (METEOSTAT) provided weather imagery for the Gulf War. Satellites in polar orbit from the United States (TIROS) and the U.S.S.R. (Meteor) provided real-time television and infrared imagery as they passed over the Persian Gulf area.
[33]Ten-kilometer resolution resembles the views seen on TV weather forecasts while 0.6-kilometer resolution allows one to identify phenomena like fog and sandstorms.
[34] Space Warriors, p. 1-18.
[35]CALL Newsletter No. 91-3, Chapter 2, "The Battlefield Environment, Section A--Weather" http://call.army.mil/products/newsltrs/91-3/chap2.htm accessed on 10 January 2003.
[36]CALL Newsletter No. 91-3, Chapter 2, "The Battlefield Environment, Section B-Terrain" http://call.army.mil/products/newsltrs/91-3/chap2.htm accessed on 10 January 2003. During Desert Shield and Desert Storm Iraq did not purchase images.
[37]CALL Newsletter No. 91-3, Chapter 2, "The Battlefield Environment, Section B-Terrain" http://call.army.mil/products/newsltrs/91-3/chap2.htm accessed on 10 January 2003.
[38] Major Korpsel and Mr. Freeman, Input to LAMP: Space Support for Desert Shield (Fort Leavenworth: Army Space Institute, September 1992).
[39]Space Warriors, 1-19.
[40]"Memorandum from Colonel Michael Keaveney on Assuming Command, 1 April 1991." He went on to state, "We trained TOPO unit personnel and acquired proper equipment for them; provided upgrades into Saudi Arabia; and initiated and completed a very significant MSI intel or IMINT project, merging classified imagery and unclassified LANDSAT and SPOT imagery."
[41]Voice communications were needed for command and control purposes on all different levels. Data circuits were used to transmit logistics information, imagery and other messages. The units in the field also had to communicate with other theaters and bases in the continental United States that were supporting them. The military communications satellite systems were supplemented by INMARSAT (international maritime satellite system) and INTELSAT (international telecommunications satellite system) satellites.
[42]Unless otherwise cited, the material about communications is based on CALL Newsletter No. 91-3, Chapter 4, "Communications" http://call.army.mil/products/newsltrs/91-3/chap4.htm accessed on 10 January 2003.
[43]Material about tactical missile warning systems is based on CALL Newsletter No. 91-3, Chapter 3, "Tactical Early Missile Warning" http://call.army.mil/products/newsltrs/91-3/chap3.htm accessed on 10 January 2003 and Space Warriors, 1-19-20.
[44]Space Warriors, 1-20.
[45]Craig Covault, "USAF Missile Warning Satellites Providing 90 Sec. Scud Attack Alert," *Aviation Week and Space Technology* 134.3 January 21, 1991: 60.
[46]"Spacecraft Played A Vital Role in Gulf War Victory," *Aviation Week and Space Technology* 134.16 April 22, 1991: 91.
[47]CALL Newsletter No. 91-3, Chapter 3, "Tactical Early Missile Warning" http://call.army.mil/products/newsltrs/91-3/chap3.htm accessed on 10 January 2003.
[48]CALL Newsletter No. 91-3, "Chapter 1, Position/Navigation (POS/NAV)," http://call.army.mil/products/newsltrs/91-3/chap1.htm accessed on 10 January 2003 and Ricky B. Kelly, Centralized Control of Space: The Use of Space Forces by a Joint Force Commander (Masters Thesis, School of Advanced Airpower Studies, Air University, Maxwell Air Force Base, Alabama, September 22, 1994), pp. 24-25.
[49]"Iraqis Still Receive Weather Data from U.S. Satellites," *Aviation Week and Space Technology* 134.4 (January 21, 1991):26.
[50]See CALL Newsletter No. 91-3, Chapter 2, "The Battlefield Environment, Section B-Terrain" http://call.army.mil/products/newsltrs/91-3/chap2.htm accessed on 10 January 2003.
[51]The establishment of a coalition force representing 28 nations, highlighted deficiencies in U.N. procedures and structures developed over nearly forty years of operations. The U.N. civilian operation was never fully staffed and could not provide effective humanitarian, political, and security leadership needed for the peace enforcement operations to succeed. The elaborate administrative procedures developed over forty years to support peacekeeping operations were unsuitable for the fast-moving political, economic, humanitarian and military operations in this frequently hostile environment.

[52]The following section on Somalia is based on summaries from the following sources unless otherwise specified: Army Space Reference Text, 2-21 Operation Restore Hope and 10th Mountain Division (LI), "Somalia After Action Report for the Secretary of Defense, Complete Lessons Learned;" 10th Mountain Division (LI), "After Action Report: Executive Summary." The latter two documents are available at the CALL Restricted Database, Central Command, Somalia, Operation Restore Hope (1992), Operation Continue Hope (1993). Also, see Daniel G. Dupont, "Army Space Command Sent to Train Ft. Drum Soldiers: Somalia Operation May Be Army's Largest Use of Space Technology to Date," *Inside the Army* 14 December 1992:3.

[53]The funds for purchasing LANDSAT imagery were not immediately available.

[54]Also see, Space Warriors, 1-14, 1-15. The Army also used space-based systems in Bosnia during Operation Restore Promise. See "Army Space Command Provides Mission Rehearsal System for Use in Bosnia," *Inside the Army* 15 March 1993; Genevieve Anton, "Troops Trained to Use Space-Age Tools," *Colorado Springs Gazette-Telegraph* 21 May 1993:B-2 and Sue McMillis, "Army Space Command Helps Plan for Possible Pilot Rescues in Bosnia," *Colorado Springs Gazette-Telegraph* 3 March 1993.

[55]The DMSP consists of sun-synchronous, polar orbiting, low earth orbiting satellites that provide daily world wide coverage with higher resolution images than those available from geosynchronous satellites. The Joint Task Force Headquarters received its DMSP from its supporting Staff Weather Officers.

[56]See Colonel Jan V. Harvey and Colonel (ret) Alwyn H. King, "Space: The Army's New High Ground," *Military Review* 65.7 (July 1985):38-51; Major Linas A. Roe and Major Douglas H. Wise, "Space Power is Land Power: The Army's Role in Space," *Military Review* 66.1 (January 1986):4-17; Lieutenant Colonel Clayton R. Newell, "The Army and Space," *Army* 37.9 (September 1987):59-61; Paul A. Robblee, Jr., "The Army's Stake in Emerging Space Technologies," *Parameters* 18.4 (December 1988):113-119; and Igor D. Gerhardt, "Space the Air Land Battle, *Army* 40.6 (June 1990):43-47.

Chapter 5
New Ideas about Space and Missile Defense After the War, 1991-1997

ARSPACE After The Gulf War

As the president unveiled a new SDI and the Soviet Union began to wither and disappear, American forces were engaged in conflicts in Southwest Asia that underscored the utility of space-based systems. ARSPACE and ASI both passed the tests presented by the Gulf War and Somalia, although this success may have sealed the fate of the latter organization. ARSPACE thrived because

> Desert Storm provided a real test for the command....ARSPACE didn't fight the war in the traditional Army sense of fighting and we sure didn't win the war. However, we believe we exposed the Army to the potential of space applications early on, prior to the war, and that exposure assisted the fighters to do their jobs better and easier.[1]

The challenge ARSPACE then faced was using space-related systems and products in later operations and weaving space into the Army's consciousness. Otherwise, old difficulties would re-emerge and the Army would again "have a problem getting back into space [because] not many of our people understand space assets and what we can do."[2] The need to meet this challenge and fix the shortcomings exposed by the test of combat in the Gulf War led to creating two new organizations: the Army Space Support Team (ARSST) and the Joint Tactical Ground Station (JTAGS).

As with the end of every major conflict, the end of the Gulf War saw a renewal of the roles and missions debate. It was preceded by an internal Army discussion of the future administrative location of ARSPACE. The Vanguard Study considered whether it made more administrative sense to continue to keep ARSPACE as a FOA reporting directly to the Department of the Army or to make ARSPACE a subordinate entity to a major command.[3] The Army chose to bring its "strategic and space assets together in a single MACOM." A single organization would be responsible for managing "strategic defense, development and use of strategic space assets to support the AirLand Battle Future concept." In addition, it would be "streamlined, cost-effective management."[4]

In a memo to the Vice Chief of Staff of the Army, a writer dissented from the study's conclusion. He observed that the study investigated two approaches: to integrate space responsibilities throughout the Army's structure, or to consolidate space operations into a focused command. If the Army followed the former path, it would embed space expertise in the places where the problems and requirements would be first identified. However, such an approach would need a careful long-term management and budget strategy in light of the "budget and force structure cuts." Following the latter path would guarantee the Army would have a critical mass of expertise, interest and responsibility in a single location but such a command could become isolated from the Army's over-all needs and responsibilities. Thus, the "VANGUARD recommendations would significantly weaken Army space capabilities over the long-term."[5]

Despite this dissenting opinion, the Army chose to follow the study's recommendation to "Reduce the size of ARSPACE Headquarters by 10 percent and consolidate [it] with the Strategic Defense Command." The rationale was direct, noting that consolidating the two entities "establishes a single Army organization for strategic and space assets. The SDC commander would be dual-hatted as CG ARSPACE, thereby ensuring senior Army representation at the U.S. Space Command." In addition, the consolidation would realize 10 percent cost savings as "the result of streamlining minimum essential functions." At the same time, "Retaining ASPO as a FOA recognizes two important features" of that organization. First was "the importance of its current mission and functions" and second, a recognition of the fact that ASPO was "predominantly involved in tactical as opposed to strategic missions that SDC and ARSPACE perform."[6] According to General Order 12, ARSPACE was "discontinued as a field operating agency of the Office of the Deputy Chief of Staff for Operations and Plans, Headquarters, Department of the Army" and "was established as a subordinate command of the United States Army Space and Strategic Defense Command."[7]

New Discussions of Roles and Missions Regarding Space

At about the same time, a roles and missions struggle began over which service would have primary responsibility for space assets. The calls for consolidation came from several quarters, beginning in February 1993 with the recommendations of the Chairman of the Joint Chiefs of Staff, General Colin Powell. He recommended eliminating USSPACECOM and creating a combined element of USSTRATCOM. The Army and Navy functions of the new command would be scaled back. The commander of the Army Space and Strategic Defense Command, Lieutenant General Donald Lionetti, responded that since the Army is the largest consumer of space products it should have a role in developing them. In May 1993, a new edition of *Joint Doctrine: Tactics, Techniques and Procedures for Space Doctrine* noted the lessons learned from the Gulf War and urged the services to make greater use of space assets. The Air Force also made repeated attempts to consolidate or transfer the Army's space mission to itself. The Air Force Chief of Staff, General Merrill McPeak, advocated transferring all military space operations to the Air Force "to avoid overlapping functions in this time of shrinking budgets." However, in September the DoD decided to leave the commands as they were because of the limited cost savings and "the need to stimulate space operations."[8]

The Air Force continued to advocate consolidation to save money. The Army contended that losing control of its space assets would have a two-fold effect: it would hamper efforts to use digital information on the battlefield and silence Army participation in joint space operations. As the Army grew smaller, space became increasingly critical for power projection. In addition, without direct links to field commanders, use of space-based capabilities would be jeopardized.[9] Defending the Army role in USSPACECOM, retired General Frederick Kroeson wrote that the only alternatives to a joint command are either "a defense space agency or assigning space activities to a single service." He pointed out that agencies only add bureaucratic layers but do not improve service to forces in the field and that while "single service assignments worked in the short term," over the long term the "other services find that their needs are not precisely met." However, joint commands have proven their worth through experience.[10]

Although an Air Force Association report called for creating a Joint Space Management Board to recommend the ways in which resources would be divided according to joint or single service requirements, the venue of the dispute moved to a congressionally mandated Commission on Roles and Missions.[11] The Air Force continued to insist that the real standard of decision-making "is whether a different organization offers opportunities for increased efficiencies, reduced costs and expanded combat capability." The USASSDC Commander, Lieutenant General Jay Garner, and the Secretary of the Army, Togo West disputed this assertion. General Garner pointed out, "Because the Army is the biggest user of space, it needs to ensure continued and significant involvement in space matters." Secretary West argued, "Space is a place, not a role, function or mission. All forces must be able to leverage the tremendous potential that free access to space offers. To ensure continued success in what is still a new frontier, we should look for efficiencies in what we have, rather than centralizing responsibilities."[12] Senior Army leaders were joined by senior Navy and Marine Corps leaders in opposing the move to give the Air Force central control over space. The main issues were defined as who will manage military space assets, how will future space requirements be addressed and to what extent the service's space commands will be organized?[13] This dispute over roles and missions continued over the next year with the Army and Navy holding onto their own space commands. As the Roles and Missions Commission of the Armed Forces began its deliberations, the Air Force continued to advocate centralizing space activities under its purview. This purview was expanded when the Air Force staked a claim to be the lead service in Theater Missile Defense, previously an Army mission. In response, the Army's Deputy Chief of Operations and Plans, Lieutenant General Paul Blackwell replied, "Simply stated, Theater Missile Defense should be directed by the man in charge (the joint task force commander). It can't be a sequential transition from ship-to-shore. It has to be seamless."[14]

The argument soon shifted to a discussion over creating a Space Architect in the Department of Defense. The crux of the argument was the role of an oversight board. The original Air Force proposal contained no provision for an oversight board. The Army urged that a Defense Space Management Board serve as a Board of Directors. The Army Chief of Staff, General Gordon Sullivan, expressed concern about "the lack of a space board of directors." He believed this board would serve as a multi-service forum for "senior level leadership involvement" in approving space 'blueprints,' policy, acquisition matters, management process and organization. "A distinct space board of directors will provide service leadership with the requisite insight into the architectural and budgetary trades while assessing the impact of final guidance on programs at all levels."[15] *Inside the Army* reported that "Army officials would be more comfortable with the Air Force proposal if it included...a joint service oversight body... that could serve as an 'appeals court' for issues" that needed to be resolved. The fears of the other services concerned the Space Architect's ability "to circumvent the service staffs and the effective elimination of an element of the coordination process."[16] The Army repeatedly expressed its reservations about the plan and worked to change it. The chief fear was losing "responsibility and authority for ground equipment that leverages space products." The Army also feared the consequences of a single service being "the executive agent for space."[17] The Army had come a long way since it returned to space in earnest during the 1980s.[18] Consolidating Army space assets and functions also continued. In 1994, ASTRO's space technology functions were transferred to the USASSDC. In 1996, ASPO (responsible for the Army's TENCAP) was also transferred to USASSDC.

This debate took place against the background of a Congress increasingly critical of the way the DoD managed its space efforts. In 1992, a Joint House-Senate conference committee asserted that the Secretary of Defense should develop a comprehensive and centralized space acquisition strategy to improve efficiency and decrease costs. In 1993, the House Appropriations Committee noted that the existing space management structures were inadequate and that a coherent management structure for space programs should be created.[19] In 1994, the Defense Department broadly reviewed its space management practices and began restructuring several of the offices and directorates in order to improve integration and coordination of Defense Department space activities.[20]

Three organizational changes took place in 1994-1995.[21] The first was the Secretary of Defense's creation of the office of the Deputy Undersecretary of Defense for Space (DUSD/Space). The office would serve as the principal contact point within the Office of the Secretary of Defense for space matters and develop, coordinate, and oversee implementing the department's space policy and oversee all space architectures and the acquisition of space programs. The DUSD Space worked under the direct supervision of the Under Secretary of Defense for Acquisition and Technology.

Second, was the establishment of a Space Architect in the Department of Defense in March 1995. The office consolidated the responsibility for space missions and system architecture in the Defense Department to eliminate overlapping and redundant programs and make acquisition and future military operations more efficient. The Space Architect worked with the DUSD Space to develop and maintain an overall space system master plan specifying how mission support would be provided by space systems to combatant commanders and deployed operational forces. The Space Architect was a major general who reported through the Air Force Acquisition Executive to the Defense Acquisition Executive and who received policy guidance from the DUSD Space.

Third, the Secretary of Defense and the Director of Central Intelligence formed the Joint Space Management Board (JSMB) in December 1995. The board would act to consolidate defense and intelligence space architecture functions into a single national space architecture that would be designed to ensure they were integrated to the greatest extent possible.

These reforms were short lived as the Clinton Administration began to streamline the DoD's organization by introducing business practices into this bureaucracy. Two Defense Reform Initiative Directives (DRID) reorganized the department's space management responsibilities. In December 1997, DRID 11 abolished DUSD Space and transferred its policy functions to the Office of the Under Secretary of Defense for Acquisition and Technology and the Office of the Under Secretary of Defense for Policy. A later amendment transferred the DUSD Space's policy systems architectures, acquisition, management and integration functions to the Office of the Assistant Secretary of Defense (ASD) for Command, Control, Communications and Intelligence (C3I). In May 1998, DRID 42 ordered the ASD C3I to work with the Under Secretary of Defense for Policy to ensure that former's decisions were integrated into overall national policy decisions.

In July 1998, an amendment to DRID 11 abolished the Space Architect's office and replaced it with an office of the National Security Space Architect, who would be responsible for maintaining, disseminating and developing the National Space Security Master Plan, developing transition strategies for future space strategies, integrating requirements into future space system architectures and advising the ASD C3I and the Deputy Director of Central Intelligence for Community Management and their staffs of appropriate budget documents. The office was created to address the needs of the warfighter directly. DRID 11 also abolished the JSMB and replaced it with the National Security Space Senior Steering Group. It addresses broad national security space management and integration issues in the Defense Department and the intelligence community.

The End of the Cold War and a New Security Environment

There had been many changes to world politics since President Reagan's 1983 SDI announcement. The most revolutionary was the end of the Cold War, signaled by the end of the Soviet bloc in Eastern Europe. The Berlin Wall fell in 1989, which ended the division of the city and led to German reunification. As Russian troops were withdrawn from Eastern European countries, their communist governments fell and were replaced by freely elected noncommunist leaders. Even President Reagan's "Evil Empire," the Soviet Union, disintegrated into its component parts in 1991, leaving the United States as the sole superpower.[22]

Despite the disappearance of the traditional Cold War enemies, it was soon evident that threats still existed. In 1990 and 1991, the world focused its attention on the activities of Saddam Hussein, President of Iraq. During the subsequent Persian Gulf War, the Scud missile, although not a new weapon, was recognized as a new threat in the ballistic missile arsenal. Analysts observed that ballistic missiles "[appealed] to leaders of developing countries."[23] They were and still are valued for their long range, short flight time, payload flexibility, and relatively low cost. A 1992 study on BMD proliferation, for example, located the 300-km Soviet Scud in the arsenals of 16 countries. The same study found that "thirteen countries have produced or [are] in the process of producing" long range ballistic missiles. As Lieutenant General Donald

Lionetti observed, "The tactical ballistic missile genie is out of the bottle and can never be put back. There won't ever again be a mid to high intensity armed conflict without tactical missiles."[24]

A New Approach – President Bush and Star Wars

In 1989 there was a new president in the White House, but there were no anticipated changes to the Strategic Defense System. In fact, on 9 February 1989, President George H.W. Bush announced in an address to a Joint Session of Congress, that he would "vigorously pursue" the Strategic Defense Initiative. Following a review of the national defense strategy, Bush "concluded that the goals of the SDI program were generally sound." In addition, the program had the potential for a deployment decision in the next few years. Bush decided that "emphasis in this effort was to be directed toward perfecting boost-phase kill technologies such as Brilliant Pebbles."[25]

In December 1989, President Bush commissioned an independent review to examine the strategic requirements for a "new world order." Conducted by Ambassador Henry Cooper, the study concurred with this assessment of Brilliant Pebbles and its potential in the Strategic Defense Architecture.[26] It would ultimately define the concept for a new missile defense system known as Global Protection Against Limited Strikes (GPALS).

Global Protection Against Limited Strikes (GPALS)

Responding to these events, President George Bush presented a revised version of the SDI concept in his 1991 State of the Union address. Rather than the massive threat posed by the Soviet nuclear arsenal, the program was redirected to "emphasize defense against limited attacks of up to two hundred warheads."[27] Specifically, President Bush announced

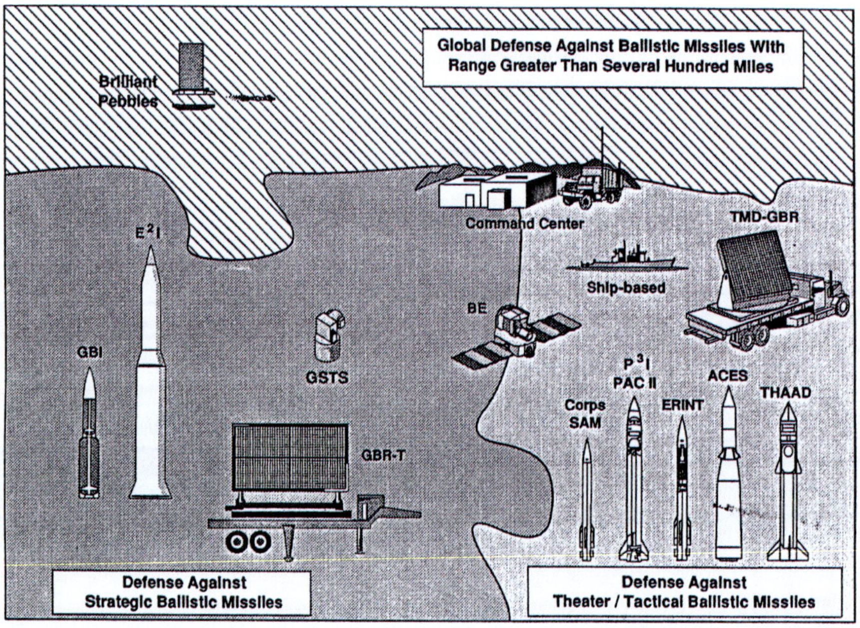

Fig. 5-1. The three pieces of the Global Protection Against Limited Strikes puzzle provide a global defense against strategic

and tactical ballistic missiles.

> I have directed that the Strategic Defense Initiative program be refocused on providing protection from limited ballistic missile strikes, whatever their source. Let us pursue an SDI program that can deal with any future threat to the United States, to our forces overseas and to our friends and allies.[28]

A smaller version of Reagan's SDI, the GPALS would provide a defense against "purposeful strikes by the various Third World powers developing ballistic missiles, or accidental or unauthorized launches from the U.S.S.R."[29] The GPALS architecture focused on three elements. The first facet was a ground-based National Missile Defense (NMD) system. The second was a ground and sea-based Theater Missile Defense (TMD) system that would protect friendly nations, allies, and deployed American forces. The third and final element was a space-based global defense system "that could stop a small attack against virtually any point on the globe."[30] These goals would be accomplished with a tiered deployment of 1,000 space-based Brilliant Pebbles interceptors, 750-1,000 long-range ground-based interceptors located at six sites, space-based and mobile sensors, and transportable theater ballistic missile defenses.

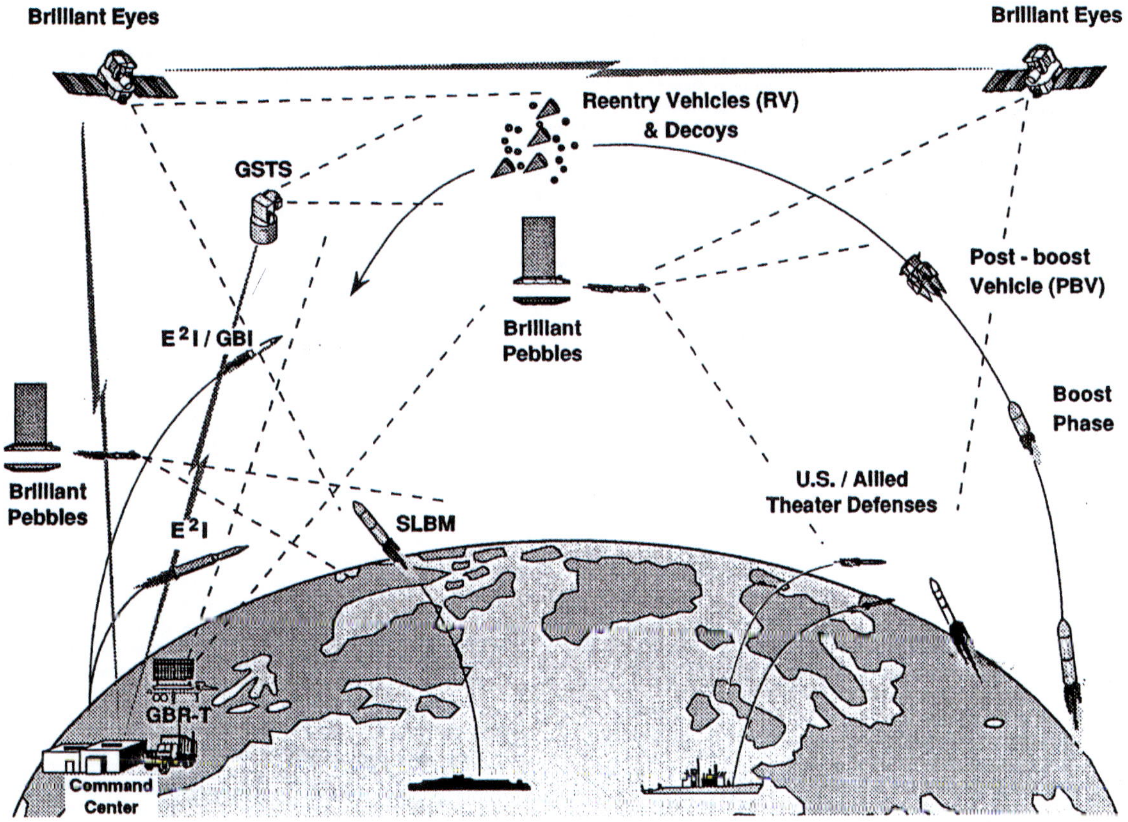

Fig. 5-2. The Global Protection Against Limited Strikes incorporated a new element - Brilliant Pebbles - in the global strategic defense scenario.

The events of Desert Storm which vividly illustrated the need for theater missile defenses had a strong impact on the American people. The Brilliant Pebbles space-based element of the GPALS System however renewed concerns about the militarization of space. The Missile Defense Act of 1991, signed into law on 5 December 1991, further defined the new initiative.[31] The legislation directed the Department of Defense to "aggressively pursue the development of advanced theater missile defense systems, with the objective of down selecting and deploying such systems by the mid-1990s."[32] With regard to National Missile Defense, DoD was to "develop for deployment by the earliest date allowed by … technology [development] or by fiscal year 1996 a cost effective, operationally effective, and ABM Treaty-compliant antiballistic missile system at a single site as the initial step toward deployment of an antiballistic missile system." Congress also supplied funding for Brilliant Pebbles and other innovative technologies, but these were not to be a part of any initial deployment. At the same time, Congress directed the President to pursue negotiations with the Soviet Union to allow the expansion of a deployed NMD system beyond the one location permitted by the ABM Treaty.

New Organization: Program Executive Office-GPALS

Fig. 5-3. Emblem of the GPALS Program Executive Office

As the roles and missions discussions over space continued, a newly revised strategic defense system was emerging. Freed from deterring a defunct Soviet Union, planners slowly began to create a new structure. The SDIO retained primary responsibility for this revised strategic defense system. With this new guidance and directives, however, on 29 July 1992 a newly created Program Executive Office (PEO) for GPALS replaced the PEO for Strategic Defense, established under the leadership of the U.S. Army Strategic Defense Command (USASDC) Commander in 1988.[33] Established by Memoranda of Agreement with the military services, the PEO provided a centralized organizational structure for the acquisition and deployment of missile defenses. Initially headed by a Major General, the U.S. Army GPALS PEO reported to the Army Acquisition Executive.

The PEO GPALS, subsequently renamed PEO for Missile Defense in 1993, was composed of elements of the USASDC and the U.S. Army Missile Command's PEO – Air Defense. The resulting organization was divided into two Program Offices – Army National Missile Defense and Army Theater Missile Defense. The NMD Program Office included the GBI, GBR, GSTS, Site Development and Regional Operations Center/Communications Project Offices, formerly of the USASDC. The TMD Office was composed of the Theater High Altitude Area Defense and Extended Range Interceptor Project Offices and the Adjunct Sensors, Arrow and Testbed Product Officers from the USASDC and the Corps SAM and Patriot Project Offices from the former Missile Command.

On 18 July 1996, the PEO Missile Defense officially became the PEO Air and Missile Defense (PEO-AMD).[34] As Colonel (P) Daniel Montgomery, PEO-AMD explained "air defense has historically included all threat platforms in the air or space – whether they are air breathing or not." The PEO's TMD systems, with the exception of THAAD, are also aircraft, cruise missile and helicopter killers.

New Organization: U. S. Army Space and Strategic Defense Command (USASSDC)

Fig. 5-4. The U.S. Army Space and Strategic Defense Command adopted this command logo in 1995

As part of the reorganization that created the PEO-GPALS, the USASDC became the U.S. Army Space and Strategic Defense Command, a field-operating agency of the Chief of Staff.[35] The new organization retained an affiliation with the SDIO, but would also provide an Army focal point for space and missile defense matters. The USASSDC continued to perform research and development for strategic and theater missile defense technologies and anti-satellite efforts, providing research and technological support to SDIO missions and matrix support to the PEO-GPALS. The command also retained operational responsibility for the Kwajalein Missile Range and the High Energy Laser Systems Test Facility.[36]

At the same time, General Orders 13 designated USASSDC as the Army's focal point for space. The creation of the USASSDC began the process, initiated by the Chief of Staff of the Army, to centralize research and development of space and strategic assets for the benefit of the soldier in the field. In the first step the U.S. Army Space Command, formerly a field operating agency of the office of the Deputy Chief of Staff, Operations and Plans, became a subordinate command of the USASSDC in August 1992.[37] Just two months earlier, the Chief of Staff of the Army had approved a realignment proposal which made the ARSPACE the "user" for deployed ground-based elements of the NMD program.

The next step in the creation of a united Army space program came in March 1993. On 3 March, Lieutenant General William H. Forster, Military Deputy to the Assistant Secretary of the Army (Research, Development, and Acquisition) ordered the transfer of the Army Space Technology Research Office (ASTRO) from the Communications-Electronics Command to the USASSDC. Created in 1988 by the Army Materiel Command, the ASTRO managed near and possible far-term space R&D programs and provided a developer focus both within the Army and with outside agencies. As part of the USASSDC, ASTRO became the Space Applications Technology Program.

The final step in consolidating the Army Space program came on 1 July 1994. On that date, the Army Space Program Office (ASPO), a field agency of the Office of the Deputy Chief of Staff for Operations and Plans transferred to the USASSDC.[38] The ASPO, which was established in 1973, has responsibility for the Tactical Exploitation of National Capabilities Program (TENCAP).

A New Priority – Theater Missile Defense

Although the emphasis upon Theater Missile Defense (TMD) began with the GPALS initiative, the command began exploring theater concepts in the mid-1980s. In December 1985, the SDIO assigned to the USASDC the task of developing TMD architectures.[39] Six months later Secretary of Defense Caspar Weinberger relayed the increasing concern in Europe of the "growing threat posed in the chemical, nuclear and especially conventional areas by increasingly accurate Soviet shorter-range missiles."[40] Secretary Weinberger directed SDIO to explore "specific ways in which the U.S.-led SDI research program [could] assist the NATO extended air defense effort in which the Europeans are taking a leading role." At a NATO Defense Ministers conference in Brussels in December 1986, Weinberger announced the first seven contracts devoted to TMD. Contractor teams from Germany, France, Italy, Great Britain, Israel and the United States participated in the first phase of the TMD Architecture Study.

Two years later, the Joint Chiefs of Staff approved the Army Tactical Missile Defense Operational Concept, which outlined the capabilities required to counter the tactical missile threat of the future. By 1990, the programs had progressed to the extent that the SDIO received a new program element, entitled Theater Missile Defense, in the appropriations legislation. The Appropriations Conference Committee also recommended that the Defense Department accelerate research on theater and tactical ballistic missile defense systems. Two Army programs, the Extended Range Interceptor (ERINT) and the Arrow, were specifically mentioned at this time. The SDIO was subsequently assigned responsibility, on 9 November, for the centrally managed DoD Theater and Tactical Ballistic Missile Defense program.[41] In January 1991, all Army TMD functions would be assigned to the USASDC in the Theater Missile Defense Applications Project Office.[42]

The events of Operation Desert Storm would prove the significance of Theater Missile Defenses. Although later studies would question its effectiveness, as Scud missiles[43] rained upon coalition forces and allied nations, the only defense was the modified Patriot anti-aircraft missile system.[44] The worst event of the war for American forces was not in battle, but rather the 25 February 1991 Scud attack on an Army barracks near Dhahran, Saudi Arabia, which killed 28 soldiers and wounded 100 others. As a *Los Angeles Times* reporter observed: "The age of Star Wars had arrived."

TMD and the End of SDI

With the arrival of the new administration of President William Clinton primary emphasis remained on TMD efforts. On 13 May 1993, Secretary of Defense Les Aspin announced that with the end of the Cold War, the United States was no longer threatened by a massive attack from the Soviet Union. Instead, the new threat was theater ballistic missiles controlled by Third World dictators, or "hostile or irrational states that have both nuclear warheads and ballistic missile technology that could reach the United States."[45]

Thus the first priority became the deployment of a TMD system with space-based sensors.[46] The second priority was the NMD program with deployment timed to meet the threat posed by rogue nations. Further research and development, follow-on technologies such as directed energy efforts, received the lowest priority rating. To reflect the new priority structure and its wider mission, Secretary Aspin reorganized and renamed the SDIO to create the Ballistic Missile Defense Organization (BMDO).[47] With this shift from research to development and acquisition of systems, the BMDO now reported to the Under Secretary of Defense for Acquisition, rather than directly to the Secretary.

Released in September 1993, the Bottom-Up Review of the Military, initiated by the Clinton Administration, outlined the national security plans for the five-year period between 1995 and 1999. The goal was to field effective TMD systems in the shortest time possible, while also "providing a basis for a speedy decision to deploy national missile defenses should a serious threat … suddenly materialize."[48] Thus in the field of BMD, the review laid out a three-tiered program with primary emphasis given to TMD, in particular the follow-on to the Patriot system, modifications to the Navy's Aegis air defense system, and the Army's Theater High Altitude Area Defense system. The TMD program would receive a budget of $12 billion over that five-year period.[49] In contrast the NMD would only be allocated $3 billion and Follow-On Technology and Research and Strategy would share an allotment of $3 billion for the same time-frame.

The ABM Treaty and TMD Demarcation

In September 1994, President Clinton and Russian President Boris Yeltsin agreed that "[b]oth sides have an interest in developing and fielding effective theater missile defense systems on a cooperative basis."[50] The issue became the definition of a TMD system, in particular with reference to the Theater High Altitude Area Defense.[51] The Clinton administration proposed that the boundary between tactical and strategic ballistic missiles be "the ability to intercept a missile traveling at 5 kilometers per second."[52] They added that this determination should be based on demonstrated capability and not theoretical ability. Following two years of negotiations, officials agreed to the Russian proposal that TMD systems with a demonstrated interceptor velocity of 3 kilometers/second would comply with the ABM Treaty. The proviso was that these systems were not to be tested against target missiles with a range in excess of 3,500 kilometers and a maximum flight velocity of no more than 5 kilometers/second.[53] The governments of the United States, Russia, Belarus, Kazakhstan, and Ukraine signed the final agreement on 26 September 1997.

Reconfiguring the Post-Cold War Army

The evolution of missile defense systems and organizations was only one series of events that made up the task of reconfiguring the Army after the Cold War. The Chief of Staff of the Army, General Gordon Sullivan, established a new vehicle to investigate and support necessary change, the Louisiana Maneuvers (LAM) process. General Sullivan consciously modeled his LAM on the series of maneuvers the Army conducted in Louisiana and the Carolinas in 1940-1941. These maneuvers were a culmination of a series of corps- and field army-level exercises to train troops, test new doctrinal and organizational concepts, identify equipment requirements and evaluate future senior Army leaders that began in 1938. In using this term General Sullivan hoped to signal that this would not be business as usual but that the results would not be foreign to the Army. His idea was "to conduct experiments that would be the basis for designing new units." He also made it plain "that I – not merely my staff – was going to be personally involved."[54] The process General Sullivan set in motion gave ARSPACE greater impetus and outside support at the highest levels to make the changes indicated by the lessons of the Gulf War.[55]

In 1993 as part of the LAM process, the Army investigated the organization and equipment necessary to establish a deployable space support team and the following year established the Contingency Operations-Space (COPS) at ARSPACE. As the Army's senior leadership was deciding on the merits of the case for permanent space support teams, the Army Audit Agency released a report confirming a need for an organization that would provide space support to warfighting commanders and their staffs. After noting ARSPACE's successful support of field units in Bosnia and Somalia as well as relief efforts in the wake of Hurricanes Iniki and Andrew, it pointed out that providing operational support to commanders and their staffs was not part of ARSPACE's mission. It did not have the resources to provide sustained, operational support to units in the field. As a result, the commanders of these units had to go to many sources to obtain the support they needed. This was probably the final push needed to bring the required level of support for this mission. Later in 1994, the COPS became the ARSST.[56]

Given the lessons learned in TMD from the Gulf War, the need for early warning capabilities was unquestioned.[57] The result was fielding a unit, the Joint Tactical Ground Station, the JTAGS, in a relatively short period of time. This unit has demonstrated its ability to fulfill Army, Joint and coalition requirements for TMD. The process of establishing and training the unit and acquiring the appropriate equipment shows how rapid the process can be when an urgent need is presented. The same may be said for the organization and deployment of the ARSST. The JTAGS supports all aspects of TMD: passive defense, attack operations, active defense and command, control, communications, computers and intelligence (C4I) and is flexible enough to be placed in any theater of operations. The JTAGS is not merely an example of the Army's versatility; it is a multi-service system and drew on multi-service research and development, acquisition, training and unit operations. As the American military slowly evolves toward joint capabilities and joint operations, the lessons learned from the JTAGS could provide important insights for all the Services.[58]

ARSPACE and Contingency and Training Operations in the 1990s

While roles and missions were being debated in Washington, ARSPACE continued to support Army contingency operations and exercises. The areas to which American troops were deployed had minimal or nonexistent national communications infrastructures and space-based systems proved their worth again.

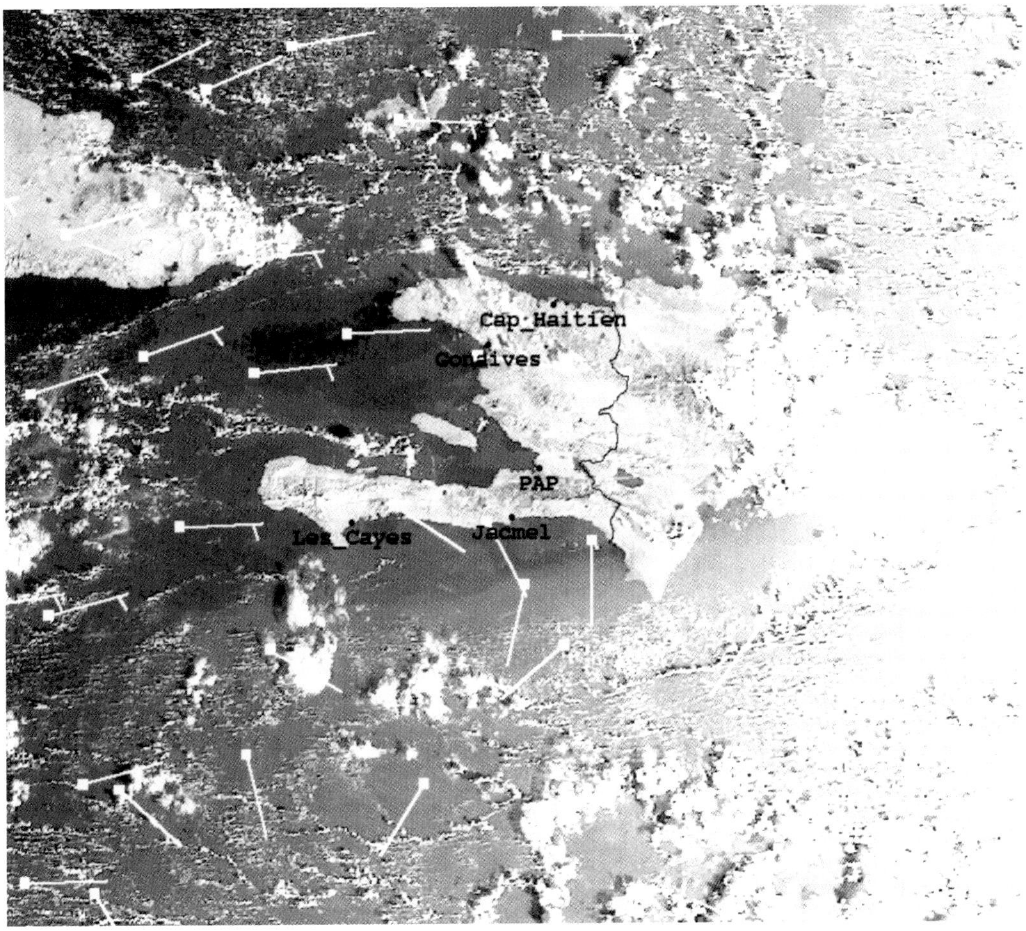

Fig. 5-5. Weather map of Haiti.

In the mid-1990s, the Army Space Command provided space products to troops involved in operations in Haiti and the Balkans, supplied material for planning an operation to evacuate noncombatants in Liberia and participated in major exercises.[59] In Operation Uphold Democracy (Haiti), ARSPACE supported Joint Task Force (JTF) 190, primarily the 10th Mountain Division and the XVIII Airborne Corps. At first the satellite communications systems were used to connect the forces ashore and afloat with decision-makers in Washington.[60] The systems employed included the Mission Planning Rehearsal System (MPRS), multispectral imagery (MSI), the Terrain Reconnaissance Tool (TRT), the Advanced Communication Satellite (ACTS) and INMARSAT. In addition, the Continental United States Regional Space Support Center supported the Atlantic Command, the XVIII Airborne Corps and the 10th Mountain Division with Defense Satellite Communications System planning support from Fort Bragg.

Fig. 5-6. Communicating in the aftermath of Hurricane Iniki in Hawaii.

As the situation in Haiti stabilized, the ACTS system was used for "morale video teleconferencing between soldiers in Haiti and their families" at their home stations.[61] In November 1995, ARSPACE personnel briefed NASA on the ways the ACTS satellite was used "in Haiti and impressed NASA with their use of the satellite and ground terminals."[62]

Space-based systems proved their worth. In fact, ARSPACE used three ACTS terminals in Haiti, two in Port au Prince and one in Cap Haitien. For the first thirty days of the deployment, "the ACTS VTC was the primary command and control system used by the JTF commander and staff." It was not until other "secure systems were brought on line" that ACTS was placed in a "secondary role of providing 'morale conferences' between soldiers in Haiti and their families." In mid-November, "a High Resolution Weather Receiver was sent to Haiti" along with trainers to instruct Air Force Staff Weather Officers in its use.[63] The joint task force continued to use this equipment until April 1995, when an ARSST brought it back, leaving the multinational force with a single INMARSAT terminal and pictel equipment for VTCs.[64]

Fig. 5-7. An Army Space Support Team in Albania supporting operations in Kosovo.

Army peacekeeping operations in the Balkans were supported in a similar fashion. The command supported the 1st Armored Division's planned entry into Macedonia with various MSI products, including three dimensional perspectives and "fly throughs" of Macedonia. It also supplied LANDSAT and SPOT maps to the division.[65] Later that year, through its 1st Satellite Control Battalion (SATCON BN) and ARSSTs, the command supported operations in Bosnia.[66]

The ARSPACE supported Allied Command Europe Rapid Reaction Corps, Task Force Eagle's and the 10th Special Forces Group's planning and preparation for Bosnia operations by providing them with a Multispectral Imagery Processor (MSIP). In addition, a single soldier from the Regional Satellite Support Center (RSSC)-Europe "was deployed to Zagreb, Croatia, as a member of the International Force Combined Joint Communications Coordination Center" and soldiers from the other two RSSCs were sent to Europe so that RSSC could operate 24 hours a day.[67] Later that month, two of the 1st SATCON BN's companies provided super high frequency (SHF) satellite communications (SATCOM) support for Operation Joint Endeavor using both DCSC satellites and the NATO IV series of SHF MILCOMSAT satellites.[68] Later reports detailed the support ARSPACE provided to this operation.[69] In this same time period, "USARSPACE RSSC-EUR provided Ground Mobile Forces (GMF) TACSAT planning" to support "EUCOM for Operation Assured Access," a noncombatant evacuation operation for Liberia. For the potential mission, the RSSC reconfigured the West Atlantic DSCS III Satellite. "The GMF terminals were operated by the 112th Signal Battalion and the 1st Combat Communications Squadron."[70]

The types of support provided for various contingency operations are summarized in the following table.

Table 5.1: Equipment Support for Contingency Operations, 1990-1994

Equipment/ Contingency	MSI	MPRS	INMARSAT	Weather
Desert Storm	X		X	X
Provide Comfort	X			
Zaire NEO Non-combatant Evacuation Operation	X			
Hurricane Relief Iniki Andrew			X	
Somalia	X	X		X
Bosnia	X	X	X	
Macedonia	X	X	X	
Rwanda	X			X
Haiti	X	X	X	X

Source: USASMDC Archives

The command also participated in major training exercises. In 1995, for example, these included Atlantic Resolve, Roving Sands, Ulchi Focus Lens and Cobra Gold. It frequently used the exercises as part of the ASEDP process. In Atlantic Resolve that year, ARSPACE deployed twenty personnel to use all the ARSST equipment (for mapping, weather, intelligence and command and control capabilities as well as selected intelligence assets).[71] Command and control capabilities included the Space Enhanced Command and Control System.[72]

In Roving Sands that year an Army Tactical Missile Defense Element (ATMDE) Force Projection Tactical Operations Center (TOC) was airlifted to Fort Bliss. The Vehicular Data Communications and Positional Awareness Demonstration was brought to Forts Hood and Bliss for the exercise.[73] At Roving Sands, the ATMDE "generated significant interest." It was deemed a success, "demonstrating its capabilities in a dynamic environment." It was hoped it would provide "initial baseline data for the Army's War Fighting Experiment."[74] The ARSST-PAC and the JTAGS at Osan AFB, Republic of Korea participated in Ulchi Focus Lens with weather (the 607th Squadron) and intelligence support (2nd Infantry Division).[75] For Cobra Gold, the Pacific Command supplied an ARSST and its associated equipment[76] to support Army units from I Corps, 25th Infantry Division and the 1st Special Forces Group that participated.[77]

Fig. 5-8. In Albania supporting Task Force Hawk.

On 20 June 1994, Vice Chief of Staff of the Army General J.H. Binford Peay III signed the Charter for the Theater Missile Defense Advocate. This charter made the Commanding General of USASSDC the Army's Theater Missile Defense Advocate. As such, the Commander was tasked to serve as the Department's focal point and coordinator for systems requirements and operational aspects of TMD. He would also conduct department level studies of all four elements of TMD – Active Defense, Attack Operations, BM/C^{4}I and Passive Defense. Some of these key TMD missile programs developed by the command are outlined below.

Extended Range Interceptor (ERINT)/ Patriot Advanced Capability-3 (PAC-3)

The Extended Range Interceptor (ERINT) was a follow-on to the 1980s FLAGE experiment.[78] To create the ERINT engineers upgraded the design of the FLAGE adding, for example, aerodynamic maneuvering fins and attitude control motors, thereby extending the range of the system.[79] Despite funding cuts, ERINT passed its final design review in December 1989. Under the new guidance, this high velocity, hit-to-kill missile was to be used primarily against maneuvering tactical missiles and secondly, against air-breathing aircraft and cruise missiles.[80]

In 1992, LTV Aerospace & Defense Company demonstrated the ERINT's flight capabilities. Later in that same year, the policy redirection towards theater/tactical defense resulted in the upgrading of the ERINT to the status of Project Office. During the third flight test, in June 1993, the ERINT tracked its target but failed to intercept the Lance missile.[81] The problems were soon rectified and the ERINT had several successful intercept tests in fiscal year 1994. These tests pitted the ERINT against two target theater ballistic missiles with simulated bulk chemical warheads and an air-breathing drone.[82]

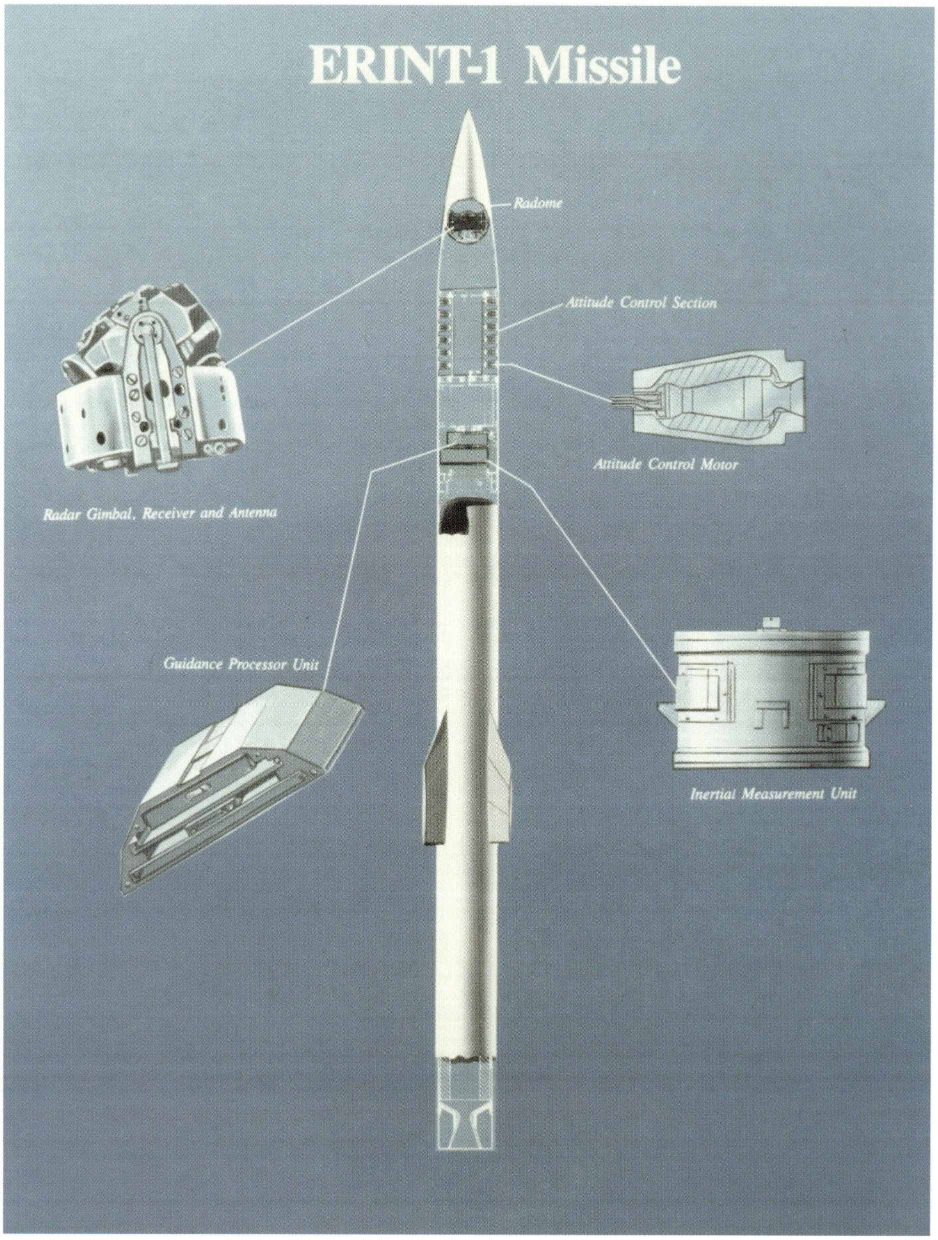

Fig. 5-9. The Extended Range Interceptor incorporated several technological advances creating a smaller, more effective interceptor.

Fig. 5-10, 5-11 and 5-12. Extended Range Interceptor 1 destroys a Storm target in this sequence of photographs from GTF-2 on 30 November 1993.

Fig. 5-13. The PAC-3 uses an attitude control system made of a ring of small solid rocket thrusters which provide "the extremely rapid air frame response accuracy to achieve hit-to-kill performance." Compared to its predecessor, the PAC-2, the PAC-3 can protect an area seven times greater and is effective against chemical and biological warheads. Photograph taken on 5 February 2000.

Budget constraints put the ERINT in direct competition with another short-range theater missile defense system, the Patriot missile developed by Raytheon Corporation for the U.S. Army Missile Command.[83] The Army evaluated the ERINT and a revised Patriot as part of the pre-planned upgrades to the Patriot system, the Patriot Advanced Capability – 3 (PAC-3). The Army System Acquisition Review Council determined the ERINT, which is half the size of a Patriot missile, "[offered] increased range, accuracy and lethality."[84] The official decision came on 19 May 1994, when the Defense Acquisition Board endorsed the Army's decision to select the ERINT missile. In July, Principal Deputy Under Secretary of Defense R. Noel Longuemare authorized the ERINT project to enter the engineering and manufacturing development phase. Following these decisions, the ERINT Project Office merged with the Patriot Project Office, within the PEO for Missile Defense, and the ERINT missile became known as the new interceptor for PAC-3. In this capacity, the PAC-3 will be the lower tier of a two-tier active theater missile defense.[85]

The Army conducted the PAC-3 deployment in a three phased configuration. The first units received the PAC-3 Configuration 1 in December 1995. This system incorporated the guidance enhanced missile or Patriot GEM, and improvements to the BMC^3I. The PAC-3 Configuration 2 system, fielded in fiscal year 1998, used both the PAC-2 and GEM missiles and made upgrades to the radar, communications and other systems. The PAC-3 Configuration 3 meanwhile introduced the new PAC-3 hit-to-kill interceptor and made additional improvements to the communications, radar and ground support systems. Plans originally called for the PAC-3 Configuration 3 to be fielded in the year 2000. However, the situation in the tense Persian Gulf region in early 1998, led to a Pentagon decision to deploy the relatively untested prototype PAC-3 missiles.[86]

Testing resumed in 1999 with a seeker characterization flight in March and the first official intercept test of a PAC-3 in September. Both tests achieved successful intercepts and led to a government decision to enter low-rate initial production phase. At the same time, however, the program experienced budget overruns and set-backs that put the program more than a year behind schedule.[87] Despite these financial controversies, the test program proved successful. Integrated tests conducted in 2000 demonstrated the system's capabilities against several types of targets – tactical ballistic missiles, cruise missiles and an aircraft.[88] The developmental test phase ended in March 2001 with the first tactical ripple mode test. The first missile destroyed the target and the second self-destructed as expected. After completing seven intercepts in as many attempts, however, operational tests conducted in 2002 proved less successful. Although targets were intercepted, one or more of the missiles failed to perform as expected in four successive ripple tests. The anomalies were identified and addressed and, as Colonel Tom Newberry observed, "Nothing that we've encountered so far would indicate that we've got some sort of systemic problem."[89]

The PAC-3 system consists of the launcher with up to 16 missiles, a radar, fire control station, power supply and communication relays. Configuration 3 deployment began in March 2000, when batteries in the 108th Air Defense Artillery Brigade received the first PAC-3 radars. The first new missiles were delivered in September 2001. In spite of the operational test issues, by August 2002, the Pentagon declared the PAC-3 combat ready.[90]

Arrow

Another element in the lower tier for the theater architecture is the Arrow missile system, developed jointly by the governments of Israel and the United States.[91] Initiated in July 1988 with a Memorandum of Understanding, the Arrow is an anti-tactical ballistic missile for specific use in Israel but capable of operating with American TMD systems. A successful first launch in August 1990 was followed by several failed tests, which resulted in a redesign of the Arrow System.[92] Nevertheless the two governments signed a Memorandum of Agreement on 7 June 1991 for the Arrow Continuation Experiments (ACES) to develop an Arrow-2 missile and launcher. Also with the new emphasis on TMD, on 29 July 1992, the Arrow Office transferred to the PEO-GPALS and became the Arrow Project Office.[93]

Fig. 5-14. Unlike other theater missile defense systems, the Arrow, which travels at Mach 9, employs a warhead to intercept its targets.

Despite two partially successful launches, in September 1992 and February 1993, concerns about the feasibility of the Arrow continued. These doubts were raised again when a planned ship-based launch against a simulated chemical warhead, the first attempt by a "western-developed missile" to intercept a target with a non-conventional warhead, was canceled.[94] Given its test record, some members of Congress expressed a reluctance to continue funding the Israeli program. The Arrow test program continued to be plagued by mechanical problems until June 1994. In two previous intercept tests the warhead had failed to detonate, although the Arrow came close to the target. On 12 June 1994, however, the Arrow successfully intercepted a surrogate tactical ballistic missile. This was the seventh and final test before the initiation of the ACES and the Arrow-2 system.[95]

The ACES program began with the initial flight test of the two-staged Arrow-2 missile in July 1995.[96] This test was followed by a successful intercept of a simulated Scud missile on 20 August 1996. With the completion of two more successful intercepts, the ACES program ended in 1998 to be replaced by the Arrow Deployability Program. The goal of this initiative was to integrate the Arrow missile with its various system components and determine the Arrow's ability to operate with American TMD systems.

The Arrow program completed its first integration test on 14 September 1998. During this test, the Arrow 2 interceptor was controlled throughout its flight by the various components of the Arrow Weapon System, specifically the surveillance/fire control radar (Green Pine), the fire control center (Citron Tree) and the launcher control center (Hazel Nut Tree). A second full system test conducted in November 1999 again demonstrated the system's ability to acquire and intercept targets.

With the completion of these tests officials declared the Arrow Weapon System to be initially operational, as a limited contingency capability. The Israeli government deployed its first battery of 14 Arrow missiles on 14 March 2000. With the first delivery to the Israeli Air Force, Major General Eitan Ben-Eliahu declared, "As of today we complete the acceptance of the only weapon system of its kind in the entire world. We are the first to succeed in developing, building and operating, a defense system against ballistic missiles."[97] In a 14 September 2000 test, the Arrow Weapon System successfully intercepted an air-launched Black Sparrow target in an in-bound trajectory. As a result of this test, Israel declared the first battery, located near Tel Aviv, operational on 17 October 2000. A second battery has since been added at Hadera, with plans for a third battery.

Since this time, officials have expanded the tests of production missiles to include new challenges. In August 2001, the Arrow-2 achieved an intercept at approximately 100 kilometers from the coastline at a distance "higher and farther than in any previous tests." In January 2003, four Arrow interceptors were launched almost simultaneously against a simulated barrage of target missiles. Israeli officials stated that "the Arrow should be able to intercept an incoming missile in less than three minutes at altitudes of more than 30 miles." Israel developed the system in preparation for a possible war against Iraq.

Theater High Altitude Area Defense (THAAD)

For longer-range protection, the USASDC and SDIO introduced the THAAD missile system in 1988, the first weapon system developed specifically to defend U.S. and allied soldiers, military assets and population centers from the threat of theater ballistic missile attack.[98] Designed to counter tactical ballistic missiles, such as the Scud, the THAAD system uses truck mounted launchers and a ground-based radar. According to plans THAAD missiles, "smaller, faster and smarter" than existing systems, would be able to defend an area "dozens of times wider" than a Patriot battery.[99]

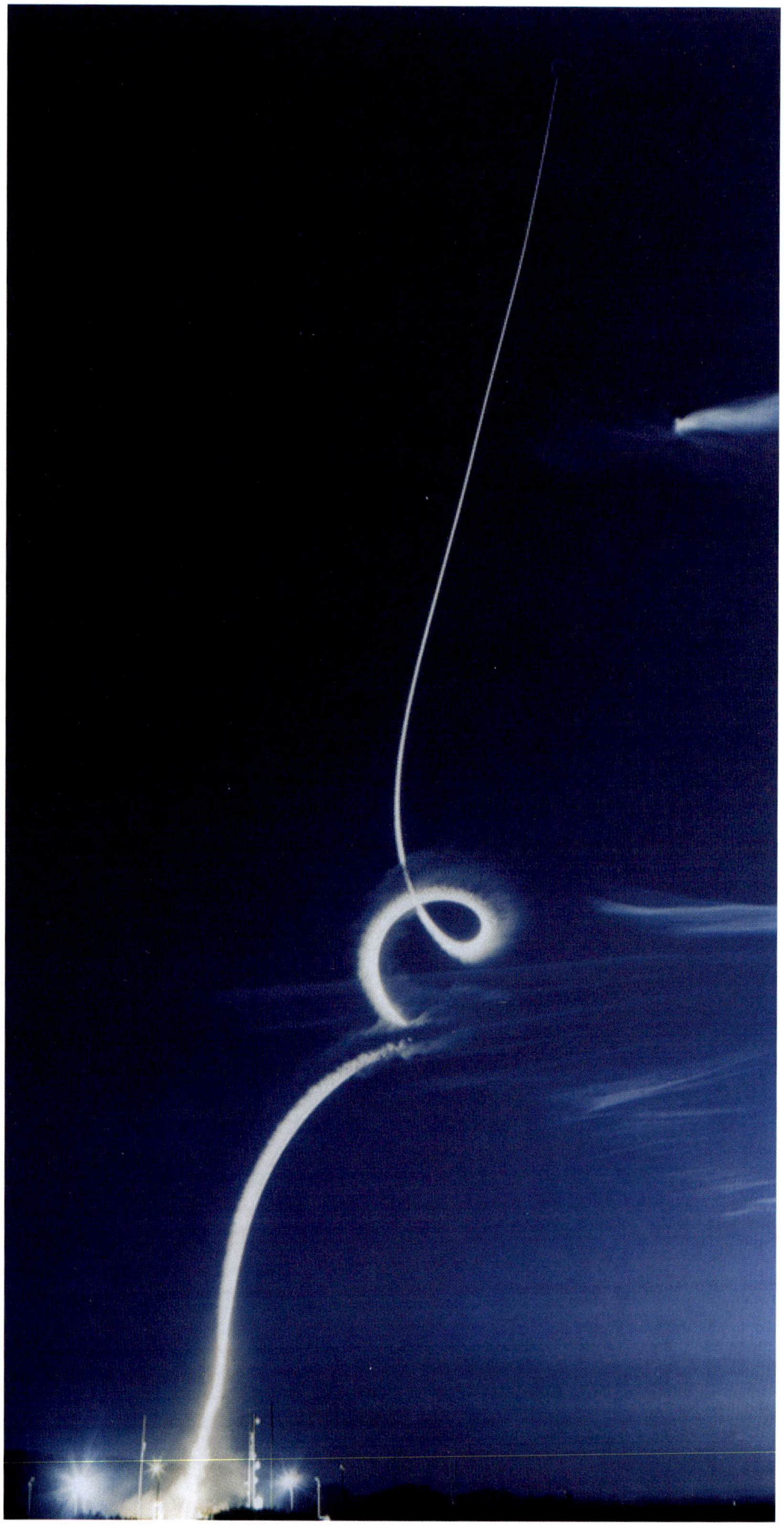

Fig. 5-15. The curly-cue in the Theater High Altitude Area Defense (THAAD) contrail is part of a purposeful maneuver to burn off excess fuel before the missile proceeds down range. The first THAAD intercept occurred during the tenth flight on 10 June 1999.

Fig. 5-16. The Theater High Altitude Area Defense, launched from a palletized truck.

Fig. 5-17. During the fifth THAAD test in March 1996, the metric accuracy of the THAAD DEM/VAL radar achieved a mark 4.6 times greater than was required, by the ninth test, the accuracy rate exceeded the baseline by 12.0 times. Essentially, if the radar was in Huntsville, Alabama, it could see an object smaller than a basketball sitting above the Washington Monument.

Fig. 5-18. Activation Ceremony of the Bravo Battery of the Theater High Altitude Area Defense Battalion.

The THAAD request for proposals was delayed several months as the SDIO and the Army debated the appropriate acquisition strategy.[100] The demonstration/validation contract, however, was awarded to Lockheed Missiles and Space Company in September 1992. In March 1993, the design underwent a revision, producing a "larger kinetic-kill interceptor and a more powerful rocket booster," to accommodate the flight termination system and ensure the system's ability to intercept tactical missiles "above and just within the Earth's atmosphere."[101]

In addition to the treaty woes, cost growths, budget cuts, management problems, and technical concerns combined to delay THAAD testing.[102] Nevertheless, with three flight tests beginning in April 1995, the THAAD project achieved its objectives and made preparations for the next phase of the demonstration/validation program. The first intercept attempt occurred on 13 December 1995. The THAAD demonstration/validation radar performed as planned, tracking and detecting all objects. The overall test, however, did not achieve an intercept due to software problems.

Despite this setback, the Army continued to move forward with plans to establish a THAAD battalion and deploy a prototype system. The Total Army Analysis 2001 validated the requirement for the battalion in 1993.[103] The Air Defense Command would be composed of a brigade, three Patriot battalions, a THAAD battalion and two Avenger battalions for each of the three major regional contingencies. Bravo battery, with 81 soldiers, was established after the first successful flight test. The Army activated the second battery, Alpha, on 23 February 1996 at Fort Bliss, Texas. Together they comprise the core of the THAAD User Operational Evaluation System battalion - 1st Battalion, 6th Air Defense Artillery. A THAAD battery consists of a THAAD Radar, a BM/C4I element, and nine launchers with a basic load of eight missiles on each launcher.

Between December 1995 and May 1998, the THAAD test program made five intercept attempts. Although the tests illustrated the exceptional qualities of the radar, proved the communications links, and demonstrated the palletized launcher system, no intercept was achieved. With this test record and cost increases, the program repeatedly faced opposition from OSD. The Army, supported by the Navy and the Ballistic Missile Defense Organization, remained dedicated to the THAAD program. The resulting investigation attributed the test failures to quality control issues in the manufacturing process and prompted program revisions.[104] In a cost-sharing agreement between Lockheed Martin and the Army, the contractor would pay up to $75 million if they failed to achieve three hit-to-kill intercepts over the remainder of the Program Definition and Risk Reduction phase.[105] Five tests remained in this phase of the test program. In many respects, however, as one Democratic Senate aide remarked, "In reality, if there's one more failure there is no more THAAD."[106]

The year 1999, then, was crucial in the evolution of the THAAD program. The first intercept test of the year though ended in yet another failure. An attitude control thruster failed and the interceptor missed the target by 12 meters. Per the cost-sharing agreement, Lockheed was penalized $15 million. Two subsequent tests demonstrated the THAAD's capabilities.[107] On 10 June 1999, in its seventh intercept attempt, the THAAD weapon system successfully intercepted a Hera target missile in the upper atmosphere over WSMR.[108] The THAAD scored its second consecutive hit on 2 August 1999. In contrast to the June test, this intercept occurred outside the Earth's atmosphere. As Ed Squires, Lockheed Martin's THAAD Vice President, explained "By achieving a target intercept under a more stressing flight test scenario, we have been able to obtain the final missile design information required to move this program forward."[109]

Following the second successful THAAD intercept, Pentagon officials instructed the Army to cancel the remaining Program Definition and Risk Reduction flight tests and begin preparations for Engineering and Manufacturing Development. A 98-month EMD contract was signed with Lockheed Martin on 28 June 2000. Designed in two phases, the primary focus of the first phase of the contract is the demonstration of the redesigned system's capabilities in a series of ground and flight tests.[110] During this phase of low rate initial production, the team will also validate the production process. The second phase calls for a battle management and other software enhancements to provide full operational requirements compliance.

The redesigned THAAD will incorporate recommendations of the soldiers of the THAAD battalion, which address everything from ergonomic changes, improvements to software operation and doctrinal issues.[111] In addition, the redesign will create a more testable missile, according to Hans Mark, Pentagon Director of Research and Engineering.[112] Officials removed the requirement to intercept targets at altitudes ranging from 15-20 kilometers to the vacuum of space, determining that design complications did out weigh the benefits of these low altitude maneuvers. The new requirement of 40 kms and up will reduce the stress on the system's seeker and guidance system. In 2002, Colonel Patrick O'Reilly, THAAD Project Manager reported that the project is "going great" – slightly ahead of schedule and under cost.[113] In fact, the THAAD Project Office received the David Packard Excellence in Acquisition Award in 2002 for developing innovative logistics concepts, based on "pit-stop technology," that potentially reduce operation and support costs throughout the life of the system.[114]

Deployment has been an issue throughout the THAAD test program. In 1996, the Pentagon explored the possibility of deploying a prototype system to South Korea due to the North Korean missile test program and the rising tensions in the region.[115] Congress even mandated, in the 1996 defense bill that a system be in place by 1998. A GAO study conducted at the same time recommended not fielding a prototype until late 2000, until the THAAD was fully tested. The current goal is to field the system to operational units in 2007, and an entire battery by 2008. Full deployment should be attained by 2013.

National Missile Defense Redefined

The national missile defense system initially defined in the GPALS concept incorporated the various elements of the Strategic Defense Initiative. Budget constraints, due in part to the redirection toward theater programs, resulted in the termination of the HEDI program at the end of fiscal year 1992. The Bush Administration's interest in the Air Force's Brilliant Pebbles program led to the cancellation of the GSTS project, as the Brilliant Pebbles could serve both as boost and midcourse sensors.

In October 1992, Congress passed the National Defense Authorization Act for Fiscal Year 1993, which amended the Missile Defense Act of 1991. It placed greater emphasis on treaty compliance for any NMD system that the United States might deploy and eliminated the 1996 target date for deployment of an NMD site.[116] Five months later, the Total Army Analysis 2001 validated the requirement for a National Missile Defense for the continental United States.[117] Meanwhile, the Bottom-Up Review released in September 1993, recommended that the NMD program be reduced to a System Technology Demonstration. Funding for the program was reduced accordingly. The BMDO leadership negotiated to create a restructured NMD "Technology Readiness Program."[118]

The Ground-Based Interceptor remained a viable element of the NMD system and plans called for one hundred interceptors to be deployed the former SAFEGUARD site near Grand Forks, North Dakota. Despite continued progress, by the end of 1993, the future for the ERIS/GBI project did not look positive. Officials deferred acquisition efforts to await future directions following a DoD review of the Strategic Defense Initiative and the release of the National Missile Defense Acquisition strategy.

The program, renamed the Exoatmospheric Kill Vehicle (EKV) program, seemed to rally in 1994. On 26 May, the USASSDC announced a downselect in the EKV contractors, from three to two.[119] Although funding was significantly reduced, the SDIO Director, Lieutenant General Malcolm O'Neill (USAF) wrote in his response to Congress that he "[envisioned] this program as a series of epochs designed to incrementally mature the technology necessary to provide defense of the United States."[120] He added that the final EKV contractor would design, fabricate and test the system, with tests scheduled for FY97. Nevertheless, funding was again cut in subsequent years, as the Congress, the administration of President William Clinton and the military continued to argue the merits of a national missile defense system.

Fig. 5-19. The Ground-Based Radar Family of Radars included the truck mobile Theater Missile Defense system, the planned National Missile Defense radar and the experimental complex to be constructed at Kwajalein.

The Ground-Based Radar Experimental program faced similar obstacles.[121] Approved for the DEM/VAL phase in 1990, SDIO ordered the cancellation of the GBR-X program in 1991 following the Midcourse and Terminal Tier Review Architecture Study. Primary attention would instead be placed upon creation of a "Family of Radars" which employed a modular antenna component concept.[122] These radars would be used in support of TMD and GPALS, or Strategic Missile Defense. The TMD radar should be integrated with a variety of theater systems, such as the Patriot and ERINT. The GPALS facility would be able to operate in both the endo and exoatmosphere modes. The family of radars included four functional systems, three for theater defense and one for strategic operations. Raytheon Corporation was selected to perform the demonstration/validation for the radars in 1992.

In the next year, the family of radars became the TMD GBR, in support of the Upper Tier Theater Missile Defense System and the NMD GBR for strategic defense. Although both radars share many qualities, the NMD radar was designed to have a larger antenna, thereby requiring a larger power supply. With dramatically reduced funding in November 1993, the NMD-GBR was "restructured to leverage off of TMD and… concentrate on critical technology issues."[123] Thus at the end fiscal year 1993, only the mobile TMD GBR had received approval to proceed. This approval was based upon the radar's ability to "meet an immediate requirement for a more capable wide-area-defense radar to provide surveillance and fire control support" to the THAAD missile system.[124] The TMD GBR was to provide threat attack early warning, threat type classification interceptor fire control, sensor/cueing, launch/impact point estimation, threat classification against theater/tactical ballistic missiles and kill assessment. Two years later, in 1995, the NMD Program Office established the NMD-GBR Product office and the GBR Project Office became a product office and was absorbed into the THAAD Project Office.[125]

Other Initiatives – Targets

The command and its predecessors have been actively involved in developing targets for test programs.[126] Traditionally, Minuteman I missiles served as targets for ICBM intercept tests. The surplus stock of these ICBMs, however, is nearly depleted.[127] Therefore, boosters designed and tested by the command are to fill this void and provide cost-effective payloads (targets) for both strategic and theater systems. In addition to serving as a target, the systems will assist in the development of detection procedures and technologies.

The Strategic Targets Product Office initiated the SDIO funded Strategic Target System (STARS), in 1985. Its goal was "to launch missiles with experimental payloads into near-space to simulate the reentry of ballistic missile warheads."[128] Lacking the range of a Minuteman, the STARS IRBM had to be launched from the Pacific Missile Range in Kauai, Hawaii. This move provoked considerable public opposition from environmentalists. An extensive review and subsequent court decision, however, allowed the project to proceed. Following this controversial beginning, the STARS initiated its test phase in 1993. In 1994, the USASSDC introduced the STARS II, a new configuration of the target which included the addition of the Operation and Deployment Experiments Simulator post-boost vehicle.[129] With this adaptation, the STARS II provided the ability to maneuver payloads and deploy them after the third-stage missile motor drops off, increasing the target's viability in interceptor and sensor test programs.

With several successful flight tests, plans called for the target to be incorporated into the BMDO's midcourse space experiment. On 24 April 1996, the BMDO launched the MSX satellite into near-synchronous orbit to collect data on missile signatures in the midcourse phase.[130] In this test, conducted on 3 September 1996, the STARS deployed 26 objects to be observed by the MSX with its infrared, ultraviolet, and visible-light sensors. This launch brought the STARS record to four successes out of four launches.

Fig. 5-20 and 5-21. The STARS (left) and Minuteman II MSLS (below) are among the Strategic Target Systems.

Following the shift away from NMD in 1993, the GAO initiated a study to determine the future of the STARS project – termination or temporary hold pending future NMD tests and possible TMD testing.[131] The STARS Project Office presented six arguments for the continuation of the program. The STARS is exempt from both START treaties. It can deliver payloads at a variety of speeds and trajectories. It is the only target system operating in the 1,500-3,000 km range. Finally, the STARS had a demonstrated ability to provide support for various experiments. The STARS remains on the inventory of available Strategic Targets.

A second system was later added to the arsenal of the Strategic Targets Product Office - the Minuteman II-based Multi-Service Launch System (MSLS), a joint Army-Air Force program. Introduced in 1996, the MSLS target system consists of an MSLS front section with a three-stage Minuteman II booster. Within a year, the system enjoyed a three-for-three success rate. A follow-on system, the Orbital/Suborbital Target Launch Vehicle completed its first demonstration flight on 28 May 2000.[132] The Orbital/Suborbital Program Target Launch vehicle is scheduled to replace the MSLS in future integrated flight tests for the ground-based midcourse defense system.

Fig. 5-22 and 5-23. Both the STORM (right) and HERA (above) have served as targets for the THAAD and the PAC-3 projects.

The theater target program, a product of the BMD Space Payloads Office, began in 1993 and progressed rapidly.[133] These target systems are used in tests of the THAAD, Patriot, Corps SAM and the GBR. The Storm was first developed in 1988, completed its fifth successful flight in December 1993.[134] During this flight, the Storm launched the new maneuvering target test vehicle in its first test.[135] The target missile with a range of 400km is designed to simulate the predicted maneuvers of future short and medium-range ballistic missiles. By 1995, with ten straight successes, the Storm had developed a reputation for reliability supporting ERINT and THAAD tests. A modified single-stage Storm, the Storm II Maneuvering Tactical Target Vehicle, became operational in 1997. The new version of the Storm was developed "for use in evaluating current and future theater missile defense weapon systems," such as the Patriot.[136]

A second theater target, the Hera achieved its first flight test in April 1995.[137] Developed to support THAAD interceptor and radar tests, the Hera has a longer range than the Storm and is capable of delivering a variety of payloads to include chemical weapons. The Hera is launched from the specially developed Launch Complex 94 at Mountainair, New Mexico, which provides appropriate distances to simulate realistic scenarios.[138] Following three successful flight tests, the Hera served as the target in the first THAAD intercept test in February 1996. In that next year, the Hera flew in support of the PAC-3 test program. A new version of target, the Hera modified ballistic reentry vehicle (MBRV-3), was tested in March 1998. Although the targets program experienced problems with the Hera target, its successes far outnumbered its failures and the Hera remains a viable tool in the BMD test program.

In order to simulate a target with a mobile launch capability, the Targets Office developed the Short Range Air Launched Target (SRALT). The SRALT is dropped from a C-130 cargo plane and descends by parachute before igniting its motors at the appropriate altitude. The system completed its first risk reduction flight at the Pacific Missile Range, in April 1999. With a range of up to 600km, the SRALT was developed for the Navy Area Defense and the THAAD test programs.

Responsibility for the targets project originally rested with the Test and Evaluation Directorate, later known as the Targets, Test and Evaluation Directorate, of the USASSDC. Its significance was elevated in March 1998 when the Army Acquisition Executive chartered the Ballistic Missile Targets Joint Project Office (BMTJPO) which sought to centralize the requirement held by all branches of the service to develop and launch ballistic missile targets. At the recommendation of the BMDO, the BMTJPO transferred to BMDO in October 2001.[139] The move "[was] expected to improve the effectiveness of countermeasures available to the military." The Targets Office remained with BMDO for less than a year. On 19 September 2002, Lieutenant General Ronald Kadish, Director of the Missile Defense Agency, transferred the targets management and execution to the U.S. Air Force Space and Missile Defense Center. Specifically, responsibility for managing targets development was to be put under the Rocket Systems Launch Programs at Kirtland AFB, New Mexico, "to streamline activities associated with development of the Ballistic Missile Defense System (BMDS)." The MDA/Targets and Countermeasures Directorate in Washington at MDA, however, "[would] remain the primary interface for overall program management and program integration within the BMDS."[140]

End Notes

[1]"Memorandum from Colonel Michael Keaveney on Assuming Command, 1 April 1991."
[2]From a speech made by Brigadier General Robert Stewart, Deputy Commander, U.S. Army Strategic Defense Command at the U.S. Space Foundation's Fifth National Space Symposium. General Stewart became the Army's first astronaut in 1979 and logged 289 hours in space. Although he was referring to the Army's situation after Vietnam, his comment illustrates the difficulties of introducing new methods into any institution. See Martin Berkey, "General: Satellite Secrecy No Longer Needed," *Huntsville Times* 9 April 1989.
[3]VANGUARD Study 1990.
[4]"Enclosure 3: Alternative 3, SDC Consolidation and Information Paper ARSPACE Resubordination to SDC, 28 March 1991."
[5]"Memorandum Through the DCSOPS to the VCSA regarding the VANGUARD Proposals, 16 October 1990." The counterpoint to this criticism appeared in the ARSPACE journal, *STAR*NET*, "The faulty assumption…is that the traditional proponents… will recognize the need for space applications and support such a need across proponent lines…. Only in Desert Storm did we see real space applications but their use and interest base quickly eroded in the post-conflict period." See "Commander's Corner, Regarding the Army and Space," STAR*NET 1.4 (15 January 1992):2.
[6]"VANGUARD Final Report, Conclusions and Recommendations, V-23-V-24."
[7]General Order 12, 1 July 1993.
[8]Donald G. Dupont, "Final Roles and Missions Report Says Space Lead Should Go To STRATCOM," *Inside the Army* (15 February 1993); Donna Haisiley, "Draft Joint Staff Pub Pushes for Greater Battlefield Use of Space Assets," *Inside the Army* (17 May 1993); Genevieve Anton, "Air Force Chief Urges Consolidation of All Military Space Operations," *Colorado Springs Gazette-Telegraph* (16 April 1993);" Neil Munro and Barbara Opall, "DoD Nixes Merger of U.S. Nuclear, Space Commands," *Defense News* (27 September - 3 October1993)" respectively.
[9]Jason Glastow and Steve Weber, "Air Force Moves to Push Army, Navy Out of Space," *Army Times* 24 October 1994. At the same time, a retired Air Force Colonel Kenneth A. Myers argued that while consolidation sounds persuasive the Air Force "has not always been a positive force in achieving formidable space capabilities," pointing out that the service had opposed geostationary satellites for communications, delayed and fought GPS, opposed MILSTAR and did not support LANDSAT. He called for creating a new space service "to organize, train, equip and sustain space forces to support military operations and national security requirements." See Kenneth A. Myers, "Military Space Control Reality Check," *Space News Supplement* 1-13 November 1994:15-16.
[10]General (ret.) Frederick J. Kroeson, "Thc Army's Role in Space," *Army* November 1994:28.
[11]"Air Force Association Calls for Reorganization of Space," *BMD Monitor* 18 November 1994:40. This same article noted that the report stated, "The U.S. military advantage in space could be lost if organizational, launch and budget problems "are not addressed."
[12]"DoD Debate Over Management of Military Space Assets Finds New Venue," *Inside the Army* 28 November 1994:27. The Army, Navy and Marine Corps implicitly defined space as another place to control to win the terrestrial battle. Space is those regions from, through or in which systems operate.
[13]"Space Control Debate Takes Center Stage," *Space News Supplement*, 12-18 December 1994:10. The Army was prepared to resist consolidation if it would minimize its assured space access and prevent it from voicing its requirements. The Navy asserted it needed an independent space command because of the "unique nature of its reliance upon space systems."
[14]"Army Remains Involved in Space," *AUSA News* 17.3 (January 1995).
[15]"Army Seeks Assurances on Space Architect 'Board of Directors'," *Inside the Army* 12 June 1995: 11-12; "Army Pushing for Space Oversight Management Plan," *Inside the Army* 1.12, (24 March 1994): 26-27
[16]"With Input from Services, Joint Staff…Air Force Secretary Submits Proposal for New Space Architect Position," *Inside the Army* 7.11 (20 March 1995):23-25.
[17]"Pentagon Acquisitions Chief Clears Space Architect's Office for Lift-off," *Inside the Army* 9 October 1995.
[18]As originally formed, the Office of the Space Architect had three divisions, one for requirements, concepts and operations (CONOPS), a second for analysis and a third for capabilities. The first would work with military space assets to understand and assess requirements and concepts of operations. The second division would serve as the

"in-house analytical capability." The third would check requirements from the perspective of existing and planned space capabilities and would have major responsibility for integrating and synthesizing architectural capabilities. While the Space Architect would direct planning and analysis, and promulgate standards and architectures. On the other hand, the Space Architect would not execute acquisition programs, validate requirements or direct resource allocation. See "With Input from Services, Joint Staff…Air Force Secretary Submits Proposal for New Space Architect Position," *Inside the Army* 7.11 (20 March 1995):23-25.

[19]United States General Accounting Office, *National Space Issues: Observations of Defense Space Program and Activities*, 16 August 1994, GAO/NSIAD-94-253, 10.

[20]United States, Department of Defense, *Annual Report to the President and Congress 1996* (Washington, D.C.: Government Printing Office, 1996), p. 79.

[21] Unless otherwise noted the following is based on Joshua Boehm with Craig Baker, Stanley Chan and Mel Sakazaki, *A History of United States National Security Space Management and Organization*, www.fas.org/spp/eprint/article03.html accessed on 3 July 2002.

[22]In 1990, Lithuania initially declared its independence. In the next year the other republics made their own declarations - Georgia (4 June), Estonia (20 August), Latvia and Lithuania (21 August), Ukraine (24 August), Byelorussia (25 August), Moldavia (27 August), Azerbaijan (30 August), Uzbekistan (31 August), Kirgizia (31 August) and Tadzhikistan (9 September). On 8 December 1991, the leaders of Russia, Ukraine and Byelorussia proclaimed the Soviet Union had ceased to exist and declared the creation of a Commonwealth of Independent States. President Gorbachev subsequently announced, on 17 December, that all Soviet central government structures would cease to exist effective at the end of the year. Gorbachev resigned as president on 25 December, effectively marking the demise of the Soviet Union and the end of the Cold War.

[23]System Planning Corporation at the direction of the Strategic Defense Initiative Organization, *Ballistic Missile Proliferation: An Emerging Threat 1992* (Arlington: System Planning Corporation, 1992).

[24]Lieutenant General Donald M. Lionetti, Commander USASSDC, 1992.

[25]Missile Defense Agency, "Missile Defense Milestones, 1944-2000," http://www.acq.osd.mil/bmdo/mdolink/html/milstone.html.

[26]Ambassador Cooper served as the chief American negotiator at the Defense and Space Talks in Geneva since 1987 and was appointed SDIO Director in July 1990.

[27]Baucom, *Origins*, p. 199.

[28]State of the Union Address by President George H.W. Bush, 29 January 1991, located at http://www.c-span.org/executive/transcript.asp?cat=current_event&code=bush_admin&year=1991.

[29]McMahon, p. 94-95.

[30]Baucom, *Origins*, p. 199.

[31]This act is part of H.R. 2100, the "National Defense Authorization Act for Fiscal Years 1992 and 1993."

[32]"Missile Defense Milestones, 1944-2000," http://www.acq.osd.mil/bmdo/bmdolink/html/milstone.html.

[33]Lieutenant General Robert Hammond, the USASDC Commander, was appointed the PEO Strategic Defense on 5 October 1988.

[34]The PEO-AMD was composed of the National Missile Defense Program Office, the THAAD, Patriot and Arrow Project Offices and the CORPS SAM/Medium Extended Air Defense System (MEADS) and Joint Tactical Ground Station (JTAGS) Product Offices. Memorandum for Assistant Secretary of the Army (Research, Development and Acquisition) from COL Daniel L. Montgomery, PEO-MD, Subject: Name Change of Program Executive Office (PEO), Missile Defense, 23 May 1996.

[35]General Orders 13, dated 1 July 1993.

[36]The HELSTF had transferred to the USASDC from the Army Materiel Command on 1 October 1990 as part of the Army initiative to centralize directed energy research. General Orders 12, dated 8 May 1991.

[37]General Orders 20, dated 4 April 1988.

[38]General Orders 17, dated 15 December 1995.

[39]The Multinational Programs Office oversaw this initiative.

[40]Memorandum from the Secretary of Defense to the Director, SDIO.

[41]The SDIO established on 15 March 1991 a new Deputy for TMD, equal in status to the Deputies for Technology and Strategic Defense.

[42]As a result of this decision, the U.S. Army Missile Command's Joint Tactical Missile Defense Management Office transferred to the USASDC.
[43]A direct descendant of the German V-2, the SS-1 Scud missile was first deployed by the Soviet Union in the 1960s. Designed to carry either a 100-kiloton nuclear warhead or a 2,000 pound conventional warhead over ranges from 100 to 180 miles, the weapon can also carry chemical or biological agents. During the Gulf War, the Iraqis used a modified Scud with a reduced warhead and larger fuel tank for greater range. It was, as a result, "structurally unstable and often broke up in the upper atmosphere," reducing its accuracy but also making it more difficult to intercept given the unpredictable flight path. Excerpted from Norman Friedman's *Desert Victory - The War for Kuwait* (Naval Institute Press, 1991) http://www.pbs.org/wgbh/pages/frontline/gulf/weapons/scud.html.
[44]The modified Patriot employed a proximity fuse to destroy the incoming missiles rather than a kinetic energy intercept.
[45]Announcement by Secretary of Defense Les Aspin, dated 13 May 1993; Memorandum for Correspondents, No. 159-M, 13 May 1993.
[46]K. Scott McMahon, *Pursuit of the Shield* (New York: University Press of America, 1997), pp. 246-247.
[47]Directive 5134.9 re: Ballistic Missile Defense Organization issued by Deputy Secretary of Defense John M. Deutch on 14 June 1994 specifically defined the new missions for the BMDO program.
[48]Office of the Secretary of Defense, *Report of the Bottom-Up Review*, October 1993, pp. 41-48 cited in Baucom, "The U.S. Missile Defense Program, 1944-1994," 27.
[49]DoD leaders would soon reduce this and other allocations with most of the cuts coming from TMD. Congress also indicated that it would not support the funding levels outlined in the review.
[50]The two presidents also agreed to conduct a joint TMD early warning exercise at this Washington Summit, see http://sun00781.dn.net/nuke/control/abmt/chron.htm.
[51]In addition to the missile's velocity and range, at issue was "the mobility of the THAAD batteries and the system's ability to process targeting information from ground, as well as space-based sensors." Barbara Opall, "Strategic Accord Inhibits Advances in TMD Programs," *Defense News*, 4-10 October 1993.
[52]Of particular interest for the American government was the protection of the THAAD system. "Plan may foment administration battle with Senate: Russia Considers U.S. Plan to Clear ABM Treaty for THAAD." *Inside the Pentagon*, 9 December 1993.
[53]See http://sun00871.dn.net/nuke/control/abmt/chron.htm.
[54]Gordon R. Sullivan and Michael V. Harper, *Hope Is Not A Method: What Business Leaders Can Learn From America's Army* (New York: Times Books/Random House, 1996), p. 169.
[55]General Sullivan had been considering the issues surrounding change since the end of the Cold War. His message concerning the Louisiana Maneuvers and his remarks at the TRADOC Desert Storm conference may be found in Gordon R. Sullivan, *The Collected Works of the Thirty-second Chief of Staff, United States Army: Gordon R. Sullivan, General, United States Army, Chief of Staff, June 1991-June 1995*, ed. by Jerry L. Bolzak (Washington, D.C.: U.S. Army Center of Military History, 1996), pp. 103-105 and 44-45, respectively. The standard work on General Sullivan's Louisiana Maneuvers is James L. Yarrison, *The Modern Louisiana Maneuvers* (Washington, D.C.: U.S. Army Center of Military History, 1999). A useful examination of previous large-scale Army exercises is Jean R. Moenk, *A History of Large-Scale Army Maneuvers in the United States, 1935-1964* (Fort Monroe: Historical Branch, Office of the Deputy Chief of Staff for Military Operations and Reserve Forces, 1969) while the standard work on the 1940-1941 maneuvers is Christopher R. Gabel, *The U.S. Army GHQ Maneuvers of 1941* (Washington, D.C.: U.S. Army Center of Military History, 1991).
[56]A complete treatment of the founding and early days of the ARSST may be found in U.S. Army Space and Missile Defense Command Historical Office and Science Applications International Corporation, Space Warriors: The Army Space Support Team (unpublished ms held at USASMDC Historical Office).
[57]The shortcomings of TERS are outlined in CALL Newsletter No. 91-3, Chapter 3, "Tactical Early Missile Warning" http://call.army.mil/products/newsltrs/91-3/chap3.htm accessed on 10 January 2003 as well as Space Warriors, 1-19-20. Nevertheless, it was better than no warning at all. Although postwar assessments of units in the Persian Gulf noted that the "early warning system was inadequate" or that "Divisions need improved air attack early warning capability and access to real-time early warning information," TERS did provide, within its limits some form of early warning. See U.S. Army Space and Missile Defense Command Historical Office and Science

Applications International Corporation, *The Joint Tactical Ground Station: Fielding and Operational Lessons Learned* (Huntsville: U.S. Army Space and Missile Defense Command, 2000; reprint ed., 2003), pp. 1-29-1-30.
[58]A complete account of the evolution of the JTAGS may be found in ibid.
[59]Unless otherwise noted, the material in this section on Haiti (Uphold Democracy) is based on Weekly Activities Report, 2 October 1995-8 April 1995 (hereafter referred to as ARSPACE Weekly Activities Report).
[60]For example, the ARSPACE Weekly Activities Report for 2-8 October 1995 states, "Using several USARSPACE Advanced Communications Technology (ACTS) terminals, President Clinton conducted a video teleconference (VTC) with members of the JTF staff in Haiti on 6 October. A VTC between the Army Chief of Staff and the JTF commanders and staff in Haiti was also conducted on 2 October."
[61]For example, see ARSPACE Weekly Activities Reports, 23-29 October 1995 and 30 October-5 November 1995.
[62]ARSPACE Weekly Activities Report, 13-19 November 1995.
[63]"ACTS Supports Forces in Haiti" and "Weather Satellite Supports Troops in Haiti," respectively. The High Resolution Weather Satellite Receiver received high resolution digital imagery from military and commercial polar orbiting satellites.
[64]ARSPACE Weekly Activities Report, 2-8 April 1995.
[65]ARSPACE Weekly Activities Reports, 7-14 January 1995 and 22-28 January 1995.
[66]The 1st SATCON BN was formed in 1995 and controls the DSCS, the super high frequency (SHF) communication network for American forces "anytime anywhere." The battalion provides technical support and troubleshooting for the system. It runs the Regional Space Support Centers (RSSC) and performs planning functions and authorizes warfighter use of the system. The battalion has a Headquarters Company (HHC) and five other companies spread around the world. The HHC resides at Falcon Air Force Base, Colorado. A Company is based at Fort Detrick, Maryland, Company B at Fort Meade, Maryland, C Company at Landstuhl, Germany, Company D at Camp Roberts, California, and Company E at Camp Buckner and Torii Station, Okinawa, Japan.
[67]ARSPACE Weekly Activities Report, 3-9 December 1995.
[68]ARSPACE Weekly Activities Report, 17-30 December 1995.
[69]Lieutenant Colonel Bill Bugert, "ASPO's TENCAP Supports Commanders in Bosnia," *The Eagle* January-February 1996:10; ARSPACE Weekly Activities Reports, 12-18 February, 23-29 September, 14-20 October, 11-17 November 1996.
[70]ARSPACE Weekly Activities Report 8-14 April 1996.
[71]ARSPACE Weekly Activities Reports, 16-22 October and 30 October-5 November 1994.
[72]The SPECC was designed to be used in a Joint Task Force/Joint operations Center and Combined Land Component Command G-3 Operations Cell. It allowed allied commanders to track progress of their respective forces from a single source. It gave commanders a quick common picture of the battlefield.
[73]The VDCPAD would provide a commander with real-time visibility throughout the phases of a conflict.
[74]ARSPACE Weekly Activities Reports, 9-15 April and 7-13 May 1995.
[75]ARSPACE Weekly Activities Report, 30 July-5 August 1995.
[76]Including the Mission Planning Rehearsal System, the Multispectral Imagery Processor (to develop and Intelligence Picture of the Battlefield [IPB]), the High Resolution Satellite Weather Receiver and INMARSAT.
[77]ARSPACE Weekly Activities Report, 23-29 April 1995.
[78]Reprinted and updated from Chapter 17, Walker, Martin and Watkins, *Four Decades of Progress*. The ERINT/PAC-3 missiles are 17 feet in length and 9.8 inches in diameter and weigh 707.6 pounds. Given its small size four PAC-3 missiles fit into a single Patriot canister or 16 per launch station.
[79]Kinetic Energy Weapons Directorate Historical Report for Fiscal Year 1988, dated 30 July 1993; Theater Missile Defense Applications Program Office Historical Report for Fiscal Year 1990, dated 18 November 1990, and the Joint Theater Missile Defense Program Office Historical Report for Fiscal Year 1991, dated 29 May 1992.
[80]Briefing, "ERINT-1 Program," [1991].
[81]"Data 'Drop-Out,' Mapping Errors are Possible Causes of ERINT Failure," Inside *the Army*, 28 June 1993. Note all ERINT tests were conducted at White Sands Missile Range, New Mexico.
[82]The Storm target contained a cluster of 38 pressurized, water-filled containers designed to simulate toxic chemical submunitions. "ERINT Scores successful intercept of simulated TBM," *Inside the Pentagon*, 2 December 1993; Martin Burkey, "ERINT missile test rates a success," *The Huntsville Times* 16 February 1994: B2; and "ERINT Hits Air-Breathing Target Drone," *Aviation Week and Space Technology*, 13 June 1994: 26.

[83]"Budget Constraints put ERINT and PATRIOT Interceptors on Collision Course," Inside *the Pentagon,* 26 November 1992.
[84]Unlike the ERINT, which used hit-to-kill technology, the PATRIOT multi-mode missile was equipped with a warhead. The kinetic energy intercept "releases 100 times more energy than is released by steel bomb fragments," increasing lethality explained Sid Wells, Vice President of Loral Vought Systems. Robert Langreth, "Sons of the Patriot," *Popular Science* June 1993: 1005; "Extended range interceptor wins approval for Patriot," *Redstone Rocket* 23 February 1994: 3; "Army recommends ERINT to DAB for Patriot Improvement," *BMD Monitor* 25 February 1994.
[85]Memorandum, Principal Deputy Under Secretary of Defense R. Noel Longuemare to Secretary of the Army and Director, Ballistic Missile Defense Organization, Subject: PAC-3 Acquisition Decision Memorandum, dated 7 July 1994; and "Army Weaponry and Equipment," *Army* 44 (October 1984): 296-297. The Patriot Project Office later merged with the Medium Extended Air Defense System Product Office, in October 2001, forming the Lower Tier Project Office, to "take maximum advantage from lessons learned from our legacy systems to ensure that interim and objective lower tier systems meet operational requirements at reduced cost."
[86]Daniel Dupont, "Pentagon Decides to Deploy Patriot PAC-3 Missiles to Persian Gulf Area," *Inside the Arm,* 16 February 1998: 1. Although deployed, industry and Pentagon officials did not expect to use the PAC-3 interceptors at this time.
[87]Requirements changes in January 1999 caused some delays and dry weather conditions at the test range postponed the first intercept test by six months.
[88]"The PAC-3 missile employs a lethality enhancer consisting of two rings of large fragments which slowly expand about the missile centerline to increase the effective radius of the PAC-3 missile against softer targets, such as cruise missiles, aircraft or helicopters." Lockheed Martin pamphlet entitled "PAC-3 missiles."
[89]Colonel Newberry, PAC-3 Program Manager, quoted in Bradley Graham 'Missile Defense Program Changes Course," *The Washington Post* 5 August 2002: A-6.
[90]Tony Capaccio, "Lockheed Defensive Missile Said Ready As U.S. Eyes Iraq Options," *Bloomberg.com*, 14 August 2002.
[91]Reprinted and updated from Chapter 17, Walker, Martin and Watkins, *Four Decades of Progress*.
[92]The MOU with Israel included an Israeli Testbed. Arrow Project Office Historical Report for Fiscal Year 1994, undated.
[93]Oversight of the Arrow Project Office transferred again on 1 October 2001 to the Ballistic Missile Defense Organization.
[94]Barbara Opall, "Arrow to Face Chemical Target Test," *Defense News* 31 May – 6 June 1993: 30 and Opall and Sharone Parnes, "Test Halts Cast Shadow on Arrow," *Defense News* 19-25 July 1993: 4, 50.
[95]Steven Pearlstein, "The Missile Makers' Next Big Target," *Washington Post* 5 August 1992: 81.
[96]Known as the Hetz-2 in Israel, the interceptor measures 23-feet in length and is designed to intercept missiles at altitudes between 10-40 kilometers. Unlike American missile defense systems, the Arrow is equipped with a warhead.
[97]See http://www.mod.gov.il/modh1/homa/.
[98] Reprinted and updated from Chapter 17, Walker, Martin and Watkins, *Four Decades of Progress*. The THAAD system is designed to defend an area about 200 km in radius and up to 150 km in altitude, an area approximately 10 times greater than the Patriot's. Project oversight transferred to the PEO-GPALS in 1992, the THAAD Project Office moved again on 1 October 2001 to the BMDO.
[99]Barbara Opall, "New THAAD Puts on Weight – Redesign of Larger Missile May Add Cost, Limit Availability," *Defense News*, 22-28 March 1993: 3, 28.
[100]Caleb Baker, "SDIO, Army Ends Dispute Over Theater Defense," *Defense News*, 25 November 1991: 4, 29.
[101]"Scuds blow up missile design," *Army Times* 5 April 1993.
[102]Daniel G. Dupont and Jeffrey Moag, "Lockheed Exceeds Its THAAD Cost Estimates by $83 Million," *Inside the Army* 15 August 1994; "Appropriations Slice $30 Million Intended for FY-95 THAAD Testing," *Inside the Army,* 8 August 1994; and "Independent Study Uncovers Management Problems With THAAD Program," *Inside the Army*, 26 December 1994.
[103]"TAA validates requirement for NMD – Total Army Analysis 2001 Adds THAAD Battalion to Corps Force," *Inside the Army* 5 July 1993.

[104]Among the identified problems were a malfunctioning lanyard which reset the avionics computers, a focal plane array overload preventing target identification, and the DACS motors did not operate properly. Colonel Louis Deeter, THAAD Project Manager, described the problems as "a failure of robustness" rather than a failure of design. Brett Davis and Martin Burkey, "THAAD thud may be $15M penalty," *The Huntsville Times*, 10 July 1998: A-11.
[105]Davis and Burkey, "THAAD thud".
[106]Quoted in Richard Parker, "Lockheed told to overhaul THAAD," *The Huntsville Times* 9 July 1998: A-1, A-9.
[107]LuAnne Fantasia, "Take THAAD!!," *The Eagle*, Special Edition 1999, Joe Guy Collier, "THAAD 'smokes split-off warhead," *The Huntsville Times* 2 August 1999: A-1.
[108]This was the tenth of 13 planned tests. In this test, the Hera target simulated a Scud missile.
[109]Lockheed Martin Fact Sheet – "THAAD Second Intercept, August 2, 1999."
[110]Ground tests, to include cold motor static firings are on-going and flight tests are scheduled to begin in 2004.
[111]Ann Roosevelt, "Soldiers Help Save Money on THAAD Program," *Defense Week* (26 June 2000): 3.
[112]The objective characteristics for the THAAD are weight 600 kilograms, length – 6.2 meters, with an infrared terminal guidance seeker. BMDO Fact Sheet 204-00-11. "Army, OSD Confident THAAD EMD Design will Fix Missile's Problems," *Inside Missile Defense* 20 October 1999.
[113]Sandy Riebeling, "THAAD System," *Redstone Rocket* 5 May 2002.
[114]"Race car technology puts THAAD team in winner's circle," *The Redstone Rocket* 17 July 2002: 7. The team studied the automotive racing industry's maintenance diagnostics to improve the system design and reduce repair times.
[115]Brett Davis, "Missile System needs more tests, report says," *The Huntsville Times* 10 July 1996: B-2.
[116]In addition, Congress eliminated the requirement to deploy a TMD system by the mid-1990s and replaced it with a requirement to develop advanced theater missile defense systems for deployment.
[117]The deployed system would be manned by reserve forces. "TAA validates requirement for NMD – Total Army Analysis 2001 Adds THAAD Battalion to Corps Force," *Inside the Army* 5 July 1993.
[118]The NMD program was funded at $3 billion over a five-year period. "National Missile Defense: An Overview (1993-2000)," http://www.acq.osd.mil/bmdolink/html/nmdhist.html.
[119]"Rockwell, Hughes get SSDC contract go-ahead," *Redstone Rocket* 1 June 1994: 8.
[120]Jason Glashow, "Hughes, Martin, Rockwell now in race: Contractors are Concerned GBI Downselect Could Yield Two Loses," *Inside the Army* 3 January 1994.
[121]Reprinted and updated from Chapter 15 of Walker, Martin and Watkins, *Four Decades of Progress*.
[122]Media Briefing presented by Colonel Arthur C. Meier II, GBR Project Manager, dated 20 May 1991.
[123]Ground-Based Radar Project Office Historical Report for Fiscal year 1994, undated; Barbara Opall, "Missile funds shift to small, mobile systems," *Army Times* 6 December 1993.
[124]Ground-Based Radar Project Office Historical Report for Fiscal Year 1994, undated, Ground-Based Radar Project Office Historical Report for Fiscal Year 1993, dated 23 March 1994
[125]Skip Vaughn, "National missile defense establishes radar project," Redstone *Rocket*, 12 July 1999: 16 and Vaughn, "Radar product office joins THAAD missile team," *Redstone Rocket*, 5 July 1995: 1.
[126]Reprinted and updated from Chapter 16 of Walker, Martin and Watkins, *Four Decades of Progress*.
[127]Melinda Larson, "Kauai to Kwajalein fight support by KMR: STARS mission a stellar success," *Kwajalein Hourglass*, 28 August 1993: 1, 3.
[128]Rick Lehner, "Strategic Target System and the National Environmental Policy Act: STARS Lessons to be Learned," 1994, 3; "STARS missile has successful flight," Redstone *Rocket*, 27 July 1994: 2.
[129]In contrast, the STARS I consists of a refurbished Polaris first and second stages and a commercially purchased Orbus I third stage that can deploy single and multiple payloads.
[130]"Stars Booster Deploys Objects for MSX Test," *Aviation Week and Space Technology*, 16 September 1996. The MSX satellite collected data on realistic missile-type targets and penetration aids against terrestrial, Earth and celestial backgrounds for both TMD and NMD programs. The command was involved with the MSX effort since its inception in 1987.
[131]"Ballistic Missile Defense: Current Status of Strategic Target System" Letter Report – 3 March 1995, GAO/NSIAD-95-78.
[132]The system consists of a front section atop a three-stage Minuteman II booster. Jeff Compton, "NMD Target vehicle flies demonstration," *The Eagle* October 2000: 22.

[133]Targets, Test and Evaluation Directorate Historical Report for Fiscal Year 1992, dated 21 December 1993.
[134]The Storm target is composed of a Sergeant motor in the first stage, a Minuteman I third stage motor in the second stage and a specially designed ballistic reentry vehicle.
[135]Gerda Sherrill, "New theater missile defense target demonstrated," *Redstone Rocket*, 22 December 1993: 15.
[136]Lieutenant Colonel James D. Matthewson, Jr. Storm Product Manager, quoted in "Missile target vehicle has successful test at White Sands Missile Range, N.M.," *Redstone Rocket*, 12 November 1997: 20. The Storm II target is made from a Minuteman II second stage booster and a Pershing II reentry vehicle.
[137]The Hera consists of a modified second and third stage from Minuteman II missile, a modified Pershing II guidance and control section, various interstage hardware, and an instrumented ballistic reentry vehicle. USASSDC News Release, "First Hera Target Flyout Test," 24 April 1995.
[138]A second launch site was constructed at Fort Wingate, New Mexico.
[139]With the transfer the JPO was renamed the Missile Defense Targets Joint Program Office. Memorandum from the Office of the Secretary of the Army, Acquisition, Logistics and Technology, dated 6 September 2001, Subject: Program Realignments.
[140]In 2002 the BMDO became the Missile Defense Agency. Memorandum from the Secretary of Defense, dated 2 January 2002, Subject: Missile Defense Program Direction.

Chapter 6
The Army's Newest Major Command, 1995-present

The U.S. Army Space and Missile Defense Command

In the mid-1990s, the roles and responsibilities of the U.S. Army Space and Strategic Defense Command (USASSDC) continued to evolve. In January 1995, for example, the Army named the Commanding General of the USASSDC the operational advocate and focal point for Theater Missile Defense. One year later, Vice Chief of Staff of the Army General Ronald Griffith designated the USASSDC a stand-alone Army Component Command.

General Griffith reached this decision based on the fact that the "USASSDC carries out responsibilities in scope and magnitude unlike other Army organizations." Specifically, as the Army component of U.S. Space Command, the USASSDC had an operational mission. In addition, as the Executing Agent for BMDO, USASSDC retained a "complex array of funding and tasking responsibilities." Finally, on acquisition issues the USASSDC reported directly to the Army Acquisition Executive. Nevertheless, General Griffith recognized a need for a "proponent like" Army facilitator to integrate space and missile defense solutions with the Army and Joint Warfighting forums. He tasked TRADOC and USASSDC to establish a Memorandum of Agreement (MOA) that would address these issues.[1]

Fig. 6-1. This shoulder sleeve insignia symbolizing freedom and constant vigilance in the U.S. Army Space and Missile Defense Command missions was adopted in February 1998.

On 18 February 1997, following General Griffith's directive, the USASSDC signed an MOA with TRADOC which made the command the Army Specified Proponent for Space and National Missile Defense and the overall Army integrating command for Theater Missile Defense.[2] The command would now determine space requirements for TRADOC approval and lead the integration of doctrine, training, leader development, organization, materiel and soldiers (DTLOMS) solutions across the Army and within appropriate joint agencies. The MOA also chartered the command to establish a battle lab to plan and conduct space and missile defense warfighting experiments.

In response to these new responsibilities and missions, the Army created its newest Major Army Command on 1 October 1997.[3] Effective that date, the U.S. Army Space and Strategic Defense Command, a field operating agency of the Army Chief of Staff became the U.S. Army Space and Missile Defense Command. The General Order reaffirms the new duties, responsibilities, and relationships outlined in the February 1997 MOA with TRADOC and reiterates the missions previously assigned to this organization. Essentially, the command ensures that Army warfighters have (1) access to space assets and the products they provide to win decisively with minimum casualties; and (2) effective missile defense to protect the nation

as well as deployed U.S. forces and those of its allies. The command has developed a number of innovative entities and products to achieve these goals. Space considerations dictate that only some of these are discussed below.
7

Missile Defense Battle Integration Center/ Space and Missile Defense Battle Lab

With the additional responsibility as TMD Advocate, the Army Strategic Defense Command's Commanding General, Lieutenant General Jay Garner, decided to develop a Battle Lab for TMD and space issues. General Garner saw the laboratory system as a means to move missile defense concepts into reality. Army officials granted permission for this proposal in October 1994 and the result was the Missile Defense Battle Integration Center (BIC) created on 16 January 1995.[4] The initial goal of the BIC was to connect the four elements of TMD – active defense, passive defense, attack operations, and BM/C3, enabling researchers to test concepts and allowing commanders to train soldiers. To achieve this goal, TRADOC and the USASSDC developed an MOA which established a working relationship between the two organizations with particular reference to "materiel development, analytical and/or simulation capabilities."[5] As a result of the 1997 TRADOC MOA, which expanded the command's missions, the BIC was reorganized with the Colorado Springs based Army Space Exploitation Demonstration Program to form a full-fledged Space and Missile Defense Battle Lab. Its missions were "to perform experimentation in the domains of space and missile defense and "to develop warfighting concepts, focus military science and technology research, and conduct warfighting experiments."[6] The mission expanded in October 2000 when the Army designated ARSPACE as the single Army component command to support U.S. Space Command's Computer Network Attack (CNA)/Computer Network Defense (CND) missions.

Fig. 6-2. The Uncooled Imaging Technology or UCIT device will enable soldiers to see objects through camouflage, smoke, fog and other

Fig. 6-3. The Extended Air Defense Testbed provides detailed simulations from the fire unit to the theater level and thorough analysis of system interoperability.

One goal for the Battle Lab was to develop a Synthetic Battlefield Environment (SBE) to link technology to the warfighter. The SBE would provide weapons developers, battle planners and commanders interactive realistic scenarios. The Battle Lab's SBE rested with the Extended Air Defense Testbed (EADTB).[7] Initiated in 1989, the EADTB models air, land, sea, and space-based forces and their contribution to theater-level extended air defense. With the innovative EADTB, the user can develop tailored simulations from the fire-unit up to the theater level for TMD and the global level for NMD. The first EADTB nodes opened in June 1994 at the Advanced Research Center in Huntsville, followed by the SHAPE Technical Center in The Hague, The Netherlands and Fort Bliss, Texas. Within three years, the EADTB had grown to include 30 nodes around the world.[8]

The synthetic environment established by the Battle Lab allowed simulated elements to be replaced with actual hardware, permitting a hardware-in-the-loop as well as a human-in-the-loop capability. They introduced the mobile STOW TMD system during Roving Sands exercises in May 1995, synchronizing the TMD battle for the land operations commander. Since then the SBE has continued to grow with the evaluation of new software and technologies to address many facets of the space and missile defense environment. Among the new technologies is Project Stalker which assists in locating, tracking and destroying mobile transporter erector launchers. Similarly, the Battlefield Ordnance Awareness system, introduced in 1999, collects and processes data on missile launches, artillery and tank fire. At another level, No Horizons is designed to support the integration of the Space-Based Infrared System into the Army's TMD force. These and other technology advances are brought to the soldier through traditional exercises, such as Roving Sands, Millennium Challenge, Optic Windmill, Ulchi Focus Lens, and Total Defender, as well as long-distance training and the Space and Missile Defense War Game.

Fig. 6-4. In March 1998, USASMDC achieved a new milestone in distance learning. Soldiers from the 32nd Air and Missile Defense Command, stationed in Kuwait, trained in a computer-simulated missile battle with the Battle Lab representatives in Huntsville, Alabama.

In addition to providing training opportunities and experiments, the Battle Lab brings the product to the soldier through the Army Space Exploitation Demonstration Program (ASEDP).[9] The goal is to "enhance Air-Land execution by demonstrating that space-based assets can support tactical commanders."[10] Many products could be used to illustrate the Battle Lab's successes in this arena. The Global Broadcast System - Joint In-Theater Injection, Joint Tactical Ground Station and the Force Protection Tactical Operations Center (FPTOC), for example, all trace their history to the ASEDP.

As envisioned by then Army Chief of Staff General Gordon Sullivan, t the FPTOC would provide overarching command and control capability for the theater missile defense fight. The mobile center collects and fuses data from a variety of sources including sensors, satellite communications and imagery, as well as air and missile defense units. Introduced in February 1995, the FPTOC was the first digitized command and control center.[11] It was designed to support the four elements of TMD - destroying missiles in flight (active defense); attacking their launchers and infrastructure (attack operations); missile defense warning and vulnerability reduction (passive defense); and, BMC^4I. The next generation system, the Future Operational Capability (FOC) TOC, improved the support provided and reduced the footprint for Joint Theater Air and Missile Defense.[12] With the new Windows-based Advanced Warfare

Environment or AWarE software, the FOC exercise demonstrated many improvements, including a 70 percent reduction in the in-theater footprint, while participating in Roving Sands '00. The new TOC is small enough to be deployed aboard a single C-141 aircraft and still provide the full execution of all TAMD functions.

Fig. 6-5 and 6-6. Tailored for theater-level joint operations, the Force Projection Tactical Operations Center's System of Systems was staffed by a 35-soldier cadre.

In addition to the large systems, the ASEDP has developed technologies that affect the communications available for the individual soldier or unit. The Iridium phone system, supported by a constellation of 70 satellites, provided the first truly global phone system for the soldier in the field. Early warning technology was first tested and deployed, during fiscal year 1998, with the Pager Alert Warning System (PAWS). The PAWS notifies troops in the expected impact zone of tactical ballistic missile attacks. Meanwhile, the soldier equipped with the Joint Expeditionary Digital Information (JEDI) program combines these capabilities with a laser range finder, GPS satellite positioning, and text messaging to send and receive information (troop locations, target data, special requirements) via satellite. Researchers continue to evaluate commercial off the shelf technology and government initiatives to develop innovative systems that bring the capabilities of space to the warfighter.

Force Development and Integration Center (FDIC)

The 1997 Memorandum of Agreement between the Space and Strategic Defense Command and the Training and Doctrine Command designated the USASSDC as the Army's proponent for Space and National Missile Defense (NMD). The USASSDC was given the lead on all NMD issues that required integration across TRADOC. The MOA specified that the Commanding General of USASMDC was the Army's specified proponent for space.[13] The FDIC was established on 1 October 1997 to provide the USASMDC with this capability. Its mission was to "coordinate and execute USASMDC's specified proponency and integrating responsibilities for missile defense and space." To carry out this mission it has four functions. As originally stated, it would develop Army concepts for missile defense and space. The FDIC would develop, manage and prioritize missile defense and space future operational capabilities (FOCs), as well as develop and/or integrate and validate DTLOMS solutions to missile defense and space FOCs by seeing to their inclusion in Army doctrine, FORCE XXI and Army After Next activities, training and leader development programs and methods. The FDIC would also see to their inclusion in new/upgraded materiel/systems and organizations and soldier proponency issues/programs. Finally, the FDIC would, in coordination with Headquarters, Department of the Army, develop and promote Army missile defense and space plans, policies and strategies. In order to carry out this mission and these functions, the FDIC was divided into four divisions, three concentrated on the TRADOC DTLOMS domains while the fourth served as the nexus for developing and articulating USASMDC's position on space and missile defense issues and worked to maintain liaison with external organizations and agencies.

The Concepts and Doctrine Division ensured a vertical and horizontal approach in developing, integrating and synchronizing space and missile defense warfighting concepts, doctrine and future operational capabilities. It also examined Army and Joint doctrine for space and missile defense implications and ensured consistency with associated warfighting concepts. The Training, Personnel Proponency and Leader Development Division translated space and missile defense training and leader development requirements into programs, methods or devices, assessed the adequacy of space and missile defense training and education programs throughout the Army and developed the USASMDC space literacy program. In this division, the Personnel Proponency Office was responsible for Functional Area 40 (Space Operations) and skill identifier 3Y (Space Activities) for officers and made sure that soldier proponency issues with future national missile defense organizations were addressed properly during planning and execution. The Combat Developments Division developed or integrated and synchronized Army space and missile defense materiel and organizational solutions and participated in all TRADOC combat developments processes. Finally, the Plans, Policy and Joint Coordination Division developed and coordinated Army space and missile defense strategies and policies in conjunction with the Army Staff and provided a liaison function between the command and outside organizations.

The FDIC's activities were pursued with vigor. The FDIC participated in the Army After Next Missile Defense and Space Game at Schriever AFB, Colorado in February 1999. Over the ten-day event, the Center drew the following six "emerging insights." The results of the game showed "the increasing importance of commercial space activities." The Center believed that the U.S. military "must have the means to leverage future commercial space capabilities," and urged military planners to pay attention to and understand the "rise of transnational space consortia." The Center noted that as the Army increases its reliance on GPS and other space capabilities, this "necessitates assured protection." In the future, the Army would have to confront "uninhibited surveillance from military and commercial space systems. Counter RISTA capabilities and deception measures will be critical in achieving information dominance." The Center also noted that the United States "may have to tolerate low-level attacks on space systems to avoid rapid geographic and conflict escalation." Finally, "adequate terrestrial missile defense capabilities are needed to avoid premature conflict escalation into space."[14]

From the inception of the program, the FA 40 specialty was a hot commodity and attracted many officers. In a 2000 interview, the FDIC Director, Colonel Glenn C. Collins, Jr., noted "We have a 400 per cent application rate - officers who want to be space officers versus how many we can actually accept."[15] FA 40 officers assist in managing, planning and integrating space capabilities to the benefit of the warfighter. The course of study involves both military and civilian schooling. However, despite its technical nature, this functional field draws officers from all the branches.

In the years since its founding, the FDIC in particular has been engaged in normalizing doctrine by including space and missile defense in significant Army and Joint publications.[16] As the Army continues its transformation efforts, the FDIC works to refine the Army's space and integrated missile defense requirements and prioritize them to support these efforts.

Space and Missile Defense Technology Center

In the mid-1990s, the U.S. Army Space and Strategic Defense Command underwent a series of reorganizations to better address its dual missions and the Army's priorities. New directorates replaced those originally established to align with the organization of the SDIO. The Staff

Realignment Study established a Missile Defense and Space Technology Center, to reflect more clearly the roles and missions of the Huntsville-based technical organizations.[17] The Tech Center also underscored Huntsville's reputation as a national center of excellence for missile defense[18] and realized plans to expand Huntsville's role in the Army space mission. In essence, the Tech Center serves as the command's technology developers, identifying and developing improvements to current systems and developing new materiel technologies. Recognized for leadership in missile defense technology, on 10 November 1995, Secretary of the Army Togo West designated USASSDC a Reinvention Laboratory to develop new, innovative and streamlined business practices.[19] Five years later the organization's accomplishments were again recognized as Lieutenant General Ronald Kadish, BMDO Director, appointed the USASMDC as the executive agent for ballistic missile defense science and technology.

This organization's continued achievements can be seen in the progress made by the variety of missile defense systems under development. While technology associated with interceptor systems remains its primary focus, the Tech Center continues to explore innovations. Directed Energy is once again the focus of attention and the USASMDC prepared the first Directed Energy Master Plan in 1999. Sensor technology also advanced. One example sought to improve the interceptor systems' ability to interpret what they see, while another was designed to expand the area covered. All in all, the overall goal of the Space and Missile Defense Tech Center is to be "more flexible, and [able] to respond more rapidly to new programs and marketing opportunities."[20]

Directed Energy Initiatives

In its short history, the High Energy Laser Systems Test Facility (HELSTF) has performed many tests, experiments and support work for the DoD, NASA, and other scientific communities. As one former HELSTF commander observed "lasers for shooting down missiles or aircraft are no longer something dreamed up by science fiction writers."[21] As if to underscore the commander's words, in the 1990s HELSTF overcame Army opposition and successfully demonstrated the feasibility of laser systems in anti-satellite and missile defense roles.[22]

Data Collection Exercise

In 1989, the Directed Energy portion of the Anti-Satellite Acquisition Decision Memorandum tasked the Army to develop the prime candidate for the DE ASAT weapon, based upon the Army-managed, SDIO GB-FEL TIE.[23] The HELSTF conducted the first satellite lethality experiment in August 1991. With the success of the Mid-Infrared Chemical Laser (MIRACL) in tests against rockets, Congress imposed a ban on testing the laser against satellites.[24] The ban expired in 1995 and the Army began preparations to attempt to lase an orbiting satellite in 1997.

The HELSTF took the first step towards the experiment tracking the MSTI-3 satellite with the laser in March 1997. The Data Collection Exercise (DCE) called for the MIRACL, a 1-million megawatt laser, to target an Air Force satellite to assess the ability of a laser to blind an orbiting satellite.[25] Given the increased dependence by American forces on satellite/space systems, this proposed experiment was vital to determine potential vulnerabilities in the space systems.

Fig. 6-7. At left, mounted on a 5-inch naval gun mount, the SEALITE Beam Director, with a 1.5 meter aperture, aims laser beams at moving targets. An infrared photo shows the MIRACL lasing a high altitude drone during a 1991 propagation test. At right are the results of a MIRACL beam directed against a TITAN ICBM stage.

As the time neared for the proposed test, however, the project met with controversy. Although the test did not violate existing treaties, a number of groups expressed opposition to it, arguing that it would result in the militarization of space and lead to a new arms race.[26] While the Pentagon had defined the experiment as a defensive test, opponents, including the Russian government, countered that the data could be used for offensive purposes.[27] Nevertheless on 2 October 1997, Secretary of Defense William Cohen approved the proposed laser test.

During five tests conducted between 8 and 25 October, the USASMDC successfully completed the DCE at White Sands Missile Range, New Mexico. The exercise began on 8 October when the Low Powered Chemical Laser (LPCL) acquired, tracked and illuminated the five-foot satellite orbiting approximately 260 miles above the Earth. In the next stage, on 17 October, both the MIRACL and LPCL successfully tracked and scanned the satellite. Although the satellite's systems failed to collect data, a camera on the sea-light beam director detected the laser beam on the satellite. Due to a technical malfunction, only the LPCL completed the last three phases of the experiment. The LPCL, which operates at 30 watts, dazzled or temporarily blinded the satellite on three successive nights. The tests provided data on atmospheric propagation and showed that even a low-powered laser could have a negative affect on a satellite's performance with only "a momentary or inadvertent exposure."[28]

Tactical High Energy Laser (THEL) and Mobile THEL

In the 1980s, the MIRACL system demonstrated the potential of directed energy systems to destroy targets using grounded missiles and helicopters. The next phase was to demonstrate the effectiveness of a more compact tactical laser to intercept a missile in fight - the Nautilus program. In 1995, the Army designated the Air Defense Center and School as the lead agency for the development of a tactical high-energy laser.[29] The USASSDC meanwhile oversaw the technical issues. As defined by the Technical Center, the THEL system, mounted on a five-ton truck, would have a range of one-kilometer for hard kills and up to 10 kilometers for sensor kills. With an engagement rate of 10 kills per minute, the THEL would be a cost-effective addition to the air defense arsenal.

Conducted by the USASSDC and the Israeli Ministry of Defense, the Nautilus program began testing in 1996. In its first attempt MIRACL achieved a successful intercept of an armed, short-range 120mm missile in flight on 9 February 1996, marking the first time that a laser had destroyed a rocket in flight. The success of this test generated increased interest in the Nautilus demonstration program and the THEL concept. In April 1996, President Clinton promised to support Israel to field a THEL by the end of 1997.[30] The Army committed additional funds to the effort and on 11 May Secretary of Defense William Perry elevated the THEL to a first priority as an Advanced Concept Technology Demonstration. Also in July 1996, the United States and Israel signed an MOA to explore the use of a THEL to negate the threat posed by short-range rockets, such as the Katyusha.[31]

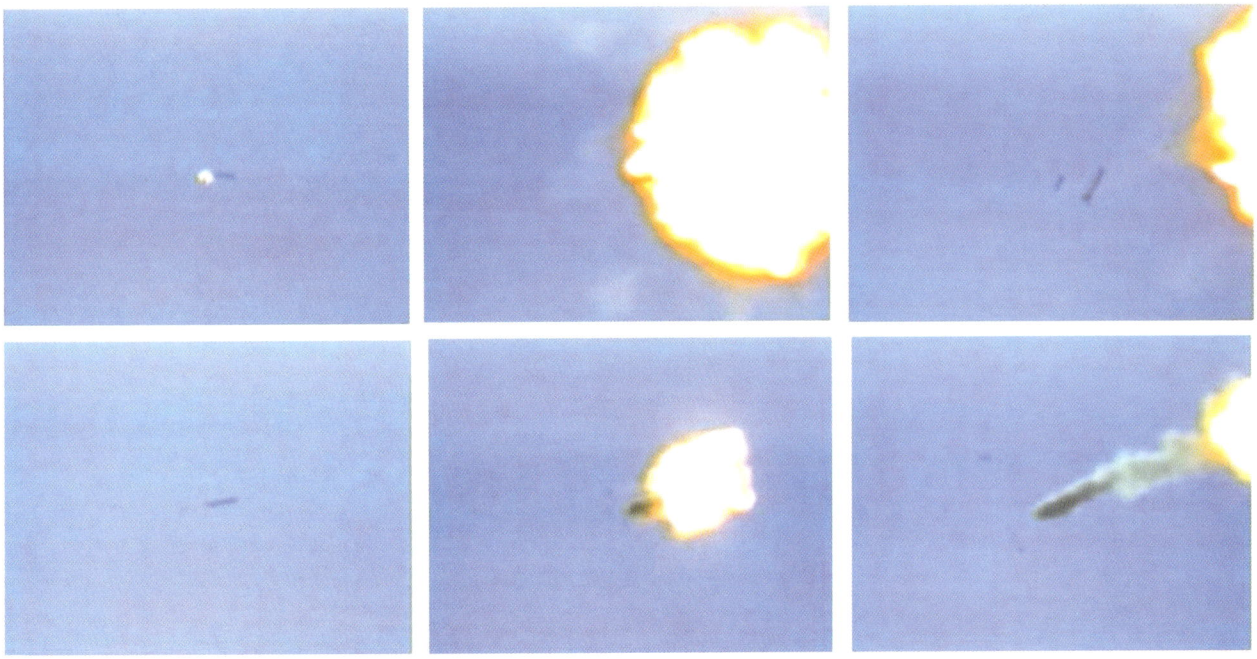

Fig. 6-8. In this series of photos the Tactical High Energy Laser heats the warheads of two missiles to detonate and neutralize them.

As work began on a prototype system, the command began to address the requirement for the future, releasing a notice for a mobile fire unit for the forward battle area capable of intercepting anything that flies at low ranges and disrupts airborne sensors.[32] Defense officials, however, were not convinced. Army Chief of Staff General Dennis Reimer, for example, testified that the truck-mounted system was "not as robust as we would like" and remarked upon its short-range limitations.[33]

With funding issues, and problems with near-term options, it initially appeared that General Reimer's assessment might be correct. Under the new agreement the design and construction phase for the THEL demonstrator was allocated 21 months with an additional 12-18 months of field testing in the United States and Israel. In that time, the contractor, TRW, would develop a transportable, tactical, deuterium fluoride chemical laser able to interface with a radar system supplied by Israel and support equipment.[34]

While the proposed 1997 field testing was delayed, testing did continue. On 14 March 1997, for example, THEL Test 8A demonstrated tracking and lasing capabilities against multiple in-flight targets. Funding however remained an issue. Despite support from Israel and members of Congress, the administration had not requested funding for the program beyond fiscal year 1999, because the Army had no formal requirement for the THEL. In 1999, the two governments, however, agreed to contribute additional funding to continue the program. They also negotiated a new contract with TRW to address schedule delays and cost overrun issues.

Fig. 6-9. Two soldiers stand beside the Tactical High Energy Laser beam director.

Despite these financial concerns, at its introduction at Roving Sands '98, the THEL demonstrated an 80-90% success rate against a variety of threats.[35] In June 1999, the THEL ACTD laser subsystem achieved first light, the first successful test of a laser, in tests at TRW's Capistrano Test Facility in California. Within the year, on 6 June 2000, the THEL demonstrator, in its first attempt, tracked and destroyed a Katyusha rocket in flight during tests at the HELSTF site. By the end of August 2000, the THEL had graduated to dual salvo tests - tracking and destroying two rockets in quick succession. Two additional dual salvo tests were successfully completed by the end of September.[36]

Between June 2000 and July 2001, the THEL destroyed 23 rockets in testing at White Sands. The next challenge, however, was to develop a more mobile version of the THEL. The Army began exploring this concept in 1999 in response to an operational needs statement from the Eighth U.S. Army in Korea.[37] As the political situation in Israel changed, they too expressed an interest in a mobile system. The resulting Mobile THEL or MTHEL system was designed to defend against a greater variety of threats - short-range rockets and mortars, aircraft, unmanned aerial vehicles, and possibly cruise missiles. In tests conducted on 5 November 2002, the MTHEL successfully demonstrated its capabilities against this set of threats. The system tracked and destroyed three 152mm projectiles fired from a howitzer.[38] The MTHEL Program transferred from the USASMDC to the Program Executive Office, Air and Missile Defense's Short Range Air Defense (SHORAD) Project Office on 28 February 2003.

Fig. 6-10. Introduced to the public in December 2002, the ZEUS laser neutralization system is a laser system designed to heat a target until the ordnance explodes. The prototype ZEUS deployed to Afghanistan in March 2003 to neutralize land mines and unexploded ordnance

TMD Critical Measurements Program

In the mid-1990s, as head of the Cooperative Targets effort, the Sensors Directorate participated in the BMDO's Midcourse Space Experiment (MSX). In one series of experiments, the satellite based MSX focused on identifying and tracking ballistic missiles and penetration aids after booster burnout and before reentry.[39] Using infrared, ultraviolet, visible light and spectrographic sensors, the MSX collected real-time data against terrestrial, earth and celestial backgrounds. The space-based sensor allowed scientists to conduct assessments not feasible in previous target data studies.

While, the MSX provided additional signature data for national missile defense system, the TMD Critical Measurements Program (TCMP) was a product of Operation Desert Storm.[40]

Fig. 6-11. The Theater Missile Defense Critical Measurements Program collects flight test data for the missile defense program's interceptor and sensor systems.

Following the war, the ability to distinguish between warheads and missile debris became a priority. In a series of campaigns beginning in 1993, the TCMP collected optical and radar data on various tactical ballistic missile target packages. The goal was to reduce TMD systems risks by characterizing "potential countermeasures and [developing and testing] computer algorithms."[41]

In the initial TCMP flights, only tested radar and sensor packages, such as the AST, the High Altitude Observatory and the USAKA based radars, participated in the data collection exercise. As the program progressed however, new products were integrated into the effort. Each test focused on the requirements of one or more TMD systems. Ultimately, all of the Army's radar systems - the GBR, the Patriot, THAAD, and Medium Extended Air Defense System, - the Navy's AEGIS and the Air Force's Space and Missile Tracking System would participate and benefit from these tests.

Joint Land Attack Cruise Missile Defense Elevated Netted Sensor System

During the 1990-1991 Persian Gulf War, U.S. forces successfully intercepted ballistic missile threats. With the systems available at the time, however, those intercepts tended to occur over friendly territory. In the mid-1990s, with the proliferation of cruise missiles, sometimes referred to as the "poor man's air force", the Defense Science Board recognized a need for a sensor that could –adapt to any terrain and essentially see over the horizon.[42] In 1995, Under Secretary of Defense for Acquisition and Technology, Dr. Paul Kaminski, directed the USASSDC to evaluate aerostats as sensor platforms for cruise missile defense.[43] The 1995 Mountain Top experiment provided positive data on the feasibility of an aerostat-based sensor. In January 1996, Dr. Kaminski and Joint Chiefs of Staff Vice Chairman Admiral William Owens directed the Army to form a joint program office and initiate an aerostat program and field two operational Aerostats by fiscal year 2002.[44] The Army assigned operational control of this first priority program to the USASSDC and established the Aerostat Project Office on 6 February 1996. By the end of the year, a million-dollar concept definition contract was awarded to H&R Co., a joint venture between Hughes Aircraft and Raytheon.

An aerostat is a tethered balloon designed with an inner ballonet. The ballonet contains air and is used to control the altitude of the system by increasing and decreasing the volume provided for the helium gas. This design and the Mylar construction provide stability for the system. A puncture from a bullet or missile would only produce a very slow helium leak. The unmanned sensors, suspended in a compartment below the aerostat, provide a 360° picture enhanced by the Identification Friend or Foe System. This data is relayed to a ground-processing center via a fiber optic tether, which would notify relevant interceptor systems. An aerostat can provide 24-hour surveillance for periods up to 30 days.

Fig. 6-12. The cost effective Joint Land Attack Cruise Missile Defense Elevated Netted Sensor System employs an aerostat, a tethered helium-filled blimp, outfitted with radar and communications equipment that operates at altitudes between 10,000 and 15,000 feet to see over the horizon.

The primary focus of the Aerostat Program, renamed Joint Land Attack Cruise Missile Defense Elevated Netted Sensor System (JLENS) in 1996, was missile surveillance, tracking and fire control for the various anti-missile systems. The program overcame availability issues and conducted its first experiment during Roving Sands '96. The system demonstrated BM/C^4 functions by successfully tracking 65 targets each hour from a distance of 200 miles and relaying data to the Force Projection Tactical Operations Center located 60 miles away. The data was then forwarded to the Air and Missile Defense Command Center which alerted Patriot and

SHORAD units.[45] This "proof of principle" test illustrated the systems "ability to significantly increase battlespace awareness."[46]

Despite repeated Congressional funding cuts, the JLENS program initiated its demonstration program and was cited as "one of the real success stories at Roving Sands '98."[47] Operating in the simulation mode, the JLENS provided the air picture for the Army Air and Missile Defense Command sending data for Patriot, Aegis and SHORAD units. In March 1999, the JLENS proved its utility in a joint operational environment. During All Service Combat Identification and Evaluation Team '99 exercises, a 15-meter aerostat served as a relay platform between an Aegis cruiser and a Patriot battery at Fort Stewart, Georgia, providing the first live real-time data exchange between the two services.[48] This exercise was a test of the JLENS processing station, which correlated data and created a single integrated air picture.[49]

Original designs called for two radar systems for the JLENS system: precision track and surveillance. A lack of funding remained a problem, however, and early in fiscal year 1999 officials opted to pursue only the precision track radar needed to relay data to the Patriot batteries.[50] With a new, slower development pace, the surveillance radar would remain an option for the future. The 2002 demonstration goal was subsequently pushed back to 2005.

The Army however remained committed to the JLENS program. In February 1999, the command submitted a proposal to convert the program from an Advanced Concept Technology Demonstration to an acquisition category II program.[51] This transition would define the program's direction and possibly solidify funding by creating a stable program. In March 1999, DA officials approved the transition. Perhaps more importantly, in May 1999 the Joint Theater Air and Missile Defense Organization identified the JLENS as a "central player in the future cruise missile defense architecture."[52] The good news continued in November 1999, when *Popular Mechanics* magazine awarded the JLENS Program Office a 2000 Design and Engineering Award. Magazine editors observed that the JLENS "represented a very clever use of existing technology to solve an extremely difficult problem."

April 2000 and the Forward Pass Mission saw the next major advance in the JLENS program. In these demonstrations, the JLENS successfully completed two target intercepts guiding a surface-launched interceptor (an Advanced Medium Range Air to Air Missile) beyond the range of its own organic radar. The concept required two types of radar, a surveillance system and a precision track and illumination radar, to identify the target and cue the system to intercept. The April test represented several firsts: the first live, over-the-horizon engagement of a cruise missile target using an elevated sensor; the first program to demonstrate the Forward Pass concept; and, the first time that control of a missile in flight was handed over to another radar (the forward pass) to intercept a low flying target.[53]

The JLENS mission includes detection and tracking of low altitude threats (cruise missiles and aircraft), tactical ballistic missiles in the boost phase, and surface moving targets; support for air-directed surface-to-air missile engagements (e.g. Forward Pass), and support for developing and displaying the single integrated air picture. By the end of 2000, the program successfully demonstrated its abilities in each of these areas. In May 2001, the JLENS program sought to demonstrate the system's versatility and a possible secondary mission of signal and intelligence support. During the Signal Symposium at Fort Gordon, Georgia, the JLENS communications package transmitted voice, video, and data from a mobile HUMVEE to the exhibit center. Following the tragic events of 11 September 2001, the Army staff and the JLENS Program Office also began to explore possible Homeland Defense missions for the elevated networked sensor.[54] On 1 October 2001, the JLENS Program Office transferred to the PEO-AMD for formal acquisition, testing and fielding.

Office of Technology Integration and Interoperability (OTII)

The significance of the Single Integrated Air Picture (SIAP), illustrated by the JLENS, was recognized by the Joint Requirements Oversight Council in March 2000 with their decision to establish a SIAP Engineer Task Force. The task force's focus was to investigate the integration and interoperability issues faced by warfighting commanders associated with emerging and legacy systems. In July 2000, the USASMDC Commander, Lieutenant General John Costello, chartered the Office of Technology Integration and Interoperability, as a Major Subordinate Element of the command, to address this issue and serve as the subject matter center for the Task Force.

Fig. 6-13. As military maneuvers become increasingly joint, the services are working together to develop a Single Integrated Air Picture.

The OTII's immediate mission is to identify and prioritize the Army's interoperability requirements for the four pillars of Joint Theater Air and Missile Defense. The goal is to link together all Army TAMD systems, and those of the Navy and the Air Force. The broader mission requires the OTII to assess and leverage technology efforts from the Department of Defense and industry with regard to TAMD as well as space and missile defense. One such initiative is the Low Cost Interceptor - a long-range interceptor costing less than $100,000 each to manufacture.[55] The program is evaluating propulsion, seeker, missile guidance and lethality components in existing and maturing technologies to develop a cost-effective counter to the proliferating threat posed by unsophisticated cruise missiles.

The USASMDC, the Army Space Master Plan and the Objective Force

In 1997, the Army established a new major command, the U.S. Army Space and Missile Defense Command, to sponsor its efforts in space and national missile defense and as overall integrator for theater missile defense.[56] Creating the command brought the Army's interest in space to a new level. The Army's earlier efforts in space have already been noted and described. They played out against a background of war and Cold War. The way space-based systems were used in the Gulf War vindicated the Army senior leadership's decision of the mid-1980s to re-enter space in order to influence the ways in which the systems it used would be developed. The challenge was to keep space-based systems in the Army's consciousness as it reorganized to face the post-Cold War world.

In 1996, the Army initiated the Army After Next (AAN) Project to craft requirements for the Army of the near future, to focus on future warfare, specifically between the years 2010 and 2025. The AAN's brief was to "explore the nature of warfare thirty years into the future and to help develop a long-term vision for the Army." Its specific mission "was to conduct broad studies of war . . . frame issues vital to the development of the U.S. Army after about 2010, and provide issues to senior Army leadership in a format suitable for integration into TRADOC combat development programs."[57] In 1997 and 1998, a series of war games initiated as a part of this project, gave the Army's senior leadership an appreciation of just how crucial space assets had become and would remain to modern land warfare.[58] The games emphasized futuristic thinking about the Army. In the first round, the AAN imagined a radically different Army - one that could self-deploy easily to anywhere in the world and one not constrained by the limits of contemporary doctrine and technologies. These virtual units enabled the players to examine notions about future warfare marginally connected to contemporary realities to stimulate unconventional thinking.

The AAN Space Game Two took place in Colorado Springs under the auspices of the USASMDC, TRADOC and the National Reconnaissance Office (NRO). The game's object was to show how space support could be integrated into a cohesive theater campaign. Its results gave the Army a better understanding of the ways in which space-based resources might affect military operations on the ground. The game also pointed out ways commercial space-based systems could amplify the commander's knowledge of the battlespace with improved position and navigation capabilities and imagery systems.[59] Many of the Army's senior leaders identified space as the battlefield's new "high ground." According to USASMDC's first commanding general, Lieutenant General Edward G. Anderson, III, "Space has become a permanent platform for capabilities whose possession or loss can decisively influence the conduct and outcome of the land battle."[60]

However, possession or loss of space is only part of the effort to learn how to use this new medium, this new area of operations. A Memorandum of Agreement between the Army's newest command and the Training and Doctrine Command explicitly enumerated the new command's role as the Army's proponent for space and national missile defense and theater defense integrator. It specifically identified USASMDC's authority and responsibility to participate in TRADOC processes and to develop DTLOMS products in the areas of space and missile defense. The MOA also authorized establishing a Space and Missile Defense Battle Laboratory. A Force Development Integration Center was also created to work with the USASMDC Battle Lab to exercise control over this process. The work of these organizations has already been described.

The USASMDC had the primary responsibility to ensure soldiers had access to space-based assets. This would be accomplished by validating space as an important part of Army and joint training operations, acting as the Army proponent for space-based systems in the military and industry in developing and testing technology to use in space-based systems and fielding and operating successful space products.[61] If the primary workhorse for achieving these goals would be ARSPACE then the vehicle would be the Army Space Master Plan (ASMP).[62] Published in March 2000, the plan concentrated on the goals of "operationalizing, institutionalizing and normalizing" space in the force structure.[63] The plan's executive summary called for the Army to integrate space into every aspect of its daily routine, including planning, training and exercises. Officers and enlisted soldiers needed to be "literate in space support," while the Army had to develop space systems that would deliver accurate and timely information directly to the battlefield.

The Army would determine the requirements, conduct the research, develop, acquire and shape the future design and application of space systems. Additionally, commanders and soldiers alike would be continually trained about space-based systems to become accustomed to using space in actual operations. Learning about space-based systems would be part of the Army schools' curricula from pre-commissioning through the advanced service schools for officers and Department of the Army civilian employees and through technical schools for non-commissioned officers and enlisted personnel. In addition, ways to use space-based systems would be placed into all Army doctrinal publications to insure that using them would become habitual and both their advantages and limitations would become known.

The ASMP itself is composed of an executive summary, an introduction, six substantive chapters and a conclusion. It starts by defining the current and future space environment, and the continued by delineating the Army space requirements determination process, non-materiel activities, current systems and modernization strategy, Army space initiatives, capabilities assessment and conclusions and challenges.[64] The ASMP provides the over-all direction and necessary guidance to implement the Army's space policy. The plan's objective is to present the necessity for embedding space systems and technologies into the Army's force structure and creating a well-trained and innovative cadre of space-literate personnel who understand the benefits space-based systems can bring to the Army. To accomplish this goal, the Army would ingrain space into its way of life, increasing understanding about the ways space-based systems can help the soldier as well as the limitations of these systems.

The ASMP begins by defining the "space environment."[65] The environment, however, is not space itself (the medium), but is the "body of policies, plans, organizations, agencies" and threats that "influence, enhance and enable the space missions, warfighting concepts, programs, initiatives and experiments." The plan reviews the documents that set the direction for future space activities and programs: the National Space Policy[66], the National Space Security Space Master Plan[67] and the United States Space Command Long Range Plan.[68]

The space requirements determination process is managed by USASMDC and coordinated with the various TRADOC branch proponents. The ASMP then explains the process specified in the 1997 TRADOC-USASMDC MOA. The plan turns to the Army's role in determining joint requirements, and outlines the national and joint policy documents that affect the determination of space requirements.[69]

The fourth chapter examines the non-materiel means to improve readiness. It outlines the three pillars forming the foundation of the institutionalized space mind-set. They are (1) leader development training and education, (2) embedding a special staff section at corps level and investigating the need at division-level and below and (3) documented space integration across the spectrum of cornerstone documents and publications. The plan's authors advocate focused integration of space throughout the Army's colleges, schools and centers as well as unit training.[70]

The fifth chapter, Current Systems and Modernization Strategy, presents an overview of the space systems and their related ground segments of most interest to the Army through 2005. It then extrapolates this overview to 2020. The modernization strategy is based on improving past capabilities while preparing for the changes that will occur when the first digitized division joins the force in 2000 and when the first digitized corps joins in 2004. The chapter assumes that the promise inherent in digitization will be realized and that the promise of success is dependent upon assured access to adequate space, related ground assets and their seamless integration.[71] The sixth chapter outlines Army Space Initiatives. The chapter defines the space initiatives; that is, the technology developments, experiments and demonstrations designed to satisfy the Army's space future operational capabilities.[72]

The seventh chapter assesses whether or not these capabilities are adequate enough to enable the Army to meet its future operational challenges in the near-term (FY 00-04), the mid-term (FY 05-10) and the far-term (FY 11-20). The capabilities are rated against the operational requirements for each time period. According to the ASMP, it appears that the future operations capability process is proceeding according to plan and will be able to attack the combat capability multipliers. Needless to say, in those areas where the Army is traditionally supported by other Services, Army space initiatives are lacking.[73] The final chapter draws conclusions from those previous and completes the ASMP methodology. It connects the goals and analyses and sets a course for the Army in the near-, mid-, and far-terms.[74]

Through the last years of the 20th century, the Army focused on modernizing its heavy mechanized units. However, in 1999 a slightly different Army transformation effort began, one that attempted to create medium weight units that could deploy swiftly and destroy an enemy with overwhelming speed. This Objective Force is built on new weapons systems, but its intellectual underpinnings for using space-based system to support it may be found in TRADOC Pamphlet 525-3-14, *Concept for Space Operations in Support of the Objective Force.*[75] It is the Army's "holistic concept" for "space and land force operations" and will be used to develop solutions across the DOTML-PF (doctrine, organization, training, materiel, leader education, personnel and facility) spectrum. The objective for TRADOC is to provide a concept that will "serve as a baseline" for developing "space-related operational capabilities and requirements."

Four space mission areas are enumerated in the Joint Doctrine for Space Operations (Joint Publication 3-14): force enhancement, space support, space control and force application.[76] The latter exists only in the minds of planners and technologists since it involves attacking forces or objects on earth from space. Space support refers to the actions taken to maintain space-based system, while space control refers to the means used to ensure access to space-based systems by friendly forces while denying access to adversaries.

Force enhancement includes what most believe is the true meaning of "support from space." This includes (1) satellite communications (SATCOM) links that ensure connectivity when terrestrial links are unavailable or nonexistent, (2) space-based and space-enabled surveillance and reconnaissance systems, (3) space-based position, velocity, navigation and timing systems, (4) space-related weather, terrain and environmental monitoring systems and (5) space-derived missile warning information. In order to achieve success, the Objective Force units must see first, understand first, act first and finish decisively. Because it will be space-enabled, the force will be able to use, as a matter of routine, the entire overhead constellation of military and commercial space platforms to accomplish these goals.

Developing the space essential operational tasks comes from wargaming and analysis and historical analyses and lessons learned derived from training exercises and actual operations. If space forces provide the necessary support for these tasks, the Objective Force will achieve operational success. There are five essential space operational tasks: (1) Supporting increased deployability and reduced theater footprint by enabling global reach to the home station operations center through 24x7 global SATCOM; (2) Enabling situational understanding of the operating environment upon arrival during entry operations. This would include space-based weather monitoring, mapping and terrain analysis that would support the intelligence preparation of the battlefield; (3) Supporting precision maneuver, fires, sustainment and information by reducing the fog, friction and uncertainty of warfare by using accurate and jam resistant GPS as well as combat identification and in-transit visibility; (4) Enabling continuous information and decision superiority to allow commanders on the scene to operate on their own terms, at times and places of their own choosing through space control protection and surveillance; and (5) Protecting the committed force during all phases of the operation including timely and accurate theater ballistic missile warning and defeating enemy attempts to use space systems. Thus, the Army and the Joint community have realized that space, an operational medium like the land, sea or air, is the new high ground and it must be seized in order to dominate the battlespace on earth.

The Army is growing more dependent upon space-based force enhancement capabilities and this means its vulnerability to disruption is also increasing. The increased use of commercial space-based systems has altered the definition of the space environment and to a certain extent represents a potential leveling of the playing field. Since the early 1990s, commercial space imagery satellite systems have improved the accuracy, quantity and timely delivery of the data they gather. Therefore, an adversary can use satellite reconnaissance photos without owning any satellites.

The Objective Force is designed to take a decisive role in joint and multinational military operations. It will be strategically responsive and immediately deployable. Units will be modular while organizations will be designed to be tactically flexible. Underpinning the new capabilities will be soldiers trained in a way that increases their mental agility and initiative.

As outlined above, control of space and space-based systems play an important role in preparing for tactical operations. Space control's contribution to the Army's Objective Force and to the joint force commander cannot be overemphasized. The Objective Force's employment of sophisticated space control capabilities should degrade or substantially diminish an adversary's military decision making process. Technology and war are interrelated, but innovative technology does not by itself win battles and wars. The doctrinal and training implications of space control technology hold the potential for changing warfare.

Underneath the story of the Army's return to space and its technological breakthroughs in the field of missile defense lay the virulent partisan political debate over national missile defense. Held temporarily in abeyance at the end of the Cold War and by the reconfiguration of national missile defense concepts, it flared up again as guided missile proliferation and nuclear proliferation continued apace.

National Missile Defense: Politics and Threat Assessment

In the early years of the Clinton administration, national missile defense was not an issue. In January 1992, the Russian government announced that it would accede to all treaties of the former Soviet Union.[77] On this date in an address to the United Nations Security Council, Russian President Boris Yeltsin described the ABM Treaty as "an important factor in maintaining strategic stability in the world." He also proposed the elimination of existing ASAT programs and suggested a ban on such weapons.

In July 1993, the Clinton administration announced its position on the ABM treaty.[78] President Clinton adhered to the "narrow" or "traditional" interpretation of the treaty. Thus the treaty prohibited the development, testing and deployment of sea-based, air-based, space-based and mobile-land based ABM systems, regardless of the technologies employed. One year later, Presidents Clinton and Yeltsin issued a joint statement that both nations "agreed on the fundamental importance of preserving the viability and integrity of the ABM Treaty."[79]

The debate over National Missile Defense reemerged in 1994 with the Republican Party's Contract with America. In this document, the 350 Republican candidates for the U.S. House of Representatives pledged to introduce and support the National Security Restoration Act. This legislation included the promise to "renew the U.S.'s commitment to an effective national missile defense by requiring DOD to deploy anti-ballistic missile systems capable of defending the U.S. against ballistic missile attacks."[80] The subsequent proposed Missile Defense Act of 1995 stated that it was the policy of the United States "to deploy at the earliest practical date highly effective theater missile defenses" and "to deploy at the earliest practical date a national missile defense system that is capable of providing a highly effective defense of the United States against limited ballistic missile attacks."[81] This document called for up to 100 ground-based interceptors at a single site or a greater number of interceptors at a number of sites as deemed necessary, fixed, ground-based radars, space-based sensors, and BM/C^3 be deployed by 2003. The language implied the abrogation of the ABM Treaty and, consequently, President Clinton vetoed this legislation in December 1995.[82] The resulting legislation advocated the deployment of an affordable and operationally effective TMD and "a cooperative, negotiated transition to a regime that does not feature an offense-only form of deterrence."[83]

In 1996, the NMD deployment question produced two conflicting proposals. Arguing that the "best defense is a good defense," the Republicans introduced the Defend America Act of 1996, which sought to deploy an NMD system by the end of 2003.[84] The stated policy was to deploy a system capable of defending the continental United States, Alaska and Hawaii, against a ballistic missile –launch, whether accidental, unauthorized or deliberate. A second criterion required DoD to develop a system that could be augmented to provide a layered defense against larger and more sophisticated missile threats. Rather than rely solely on a land-based ABM system, the proposal incorporated a variety of space-based options.[85] Congressional Budget Office cost estimates put deployment of this system, composed of 100 ground-based interceptors, ground-based radars, a constellation of 24 space and missile tracking sensors and a constellation of 500 space-based kinetic energy interceptors, at $31 to $60 billion. As a result of these estimates the bills never came to the floor for a vote.

The Clinton administration countered that a missile defense system is not required because, "No rogue nation today has ICBMs; only the established nuclear powers have ICBMs . . . [O]ur ability to retaliate with an overwhelming nuclear response [would] serve as a deterrent."[86] The administration's NMD Deployment Readiness Program, known as "3 plus 3", called for three additional years of development, followed by a review of the ballistic missile threat, to be conducted in the year 2000. If warranted the program would then proceed for three more years to deploy a system.[87] This treaty-compliant deployment focused on a single site - the former SAFEGUARD Complex in North Dakota - and included 100 ground-based interceptors, a GBR, an upgraded early warning radar, an adjunct forward based radar in Alaska, and in-flight interceptor communications for BM/C^3. As designed, this NMD system could provide a defense against a limited attack by a rogue nation or a small accidental launch.

Critics questioned the administration's commitment to its program, pointing to a lack of procurement funding in the long-range plans for defense spending.[88] Nevertheless, repeated attempts to enact legislation requiring the deployment of an NMD system by the end of 2003 failed to reach a vote in Congress. With the support of the Joint Chiefs of Staff and others, the 3 plus 3 program remained the standard throughout the Clinton Administration.[89]

Some change did appear in 1999. Responding to a new threat analysis, the Clinton administration included an additional $6.6 billion in the fiscal year 2000 budget for the development of NMD technology to be deployed by 2005.[90] Later that year, the President reversed his initial opposition and agreed to support the National Missile Defense Act of 1999. Public Law 106-38 was signed into law on 23 July 1999. The law states that it is the policy of the United States to (1) deploy as soon as technologically possible a National Missile Defense (NMD) system capable of defending U.S. territory against limited ballistic missile attack (whether accidental, unauthorized, or deliberate) and (2) seek continued negotiated reductions in Russian nuclear forces.[91]

Under the "3 plus 3" program, the year 2000 was pivotal to the NMD program. President Clinton was to decide whether or not to deploy the NMD system following a June 2000 technology review. In fact, to meet the proposed 2005 deployment date a decision would be needed no later than September, as weather conditions in the North dictated ground-breaking for construction.

A General Accounting Office report, written in May 2000, found that although DoD had taken measures to reduce program risks, performance and schedule risks remained.[92] Opponents revived the "rush to failure" criticism of the NMD program. In June, however, the NMD Independent Review Team "concluded that the technical capability to develop and field the limited system to meet the defined C1 threat is available."[93] The report added that the 2005 deployment remained "high risk" but did not propose to change the schedule. Secretary of Defense William Cohen's recommendations, however, would hinge on the 7 July test of the ground-based interceptor. Defense officials wanted two successful intercepts before making their recommendations. With two tests completed, the interceptor had one successful intercept and one failure. Due to a problem with the surrogate booster, the EKV failed to separate and did not achieve a target intercept. Later that month, administration lawyers advised the President that preliminary construction on an X-band radar on Shemya Island, Alaska, would not violate the ABM treaty. Despite the test results, Secretary Cohen recommended that the United States proceed with deployment.

In a speech at Georgetown University, on 1 September 2000, President Clinton announced his decision to defer the decision to deploy an NMD system to the next president. While Clinton recognized the existence of the threat posed by ballistic missiles and the advances made by the Defense Department, he placed greater emphasis upon the significance of treaty negotiations. With regard to NMD Clinton stated: "I simply cannot conclude with the information I have today, that we have enough confidence in the technology and the operational effectiveness of the entire NMD system, to move forward to deployment. Therefore, I have decided not to authorize deployment of a national missile defense at this time."[94]

Throughout this decade, a key distinction between the proposed missile defense systems was the threat assessment. During the mid-1990s two documents served to define this aspect of the American missile defense policy. The first document, the National Intelligence Estimate, was presented to officials in 1995. The second document, produced by the Commission to Assess the Ballistic Missile Threat to the United States, was released in 1998.

In November 1995, the National Intelligence Council presented its National Intelligence Estimate (NIE 95-19) entitled "Emerging Missile Threats to North America During the Next 15 Years." The report determined that "no country, other than the major declared nuclear powers, will develop or otherwise acquire a ballistic missile in the next 15 years that could threaten the contiguous 48 states or Canada." At the same time, it asserted that North Korea was developing a missile, the Taepo Dong 2, which could have a range sufficient to reach Alaska. The report's authors however did not expect North Korea to achieve the technological capability to develop a system able to reach the contiguous 48 states within the time parameters of the study. A third assessment was that "no other potentially hostile country has the technical capability to develop an ICBM in the next 15 years." Fourth, those nations with an indigenously developed space launch vehicle could produce an ICBM within five years, but any such activity would be detected. Finally NIE 95-19 accepted that foreign assistance could affect the rate of development of a missile program, but did not expect any country currently owning ICBMs to sell them.[95]

Republicans, such as Congressman Curt Weldon (R-PA), claimed that the report was highly politicized and downplayed the threat to the nation. Others, including Lieutenant General Malcolm O'Neill, BMDO Director, expressed concern with the manner in which uncertainties were handled.[96] As a result of these and other concerns, Congress established a bipartisan panel, headed by the former Director of Central Intelligence Robert Gates, to review the report and its findings. In his presentation to Congress, Gates testified that the report, while not politicized, was "politically naïve and not as useful as it could have been" and added that the "methodology was deeply flawed."[97] Nonetheless, the team believed that the NIE-95-19 findings were valid and no threat was anticipated within the next 15 years.

The July 1998 report of the Commission to Assess the Ballistic Missile Threat to the United States presented a radically different assessment of the ballistic missile threat. Established by the 1998 Defense Authorization Act and chaired by former Secretary of Defense Donald Rumsfeld, the Commission found that "the ballistic missile threat to the U.S. is real, credible and could appear sooner than early intelligence predictions." Specifically the Commission found:

> Concerted efforts by a number of overtly or potentially hostile nations to acquire ballistic missiles with biological or nuclear payloads pose a growing threat to the United States, its deployed forces and its friends and allies. These newer, developing threats in North Korea, Iran and Iraq are in addition to those still posed by the existing ballistic missile arsenals of

> Russia and China, nations with which the United States is not now in conflict but which remain in uncertain transitions. The newer ballistic missile-equipped nations' capabilities will not match those of U.S. systems for accuracy or reliability. However, they would be able to inflict major destruction on the U.S. within about five years of a decision to acquire such a capability (10 years in the case of Iraq). During several of those years, the U.S. might not be aware that such a decision had been made.[98]

Finally the Commission concluded that "the threat to the U.S. posed by these emerging capabilities is broader, more mature and evolving more rapidly than has been reported in estimates and reports by the Intelligence Community." They further recommended that the "U.S. analyses, practices and policies that depend on expectations of extended warning of deployment be reviewed and, as appropriate, revised to reflect the reality of an environment in which there may be little or no warning."

One month after the report was released, on 31 August, North Korea launched a three-stage ballistic missile to put a satellite into orbit. Although the launch failed, such a missile would have a range of 4-6,000 kilometers sufficient to reach Alaska and Hawaii. Citing a CIA Briefing, Representative Weldon later added that the Taepo Dong I, "depending upon the payload can hit well into the central part of the mainland."[99] At the same time another "rogue nation" Iran tested an intermediate range ballistic missile and is developing a longer-range version.[100] Also during the summer of 1998, both India and Pakistan tested nuclear weapons.

The Welch Reports

In 1998, the Pentagon also received the first report from "Task Force on Reducing Risk in Ballistic Missile Defense Flight Test Programs." Headed by retired Air Force General Larry Welch, the committee presented its findings to Congress in February 1998. The Welch report warned the government that the NMD's "3 plus 3" program was on a "rush to failure" due to an over-emphasis on compressed time schedules. As a result, tests were defeated "by poor design, test planning, and preflight testing deficiencies; poor fabrication; poor management; and lack of rigorous government oversight." The Welch panel recommended that all ballistic missile programs adopt a more realistic sequential schedule, pointing out that "accelerating schedules by simply adding risk carries a very high risk of failure." [101] Reviewers also advocated increased ground testing with simulations and test facilities to reduce the risks associated with flight testing. Ultimately the Welch panel advised the Pentagon to restructure the flight program to ensure sequential testing and allow adequate time to correct deficiencies, increase funding for flight tests and the number of planned tests, provide support for ground tests and continue the development of key technologies and follow-on system capabilities.

By 1999, the Army had awarded the contract to Boeing to serve as the lead system integrator for the NMD program and the BMDO had restructured the program. In January 1999, Secretary Cohen announced that the second phase, the deployment period, would be extended to five years.

The new schedule sought to allow developers additional time to conduct further testing and delay if necessary critical decisions on final production versions of the various system elements.

The BMDO reconvened the Welch Panel in 1999 to reassess the NMD program. They discovered that delays in test programs and the development of simulation and test facilities had already compressed the revised schedule. Panel members also found the organizational structure and lines of authority to be unclear causing further schedule delays and confusion. In general the reviewers placed less emphasis on the deployment readiness decision to be made in 2000, as the restructured program had phased the decision milestones through the year 2003. In addition, the panel recommended against focusing strictly upon the Capability 1 deployment and 2005 initial operating capability date to the neglect of future technology growth. The detailed report found that the restructured program had reduced the associated risks, yet NMD remained a high-risk initiative.[102]

The Structure of Missile Defense

During the 1992 reorganization, responsibility for National Missile Defense had transferred from this command to the Program Executive Office GPALS. With the commitment to the "3 plus 3" NMD deployment readiness program, in 1996 Under Secretary of Dense for Acquisition and Technology, Paul Kaminski ordered that NMD be designated an acquisition category 1D Major Defense Acquisition Program.[103] At the same time, Dr. Kaminski recognized that the development of an NMD system is a joint commitment involving the military services, industry and DoD agencies. As such, he directed the BMDO to create a Joint Program Office for National Missile Defense (JPO NMD) by 1 April 1997. The JPO NMD would provide management oversight for NMD program elements and is responsible for the design, development, and demonstration of an NMD system to defend the United States from ballistic missile attack by 2003.[104] To further streamline the organization, the JPO NMD commander reports directly to the BMDO Director.

Also in 1996, the Joint Requirements Oversight Council (JROC) approved the capstone requirement document, which requires the NMD system to intercept incoming ballistic missiles 95 percent of the time.[105] The Army received the task to write the draft joint operational requirements document (ORD) for NMD. The JROC validated the ORD on 10 March 1997 and designated the Army as the executive agent.

The Army's role in the joint NMD continued to grow when in September 1999, the JROC recommended that the Army be designated the lead service and user-representative for the land-based NMD system. Mr. Jacques Gansler, Under Secretary of Defense for Acquisition and Technology, accepted the recommendations and assigned these duties on 15 November 1999.[106] The Army was at the same time granted ORD approval authority for land-based NMD systems that are not a part of specific key performance parameter requirements.

The USASMDC's organizational duties during these developments were many. For example, in April 1998, the command submitted to TRADOC a force design update for the future NMD system.[107] In August of the next year, General John Abrams, TRADOC Commander, approved the charter for the National Missile Defense TRADOC Systems Manager Office. Assigned to the USASMDC, this new agency was authorized to act as the Army's representative, manager and integrator for the entire spectrum of doctrine, training, leader development, organizational, materiel, and soldier products associated with the land-based NMD system. Then, on 22 March 2000, Lieutenant General Ronald Kadish, BMDO Director, issued a memorandum appointing USASMDC as the executive agent for ballistic missile defense science and technology.

A New Direction in Missile Defense

While the BMDO and other organizations had focused primarily upon theater level systems and a limited NMD in the 1990s, the arrival of the new administration of President George W. Bush signaled renewed interest in a vigorous missile defense at the highest levels of authority. In September 2001, Mr. Kenneth Oscar, Acting Army Acquisition Executive announced that Secretary of Defense, Donald Rumsfeld, was "actively transforming the [BMDO] into an organization that focuses on strategic missile defense."[108] As a result of this directive, the BMDO gained operational control of the THAAD, Arrow and Ballistic Missile Targets Joint Project Offices from the PEO-AMD and USASMDC and returned the elements of the Lower Tier Project Office to the PEO-AMD.[109]

In January 2002, Secretary of Defense Donald Rumsfeld further restructured the BMDO and elevated it to the status of agency, in recognition of the national priority and mission emphasis on missile defense.[110] The newly renamed Missile Defense Agency reports to the Under Secretary of Defense Acquisition Technology and Logistics. In the same document, Secretary Rumsfeld identified the top four missile defense priorities and granted the MDA the means to accomplish them.[111] For example, to expedite the development process, officials devised a system of "streamlined executive oversight and reporting." Similarly, the evolution of the Ballistic Missile Defense System would be managed by a three-phased program of development, transition, and procurement and operations, guided by the MDA Director and the Defense Acquisition Board. In addition, "to encourage flexible acquisition practices," the MDA was granted the authority to use transactions other than contracts, grants, and cooperative agreements to conduct its research. The document also exempts the BMD system from the traditional requirements generation process and assigns responsibility for the Developmental Testing and Evaluation of the BMDS and its elements to the MDA itself. Although these and other decisions generated considerable controversy in both the Congress and the press for eliminating outside and Congressional oversight from the program, the MDA continues to hold these unique powers to develop and deploy effective missile defense systems in a timely manner.

Withdrawal from the ABM Treaty

Throughout his campaign, President George W. Bush had questioned the relevance of a 30-year-old treaty to the current missile defense situation. Soon after his inauguration, in a speech at the National Defense University, Bush announced that he had tasked Secretary of Defense Rumsfeld to explore all available technologies and basing options for an effective missile defense to protect the United States, our deployed forces, and our friends and allies. Beginning in May 2001, the United States sent envoys to allied leaders "to seek their input on all the issues surrounding the new strategic environment." Bush argued that the ABM treaty "does not recognize the present, or point us to the future. It enshrines the past." President Bush continued, stating "No treaty that prevents us from addressing today's threats, that prohibits us from pursuing promising technology to defend ourselves, our friends and our allies is in our interests or in the interests of world peace"[112]

In November 2001, President Bush met with Russian President Vladimir Putin at Crawford, Texas, to negotiate the ABM Treaty. No agreement was reached. One month later, on 13 December 2001, President Bush announced that he had given formal notice to Russia that the United States was going to withdraw from the ABM Treaty, exercising Article XV of the 1972 treaty.[113] As Bush explained, one of the signatories, the Soviet Union, no longer exists and neither do the hostilities that created the treaty. Terrorism, such as the attacks against the United States on 11 September 2001, now represent the greatest threat to both nations. At the same time, President Bush reiterated a pledge made earlier with President Putin to reduce the American nuclear arsenal by 1,700 and 2,200 operationally deployed strategic nuclear weapons.[114]

The President's decision was not universally welcomed. On 12 June 2002, a group of 30 Democrats filed suit against the President, Secretary of Defense Donald Rumsfeld, and Secretary of State Colin Powell in an attempt to block the American withdrawal from the ABM Treaty. The group argued that it was illegal for the president to pull out of a treaty without the approval of Congress. Nevertheless, the United States formally withdrew from the 1972 ABM Treaty on 13 June 2002.[115] In a four-paragraph statement released by the White House, President Bush remarked, "With the Treaty now behind us, our task is to develop and deploy effective defenses against limited missile attacks. As the events of September 11 made clear, we no longer live in the Cold War world for which the ABM Treaty was designed. We now face new threats from terrorists who seek to destroy our civilization by any means available to rogue states armed with weapons of mass destruction and long-range missiles....I am committed to deploying a missile defense system as soon as possible to protect the American people and our deployed forces against the growing missile threats we face. Because these threats also endanger our allies and fiends around the world, it is essential that we work together to defend against them, an important task which the ABM Treaty prohibited."[116]

A New Deployment Decision

Following the terrorist attack of September 11th, President Bush outlined a new policy or doctrine of pre-emption to the graduating class at West Point on 1 June 2002. Bush argued that deterrence and containment, the doctrines of the Cold War, have a limited role in the battle against terrorist networks and "unbalanced dictators." On the home front, both homeland defense and missile defense are "essential priorities for America." Bush explained that a proactive stance is necessary to win the war on terrorism - "the only path to safety is the path of action."[117]

The order to deploy a missile defense system came on 17 December 2002.[118] President Bush gave the Pentagon two years to deploy a system to defend American territory, troops and allies against missile attack. The President described this initial move, which builds upon the testbed at Fort Greely, as "a starting point for improved and expanded capabilities" which will be augmented as needed given developments in research and technology and changes in the threat. Ultimately the system will protect American territory, troops and allies from ballistic missiles in all stages of their flight.

The initial 2004 deployment, which plans to address the near-term threat, calls for both land and sea-based interceptors.[119] To counter the ICBM threat, up to 20 ground-based interceptors will be located at Fort Greely, Alaska (16) and Vandenberg AFB, California (4). To counter short- and medium-range ballistic missiles, the plan envisions two systems: sea-based interceptors to be deployed on existing Aegis ships, and the deployment of an unspecified number of air-transportable PAC-3s. These systems are supported by an array of land, sea and space-based radars and sensors.

President Bush's proposal was not uniformly accepted. Opponents criticized the deployment of systems that had not yet been fully tested. Nevertheless, given its modest nature and the existing threat, some Democratic leaders, such as Representative John Spratt (D-SC), have described the proposal as the "best first step to take." As Representative Curt Weldon (R-PA) has observed "It's giving us a capability that we've never had and do not have today. If a missile were launched today there would be nothing we could do to take it down - nothing."[120]

Ground-based Midcourse Defense (GMD): The System to be Deployed

On 7 December 2000, during the Association of the United States Army Symposium in El Paso, Texas, Lieutenant General John Costello, USASDMC Commander, announced a new initiative in cruise, theater and national missile defense. General Costello declared that he would develop an operational concept for globally integrated missile defense, as the line between theater and national missile was increasingly blurred. Three months later, Secretary of Defense Donald Rumsfeld, in a joint press conference with NATO Secretary-General George Robertson, observed that "tagging the missile defense effort as either theater or national is 'unuseful.'" He further stated, "What's 'national' depends on where you live, and what's 'theater' depends on where you live." "Over time," Rumsfeld added, "it's every bit as important to us to be able to defend this piece of real estate and our population in this location as it is to defend our deployed forces and to have our allies feel equally secure to the extent that's possible."[121] From this point forward, the National Missile Defense effort was redesignated the Ground-Based Midcourse Defense or GMD segment.

The current missile defense system, as defined by the Missile Defense Agency, has no final or fixed architecture. Officials adopted an evolutionary deployment concept. In the first phase, DoD will field an initial capability as defined by the President. During the next two years, 2006-2007, additional networked sensors will be added to increase the effectiveness of the interceptors. These sensors will be forward-deployed ground-, sea- and space-based systems. Additional interceptors will be added in the next phase. Then as the technology develops more advanced weapons and sensors will be added to the ballistic missile defense system.

Upgraded Early Warning Radars (UEWR)

Fig. 6-14. Part of the Ballistic Missile Early Warning System of radars, this site at Clear Air Force Station, Alaska, is a potential addition the Ground-based Midcourse Defense deployment as an Upgraded Early Warning Radar.

The UEWR system focuses on the nation's existing early warning system composed of early warning radars[122] and defense support program satellites. The satellites, which fly in a geosynchronous earth orbit, are a relatively simple system with an unalterable scan pattern. As the technology becomes available, they will be replaced with the Air Force's Space-Based Infrared System. Designed to detect incoming ballistic missiles, the radars are deployed at sites, for example, in Massachusetts, California, and the Alaskan Aleutian Islands and across the globe. The upgraded software and hardware will enable the radars to acquire, track and identify small objects near the horizon, without increasing radar outputs. At the same time, the radars will be able to detect and track ballistic missiles in their midcourse phase. In 2003, the United States received permission from Denmark and the United Kingdom to pursue the upgrades to the radars deployed in their countries in support of the GMD mission.[123]

X-Band Radar (XBR)

Fig. 6-15. The X-Band Radar can be populated with 69,632 transmit/receive modules. The massive radar stand requires an area of 7 hectares (17.46 acres) for the radar alone.

Construction began on the testbed prototype on Kwajalein Island in January 1997. Expanding upon the technologies of the GBR-prototype and the THAAD radar, the XBR is a ground-based, forward deployed phased array radar. It operates in a bandwidth that ranges from 8 to 12 gigahertz and will provide cued search, detection, track, discrimination and kill assessment. Improved target resolution and processing technology enable the system to identify closely spaced warheads, debris and penetration aids. High resolution waveforms enable the X-band radar to determine a reentry vehicle's diameter, length, spin rate, velocity, and mass, the position of other objects and the respective nose wobble patterns facilitating discrimination.[124] Systems tests began in 1998, just six days after receiving approval to operate at full power. In this test, the radar successfully tracked a satellite demonstrating the system's ability to gather data for radar calibration and validating the electro-mechanical scan technology.[125] Since then, the prototype has participated in every intercept test for the EKV and has successfully provided real-time data - acquiring the target complex, tracking the objects, discriminating the target and providing kill assessment.[126]

Fig. 6-16. The Ground- Based Radar-Prototype constructed at Kwajalein (picture taken at night).

Under the initial deployment proposal, the XBR would be constructed at an Air Force facility on Shemya Island, Alaska. In 2002, the Missile Defense Agency began to explore the possibility of a sea-based system. In August 2002, the Pentagon announced the construction of a floating X-Band radar station off the coast of Alaska.

Inflight Interceptor Communications (IFICS)

The GMD BM/C^3 network is composed of two elements: the BM/C^2 and the IFICS. The data processing capabilities of the BM/C^2 make it the "brains" of the network. In the event of an attack, this element receives and processes data from the various sensor systems and plans, selects and adjusts courses of action. The IFICS meanwhile relays target updates and status information from the BM/C^2 to the interceptor during the intercept flight. An IFICS data terminal consists of a radio transmitter receiver enclosed in a radome and an equipment shelter. These terminals would be located at possibly 14 pairs of geographically dispersed sites near NMD elements and in New England.[127] A prototype IFICS terminal was installed at Kwajalein and has been incorporated into the GMD integrated flight tests

Fig. 6-17. The unmanned In-Flight Interceptor Communications data terminals are approximately 10 feet in height to include the 3-foot radome.

Ground-Based Interceptor (GBI)/Exoatmospheric Kill Vehicle (EKV)

The GBI is composed of an EKV and a booster. The program entered its next phase in December 1998. Following very successful fly-by tests with both of the competing EKV designs, the NMD Joint Program Office decided in favor of the Raytheon sensor.[128] This sensor integrates a series of modularized subsystems that facilitate upgrades and replacements. In addition to the infrared seeker, the EKV is composed of propulsion, communications link, discrimination algorithms, guidance and control system, and computers to support target selection and interception decisions.[129]

Fig. 6-18. The 121 pound Raytheon (Hughes) Exoatmospheric Kill Vehicle is navigated by an inertial navigation system updated exoatmospherically by star sightings.

The EKV attempted its first intercept on 2 October 1999. The system successfully distinguished between the warhead and a decoy. A clogged cooling pipe and problems with a surrogate booster had a negative impact on the two subsequent tests.[130] The GBI system overcame these problems, however. Integrating all elements of the GMD system, in four consecutive tests conducted between July 2001 and October 2002, the EKV successfully identified the elements of the target complex and intercepted the warheads.[131] A booster separation problem arose again during a December 2002 intercept test, bringing the EKV test record to five successful intercepts out of eight attempts.

Fig. 6-19. During an intercept, the Exoatmospheric Kill Vehicle approaches the target at a speed of 7,000 mph. The target itself is traveling at a speed of 16,000 mph. This photo was taken during IFT-6 on 14 July 2001.

Deployment: Fort Greely, Alaska

Initial concepts for the deployment of an NMD system focused upon a single ABM treaty compliant site: the former Stanley R. Mickelsen SAFEGUARD Complex near Grand Forks, North Dakota. With the recognition of the increasing threat posed by such nations as North Korea, authorities questioned whether or not this single site could protect the entire United States. A study conducted by the BMDO determined that while the North Dakota site could address "most threats," it was "not optimal against threats to Alaska and Hawaii."[132] The BMDO proposed at this time that a second site be added to provide increased protection and extend it to all 50 states. A second site however would require an amendment to the ABM Treaty as would a system that provided protection to the entire nation.[133]

In June 1998, BMDO announced that the best site for an initial limited NMD system would be central Alaska. Based upon the type of projected threat and the state's proximity to the North Pole, officials deemed that Alaska was the "optimum" location to protect the nation.[134] A team from USASMDC and the Corps of Engineers began site surveys in Alaska in August of 1998.[135] Nevertheless, a deployment of a single NMD site in Alaska would require an amendment to the ABM Treaty which limited each nation to an ABM complex located either at the national command center or near an ICBM base. At the end of the year, sites in both Alaska and North Dakota were still under consideration.

Although no location would be announced until President Clinton made his decision on the deployment readiness, by all accounts the NMD would be constructed in Alaska. Recognizing the constraints imposed by the weather conditions, Secretary Cohen recommended a limited go-ahead for the construction phase for the X-band radar. He was supported by administration lawyers who had concluded that the initial construction work associated with radar on Shemya Island would not violate the ABM Treaty. On 1 September 2000, during his speech at Georgetown University, in which he opposed NMD deployment, President Clinton added that he would not authorize the Pentagon to award construction contracts.

Fig. 6-20. Installation of the first of six Ground-based Midcourse Defense missile silos at the GMD Testbed at Fort Greely, Alaska in 2003. The 75-foot long silo weighs 130,000 pounds.

In July 2001, the Pentagon submitted a request to Congress for funds to support a missile defense test bed at Fort Greely, Alaska.[136] This test site with its command center and five silos could if required provide a limited defense against missile attack. The test bed, meanwhile, would create a triangle with assets in Hawaii, Kwajalein, and the Alaskan and California coasts, providing the military with a means to test different trajectories and geometries for several types of missile systems. One month later, the BMDO issued a Record of Decision to conduct initial site preparation activities and a construction company began clearing the site on 27 August. None of this work violated the ABM Treaty.

In December 2001, President Bush stated that the United States would withdraw from the ABM treaty. This announcement allowed Pentagon officials to proceed with the construction plans. The Corps of Engineers awarded the first construction contract in April 2002. Two days after the official withdrawal from the ABM Treaty, on 15 June 2002, the JPO GMD oversaw a ground-breaking ceremony at Fort Greely for six underground silos, part of the GMD Testbed. At the end of 2002, officials declared that the construction efforts were on schedule.

End Notes

[1]General Ronald H. Griffith, Memorandum for Space and Strategic Defense Command, Subject: Realignment of the Space and Strategic Defense Command, 12 July 1996. The HQDA Redesign Functional Area Assessment had recommended realigning USASSDC with TRADOC.
[2]Memorandum of Agreement between the United States Army Training and Doctrine Command and the United States Army Space and Strategic Defense Command, signed by General William W. Hartzog and Lieutenant General Edward G. Anderson III on 18 February 1997. The Force Development and Integration Center and the Space and Missile Defense Battle Lab were both products of this agreement.
[3]Department of the Army General Orders 5, dated 1 March 1998.
[4]Unlike other Battle Labs, the USASSDC BIC was not associated with its own proponent or school. The other Army battle labs addressed Battle Command, Combat Service Support, Early Entry Lethality and Survivability, Mounted Battle Space, Dismounted Battle Space and Depth and Simultaneous Attack.
[5]Signed on 11 May 1995, the MOA also made the USASSDC a voting member of the Battle Lab Board of Directors.
[6]MOA between the TRADOC and USASSDC, signed by General Hartzog and Lieutenant General Anderson on 18 February 1997.
[7]The lab would also employ attack operations models developed by the Air Force for the F/A-18 and F-16 aircraft and the Navy for the Aegis ship-based TMD. Lieutenant General Garner cited in Daniel G. Dupont's "SSDC Working to Create Battle Lab for Theater Missile Defense, Space," *Inside the Army* 24 October 1994: 1.
[8]The EADTB Product Office was disestablished in June 2002, although the EADTB continues to function as part of the Office of Technology Integration and Interoperability.
[9]This Battle Lab mission can be traced to the Army Space Exploitation Demonstration Program begun in 1987. The mission transferred from the ARSPACE to the Battle Lab in 1997. The ASEDP systems employ a combination of commercial off the shelf and government technologies.
[10]Space and Missile Defense Battle Lab, "Army Space Exploitation Demonstration Program," 21 August 1998.
[11]The system transferred to the Army Air and Missile Defense Command, Fort Bliss, Texas, in November 1997.
[12]Begun in 1999, the FOC TOC was designed with guidance from the commanders of the USASMDC, the U.S. Army Air Defense Artillery School and the 32nd AAMDC.
[13]Military space mission areas were defined as force enhancement, force application, space control and space support.
[14] *FDIC Newsletter* 1.2 (April 1999): 4.
[15]Jonathan Pierce, "Functional Area 40 Continues Growth," *The Eagle* October 2000: 6.
[16]By 2001, these included Field Manual 3-14.6 *Army Space Support Team Operations*, Joint Publication 3-01.3 Joint Doctrine for Defensive Operations for Countering Air and Missile Threats and Field Manual 3-01.1 *National Missile Defense Operations*.
[17]Later renamed the Space and Missile Defense Technology Center, the Tech Center was originally composed of five directorates (Advanced Technology, Sensors, Systems, Targets, Test and Evaluation and Weapons) and the Cost Analysis Office, the Program Integration Office, the Public Affairs Office and the Staff Action Control Office. A Space Technology Directorate was added in 1997.
[18]The five areas of excellence include kinetic energy weapons, lethality, discrimination and phenomenology, targets development and range support and radar and ladar (laser radar).
[19]The command was given the authority to waive DA Regulations and DoD Initiatives, with justification and legal review. However, legislative regulations, etc. cannot be waived.
[20]Memorandum dated 4 November 2001.
[21]Jack Moore, "HELSTF puts Tularosa Basin into laser weapons age," *Alamogordo Daily News*, 8 March 1992: 2-E; *FDIC Newsletter* 1.2 (April 1999): 4; Jonathan Pierce, "Functional Area 40 Continues Growth," *The Eagle* October 2000: 6. By 2001, these included Field Manual 3-14.6 *Army Space Support Team Operations*, Joint Publication 3-01.3 Joint Doctrine for Defensive Operations for Countering Air and Missile Threats and Field Manual 3-01.1 *National Missile Defense Operations*.
[22]In January 1995, for example, Members of Congress urged Secretary of Defense William Perry to continue HELSTF operations despite Army's plans to remove funding from its budget.

[23]The Air Force was tasked to develop a candidate based on alternate technologies.
[24]Caspar Weinberger, "Testing the MIRACL" *Forbes*, 6 October 1997: 37. The monograph, by SAIC and the USASMDC Historical Office, *DCE: Army Leadership at the Cutting Edge of Space Control,* (Huntsville: USASMDC, June 1999) provides a thorough overview of the Data Collection Exercise and the history of ASAT efforts.
[25]The Air Force's Miniature Sensor Technology Integration (MSTI-3) satellite provided a suitable target. It had exceeded its planned life span and was scheduled to be switched off.
[26]The satellite's manufacturer also opposed the test stating that the $60 million satellite was still viable.
[27]The parameters set by the experiment would not destroy the satellite, produce orbital debris or pose a risk to other spacecraft. William J. Broad, "White Sands to test laser," *El Paso Herald-Post*, 3 October 1997: A-1, A-2.
[28]Robert Bell, National Security Council arms control specialist, quoted in Bill Gertz, "Shared satellite laser test weighed," *Washington Times*, 2 January 1998: 1. An unclassified summary of the DCE findings can be found in *DCE: Army Leadership at the Cutting Edge of Space Control.*
[29]The Marine Corps had developed a mission need statement for a mounted THEL system in 1994. Daniel G. Dupont, "Army to Propose ACTD for Low-Cost, Tactical High Energy Laser System," *Inside the Army*, December 1995.
[30]John Donnelly, "Perry Puts U.S.-Israeli Tactical Laser on Fast Track," *Defense Week,* 20 May 1996: 1.
[31]First announced in April 1996 by President Clinton and Israeli Prime Minister Shimon Peres, the U.S. would assist Israel in developing a defensive capability against terrorist rocket attacks with the THEL. Soviet made Katyusha rockets had been used by Hezbollah, for example, in northern Israel. The agreement had been delayed due to a technology transfer ban restricting the sale of THEL technology.
[32]Daniel G. Dupont, "While U.S., Israel Discus Tactical Laser Prototype, Army Looks to the Future," *Inside Missile Defense,* 16 May 1996.
[33]"SASC-funded Nautilus laser seen winner in conference," *Aerospace Daily,* 10 May 1996.
[34]The support equipment would include for example the development and testing of the laser/fluid supply assembly, the pointer/tracker, and a command, control and communications and fire control system.
[35]A complete overview of the role of THEL in Roving Sands '98 appears in "Tactical High Energy Laser" *ADA Magazine Homepage* Summer 1998 at http://147.71.210.21/summer98/newpage16.htm.
[36]In November 2000, *Popular Science* magazine recognized these achievements, selecting the THEL Demonstrator as the Grand Winner in the General Technology category for its "Best of What's New" awards for 2000.
[37]Daniel G. Dupont, "SMDC Eyes Tactical Laser for Korea, Touts Directed Energy Development," *Inside the Army* 3 May 1999: 1; "Space & Missile Defense Command Seeking Information on Mobile THEL," *Inside the Army*, 9 August 1999.
[38]The projectiles are eight feet shorter than the Katyushas, fly at a faster rate and emit less heat.
[39]"Midcourse Space Experiment goes into orbit," USASSDC Public Affairs Office News Release, undated [April 1996].
[40]Specific tests using Scud missiles were conducted in the Willow Dune Program.
[41]Gerda Sherrill, "TCMP Successful Again," and Cynthia Brewer, "TCMP born in Desert Storm," *The Eagle* September/October 1996: 12-13.
[42]The Defense Science Board conducted its study in 1994.
[43]The U.S. Army first used tethered balloons during the Civil War for meteorological data collection, reconnaissance, and fire control. Telegraphers relayed information to the ground. Balloons were subsequently employed by the military for anti-aircraft support (barrage balloons) and scientific exploration.
[44]The management structure for the JPO calls for an Army Program Manager, with Navy and Air Force Deputy Program Managers. Support is also provided by the Advanced Research Projects Agency.
[45]Connie M. Davis, "Aerostat Rises to the top at Roving Sands '96," *The Eagle* July 1996: 11.
[46]"Army Happy With Use of Cruise Missile Defense Aerostat in Roving Sands," *Inside the Army* 1 July 1996.
[47]In this test the JLENS aerostat measured 233 feet in length and was filled with 590,000 cubic feet of helium, two and one half times the volume of the current largest advertising blimps. JLENS Over-the-Horizon Surveillance & Tracking Fact Sheet, USASMDC Public Affairs Office Fact Sheet, dated November 1997.
[48]The 15-meter aerostat, with a cubic volume of 8,000 feet, is designed to fly at 1,000 feet. The Marine Corps has expressed an interest in this platform to relay ship to shore information. The optimum Army aerostat however is a

71-meter platform which flies at a height of 15,000 feet. LuAnne Fantasia, "Carr's Crew Ready for Roving Sands '99," *The Eagle* May 1999: 1.
[49]A Single Integrated Air Picture is defined as "the product of fused, near-real-time and real-time data from multiple sensors to allow development of common, continuous, and unambiguous tracks of all airborne objects in the surveillance area." Capstone Requirements Document (CRD) for Theater Missile Defense, dated July 1998.
[50]Michael C. Sirak, "Participation in CEC-Patriot Link Set for Next Month; Administration Requests $25 Million for JLENS in Fiscal Year 2000 Plan," *Inside Missile Defense* 10 February 1999.
[51]Stephanie G. Rosenfeld, "SMDC Requests Approval to Make JLENS An Official Acquisition Program," *Inside Missile Defense* 10 March 1999. According to Army guides, ACAT II status requires eventual research, development, test and evaluation expenditures in excess of $75 million or eventual procurement of more than $300 million.
[52]Created in 1997, the JTAMDO generates the DOD's requirements for TMD. Estimated operating costs for the JLENS are $400 per hour compared to $4,000 to $5,000 for an aircraft. David Mulholland, "Army Ties Balloon Radar to Missile Defense," *Defense News* 17 May 1999: 1.
[53]Gerda Sherrill, "JLENS hits over-the-horizon cruise missile," *The Eagle* October 2000: 19.
[54]Emily Hsu, "JLENS Officials Awaiting Army Decision on Homeland Defense Role," *Inside Missile Defense* 26 June 2002: 1, 5, 6.
[55]Debra Valine, "Low Cost Interceptor seeking ways to balance cost per kill," *The Eagle* October 2000: 6 and Low Cost Cruise Missile Defense Initiative Low Cost Interceptor Fact Sheet, USASMDC Public Affairs Office, dated April 2002.
[56]U.S. Department of the Army, General Order Number 5, 1 March 1998.
[57]Walter Perry, Bruce Pirnie, John Gordon, IV, *The Future of Warfare: Issues from the 1999 Cycle Army After Next Study Cycle* (Santa Monica: RAND Corporation, 2001), pp. xi-xii. Accessed at http://www.rand.org/publications/MR/MR1183/ on 13 March 2002.
[58]William B. Scott, "Wargames Underscore Value of Space Assets for Military Operations," *Aviation Week and Space Technology* 28 April 1997:60-61; U.S. Army Training and Doctrine Command, *Knowledge and Speed: Battle Force and the U.S. Army of 2025: The 1998 Annual Report on The Army After Next Project to the Chief of Staff of the Army* (Fort Monroe: TRADOC, 1998), pp. 11-21.
[59]Martin Burkey, "Defense Official Urges Planning for Space Warfare," *Huntsville Times*, 9 February 1998: D-2.
[60]Lieutenant General Edward G. Anderson, III, "The New High Ground," *Armed Forces Journal International* October 1997:66; Martin Sieff, "Space Dependence: A Vulnerability," *Washington Times* 2 February 1998:12; "Is Space the Pentagon's Final Battleground?" *Business Week* June 15, 1998; Frederic W. Kagan, "Star Wars in Real Life: Political Limitations on Space Warfare," *Parameters* Autumn 1998:112-130.
[61]Lieutenant General John Costello, "Space and Missile Defense," *Army* December 1998:12; "Update: SMDC Commander on Space," *BMD Monitor* 30 April 1999:2.
[62]The USARSPACE is organized to weave space into the fabric of the Army's daily life. The ARSPACE controls the 1st Space Battalion, the 1st Satellite Control Battalion, an Operations Division and Regional Space Support Centers. The Satellite Control Battalion is in charge of the day-to-day operations of the DSCS. The Operations Division develops, manages and archives remote sensing products for the ARSSTs and other DoD users and the Regional Space Support Centers, in Hawaii, Germany, and Florida, coordinate satellite communications products to the various regional combatant commanders.
[63]*United States Army Space Master Plan* (2000) (hereafter referred to as *ASMP*), ES-2.
[64]The following is based on *ASMP*, ES-2-15. Unless otherwise noted, all direct quotations come from this source.
[65]The following is based on *ASMP*, 2-1-30. Unless otherwise noted, all direct quotations come from this source.
[66]The National Space Policy, updated in 1996, outlines five goals of the U.S. space program. (1) Strengthen and maintain the national security of the United States. (2) Enhance our knowledge of the earth, the solar system and the universe through human and robotic exploration. (3) Enhance American economic competitiveness as well as its scientific and technical capabilities. (4) Encourage state, local and private investment in and use of space technologies. (5) Promote international cooperation to further American domestic national security and foreign policies. The Secretary of Defense and the Director Central Intelligence oversee national security space activities. There are eight ways national space security space activities contribute to national security. (1) Support the right to self-defense and support American defense commitments to allies and friends. (2) Deter, warn, and if necessary,

defend against enemy attack. (3) Assure that hostile forces cannot deny us use of space. (4) If necessary, counter space systems and services used for hostile purposes. (5) Enhance operations of U.S. and allied forces. (6) Ensure our own ability to carry out military and intelligence space-related activities. (7) Satisfy military and intelligence requirements during peace and crisis and through all levels of conflict. (8) Support the activities of national policy makers, the Intelligence community, the National Command Authority, combatant commanders and the military Services, other Federal officials and continuity of government operations.

[67]This is the DoD's "strategic business plan" for space programs. The plan concentrates on policies, capabilities and practices that should exist beyond the Future Years Program. The plan provides senior leadership with long-range guidelines to make the transition to the future.

[68]The principal themes dominating the USSPACECOM Vision are "dominating the space medium" and "integrating space power throughout military operations." To reach this vision, USSPACECOM created four concepts that provide the conceptual framework to change the vision into a set of capabilities. They are (1) control of space, (2) global engagement, and (3) full force integration and (4) global partnerships.

[69]See *ASMP*, 3-1-3-18.

[70]See *ASMP*, 4-1-414.

[71]See *ASMP*, 5-1-5-14.

[72]See *ASMP*, 6-1-6-22.

[73]See *ASMP*, 7-1-7-14.

[74]*ASMP*, 8-1-8-10.

[75]United States Army, Training and Doctrine Command, *Concept for Space Operations in Support of the Objective Force* (Final Coordination Draft, Fort Monroe: U.S. Army Training and Doctrine Command, 22 February 2003) Unless otherwise noted, all direct quotations are from this source.

[76]This publication is the doctrinal foundation describing military space operations.

[77]On 9 October 1992, in the Bishkek Agreement, the Commonwealth of Independent States (CIS) signed an agreement pledging to support and implement the ABM Treaty.

[78]Federation of American Scientists, Anti-Ballistic Missile Treaty Chronology, located at http://www.fas.org/nuke/control/abmt/chron.htm.

[79]The two presidents also noted that both nations "have an interest in developing and fielding effective theater missile defense systems on a cooperative basis. The presidents agreed that the two sides will conduct a joint exercise of theater missile defenses and early warning." See http://sun00781.dn.net/nuke/control/abmt/chron.htm. The two nations reaffirmed their commitment to the ABM Treaty in a joint communiqué dated 20 June 1999

[80]Republican Contract with America, http://www.house.gov/Contract/CONTRACT.html.

[81]National Defense Authorization Act for Fiscal Year 1996, Subtitle C – Ballistic Missile Defense Act of 1995, H.R. 1530 at http://www.fas.org/spp/starwars/congress/1995/h950614i.htm.

[82]Representative John Spratt, (D-SC) proposed a similar amendment which held that the U.S. should deploy a system that complied with the ABM treaty and its amendments. This amendment, which failed to pass the house, generated considerable debate on the ABM Treaty and its significance. In his editorial, Frank Gaffney interprets this decision as quite decisive stating, "The 221 and fully 21 Democrats who refused to affirm the ABM Treaty suggests that there is … a growing – and bipartisan – appreciation that America's present vulnerability to ballistic missile attack is as reckless as it is absurd." From "Shifted trajectory on anti-missile defense," *Washington Times* 20 June 1995: 16.

[83]Public Law 104-106, 10 February 1996, National Defense Authorization Act for Fiscal Year 1996, http://lcweb2.loc.gov/law/usa/us040106.pdf.

[84]Senator Kay Bailey Hutchison (R-TX) quoted in "Top Republicans offer Defend America Act of 1996," *BMD Monitor* 5 April 1996. S. 1635 and H.R. 3144 – To establish a United States policy for the deployment of a national missile defense system, and for other purposes, 21 March 1996, 104th Congress 2nd Session.

[85]Greg Coines, "Senior NSC Official Blasts Dole-Gingrich 'Defend America' Bill," *Defense Daily* 9 May 1996.

[86]Secretary of Defense William Perry in a speech at George Washington University quoted in Bill Gertz, "Perry: Missile defense unnecessary," *Washington Times* 26 April 1996: 6.

[87]If it was determined that a threat had not emerged by 2000, the program included provisions to continue to advance technology and add new elements.

[88]See for example, Rep. Curt Weldon (R-PA) Dale Eisman, "Bill Introduced in Congress to Make Anti-Missile Shield A National Policy." *The Virginian-Pilot* 6 August 1998.

[89]In 1998, for example, Chairman of the Joint Chiefs General Henry Shelton expressed some doubt about the technical capabilities of the system stating: "If the administration changed its policy tomorrow, the Joint Chief's position would remain the same because we don't think we need to pour a lot of money into something that doesn't work right this minute." Quoted in Erin Q. Winograd, "Shelton Defends Stance On NMD, Labeling Technology 'Just Not There' Yet," *Inside the Army* 14 September 1998: 1.

[90]A risk reduction effort and funding cuts in previous fiscal years delayed the initial fielding to 2005. The threat issue will be addressed in a separate section.

[91]Public Law 106-38 at http://thomas.loc.gov. This measure was criticized both by American allies in Europe as well as the leaders of Russia and China.

[92]General Accounting Office, "Missile Defense: Status of the National Missile Defense Program," GAO/NSIAD-00-131, Mary 2000.

[93]National Missile Defense Independent Review Team, "Executive Summary," 13 June 2000. The report was written by General Larry Welch (USAF-Ret.) of the Institute for Defense Analyses. The C-1 threat is defined as a threat involving a few enemy missiles and simple countermeasures. The Capability 2 and 3 systems would defend against an increased number of missiles and more sophisticated countermeasures.

[94]Speech by President William J. Clinton, Georgetown University, 1 September 2000, at Carnegie Non-Proliferation Project – http://www.ceip.org/files/projects/npp/resources/Presidentdelaysnmd.htm. According to a White House Fact Sheet entitled "White House on National Missile Defense" issued on 1 September 2000, Clinton based his decision upon "the threat, the cost, technical feasibility and the impact overall on our national security." http://www.usinfo.state.gove/topical/pol/arms/stories/00090101.htm.

[95]Emerging Missile Threats to North America During the Next 15 Years, Statement for the Record by Richard N. Cooper, Chairman, National Intelligence Council for Hearings of the 28 February 1996, House National Security Committee, Hearings on Ballistic Missile Defense. Cooper's testimony which was prepared, but not delivered can be located at http://www.ceip.org/files/Publications/NIE1995Cooper.asp?p=8&from=pubtype.

[96]The report stated that these "wild cards" could flair up and produce a threat within the 15 year window.

[97]Among the flaws identified in the report were the failure to address the economic conditions in Russia which could "provide incentives that increase the risk of leakage of hardware and expertise that could help governments aspiring to develop ballistic missiles, cruise missiles, and weapons of mass destruction." Similarly, Gates stated that NIE 95-19 "too easily dismisses missile scenarios alternative to an indigenously developed and launched intercontinental ballistic missile by countries hostile to the United States, alternatives such as a land-attack cruise missile." Robert Gates, Intelligence Analysis on the Long-Range Missile Threat to the United States, 4 December 1996, Senate Select Committee on Intelligence, http://www.fas.org/irp/congress/1996_hr/s961204p.htm.

[98]All direct quotes are from "Executive Summary of the Report of the Commission to Assess the Ballistic Missile Threat to the United States, 15 July 1998," http://www.house.gov/hasc/testimony/105thcongress/BMThreat.htm.

[99]Quoted in Eric Rosenberg, "Reagan's 'Star Wars' Vision Finds Life Under Clinton," *Legislative Focus* 13 May 1999: 26. In October 2002, the government of North Korea admitted that they were pursing a nuclear weapons program, in violation of their 1994 agreement with Washington.

[100]According to the National Air Intelligence Center, Iran is pursuing, with foreign assistance, an "ambitious ballistic missile development program" and could have a missile capable of reaching the United States before 2015. NAIC, "Ballistic and Cruise Missile Threat," Wright-Patterson AFB, Ohio, September 2000.

[101]Report of the Panel on Reducing Risk in Ballistic Missile Defense Flight Test Programs, 21 February 1998, http://www.fas.org/spp/starwars/program/welch/welch-2.htm.

[102]Briefing report located at http://www.acq.osd.mil/bmdo/bmdolink/pdf/welch.pdf.

[103]Memorandum from Paul G. Kaminski for Director, Ballistic Missile Defense Organization, Subject: National Missile Defense Joint Program Office, 9 April 1996.

[104]The NMD Program Elements were located within the U.S. Army Space and Strategic Defense Command and the PEO for Air and Missile Defense in Huntsville, Alabama; the U.S. Air Force Electronic Systems Command, Hanscom AFB, Massachusetts; the U.S. Air Force Space and Missile System Center, Los Angeles AFB, California; and the Joint National Test Facility, Colorado Springs, Colorado. Department of Defense, Memorandum for Correspondents, Memorandum: No. 049-M, 3 April 1997.

[105]U.S. Space Command held the capstone requirement document authority of all NMD systems.

[106]The BMDO remained the Ballistic Missile Defense Acquisition Executive. Memorandum for Secretary of the Army, Director, BMDO, Subject: Designation of Land-Based National Missile Defense System Lead Service, 15 November 1999.

[107]The proposed NMD unit or units would be composed of National Guardsmen and contractors.

[108]Memorandum from Mr. Kenneth J. Oscar, Acting Army Acquisition Executive to Commanders USASMDC, Acting PEO-AMD and Commander U.S. Army Aviation and Missile Command, Subject: Program Realignments, 6 September 2001.

[109]At the same time, the Army sought to restructure the PEOs "with a single integrated commodity focus." To this end, the Lower Tier Program (composed of Patriot and Medium Extended Air Defense System (MEADS)) left BMDO for the PEO-AMD; the Short Range Air Defense (SHORAD) Project Office transferred from AMCOM to PEO-AMD. The new Lower Tier organization was designed to "streamline the management of lower tier systems" and "take maximum advantage from lessons learned from our legacy systems to ensure that interim and objective lower tier systems meet operational requirements at reduced cost."

[110]Memorandum from Secretary of Defense Donald Rumsfeld, Subject: Missile Defense Program Direction, 2 January 2002, with attachment entitled "Missile Defense Program Direction".

[111]The first priority is to defend the United States, deployed forces, allies and friends from ballistic missile attack. The second calls for a Ballistic Missile Defense System (BMDS) that employs layers defenses to intercept missiles at all phases of their flight against all ranges of threats. The third priority is to enable the Services to field elements of the overall BMDS as soon as practicable. The fourth priority is to develop and test technologies, use prototype and test assets to provide early capability, if necessary, and improve the effectiveness of deployed capability by inserting new technologies as they become available or when the new threat warrants an accelerated capability.

[112]Remarks by the President to Students and Faculty at National Defense University, Washington, D.C., 1 May 2001, http://www.whitehouse.gov/news/releases/2001/05/20010501-10.html. During the same speech, the President discussed significant cuts, possibly unilateral cuts, to the American nuclear arsenal.

[113]Remarks by the President on National Missile Defense, 13 December 2001, http://www.whitehouse.gov/news/releases/2001/12/20011213-4.html.

[114]The two nations formalized this agreement on 24 May 2002 with the Treaty of Moscow. Essentially both nations would reduce the number of strategic nuclear weapons to the stated amounts by 31 December 2012, with no restraints upon the number of short-range nuclear missiles, bombers, missiles or submarines. Excerpt from Under Secretary for Arms Control and International Security John Bolton's Remarks to the Fourth RUSI Missile Defense Conference in London, 18 November 2002. The RUSI is the Royal United Services Institute. President Putin submitted the treaty to the State Dumas on 7 December 2002.

[115]In response to this decision, on 14 June 2002 Russia formally withdrew from the START II nuclear arms treaty. The Russian parliament had ratified the agreement in 2000, but tied START II to the preservation of the ABM. The two-thirds reduction in nuclear arsenals agreed to in START II, was reiterated in the Treaty of Moscow signed by Presidents Bush and Putin in May 2002. Note: On 5 May 1997, Lieutenant General Eric Shinseki, DA DCSOPS, had designated the Headquarters, USASSDC as Army Implementing Agent for the Strategic Arms Reduction Treaty (START) and START II implementation.

[116]White House Press Release, Statement by the President, 13 June 2002, located at http://www.whitehouse.gov/news/releases/2002/06/20020613-9.html.

[117]"President Bush Delivers Graduation Speech at West Point," Remarks by the President at 2002 Graduation Exercise of the United States Military Academy, 1 June 2002, located at http://www.whitehouse.gove/news/releases/2002/06/20020601-3.html.

[118]"President Announces Progress in Missile Defense Capabilities," Statement by the President, White House Press Office, 17 December 2002.

[119]News Release from the U.S. Department of Defense, "Missile Defense Operations Announcement," 17 December 2002.

[120]Associated Press, "Missile Defense Plan Draws Some Fire," *The Guardian* 18 December 2002.

[121]Transcript of the Joint Media Availability - Secretary Rumsfeld and Secretary General Robertson, 8 March 2001, http://www.defenselink.mil/news/Mar2001/t03082001_t308sd2a.html.

[122]Developed by the Air Force, these UHF, phased-array radar systems are part of the Ballistic Missile Early Warning System and PAVE PAWS.

[123]Support for the missile defense plan has been a controversial issue for American allies. Australia had already expressed its support for the NMD program in 2001.

[124]Institute for Foreign Policy Analysis, *National Missile Defense: Policy Issues and Technological Capabilities*, July 2000. This work, also available at www.ifpa.org provides a detailed description of the various elements of the GMD system.

[125]"Raytheon Tracks Satellite with Prototype Radar," *Space Daily* 18 September 1998, http://www.spacedaily.com/news/radar-98a.html.

[126]On 14 October 2002, the Missile Defense Agency incorporated a SPY-1 radar system aboard a U.S. Navy Aegis destroyer, the *USS John Paul Jones*, into the GMD intercept test. Participation by the sea-based system was previously prohibited under the ABM Treaty. The SPY-1 radar collected data but was not integrated into the test.

[127]The sites under consideration are in Alaska - Clear Air Station, Eareckson Air Station, Eielson AFB, Fort Greely, the Yukon Maneuver Area at Fort Wainwright and the western Aleutians; North Dakota – Grand Forks AFB, Minot AFB, Missile Alert Facility ECHO (near Hampden), and the SRMSC MSR site.

[128]A flyby test is designed to assess an interceptor's on-board sensor. Launched by a booster, the sensor passes by the target collecting data on the target package, discriminating the warhead from decoys. No intercepts are attempted.

[129]Ground-based Midcourse Defense Segment Exoatmospheric Kill Vehicle Fact Sheet released by Raytheon, 2001. The EKV weighs 121 pounds and is 55 inches in length and about 24 inches in diameter.

[130]A surrogate booster was used during the testing phase. Two competing designs for a three-stage, solid-rocket booster are under investigation.

[131]Despite these successes, opponents criticized the decoys used in the EKV test program. Although tests incorporated increasingly more complex countermeasures, critics held that the decoys did not reflect obstacles faced in an intercept. Some simply argued that discrimination technology simply would not work. In contrast, a recent report by the Union of Concerned Scientists stated: "While using such decoys may be appropriate for early stages of testing, the Pentagon should make clear that these tests do not provide a meaningful test of discrimination that is relevant to real-world situations." Quoted in Mike Nartker, "U.S. Plans: Activist Group Provides More Details on Decoys Used in Intercept Test," *Global Security Newswire*. In September 2000, Philip Coyle, Director of Operational Testing and Evaluation presented a series of initiatives to enhance the test flights. Paul Mann, "Next President Faces Missile Defense Knot," *Aviation Week and Space Technology* 18 September 2000: 27.

[132]Quoted in Bill Gertz, "Single-site missile defense leaves Alaska, Hawaii naked," *Washington Times* 9 May 1997.

[133]Article One of the ABM Treaty prevents the deployed system from defending the entire nation.

[134]The drawbacks to a deployment in Alaska are its decreased ability to defend against attacks from southern locations or to protect Eastern states from a launch by Libya. Michael Sirak, "U.S. Wrestles With Location, Number of NMD Sites," *Inside Missile Defense* 7 April 1999.

[135]Three sites were under review for the missile base - the Yukon Training Area, Eielson AFB, Clear Air Station and Fort Greely. Eareckson Air Station on Shemya Island was the proposed site for the XBR.

[136]Fort Greely was selected in part because it contained much of the infrastructure needed to support the test bed or a deployment. Fort Greely was closed as part of the 1995 Base Realignment and Closure decisions. On 1 October 2002, Fort Greely, Alaska, officially transferred to the USASMDC.

Conclusion

The Army's roles and missions in space and missile defense stem from its oldest mission: protecting American territory from foreign attack and invasion. While the means of performing the mission have changed from building and manning coastal defense forts in the 18th and 19th centuries, to building and manning antiaircraft and air defenses in the 20th and 21st centuries, the intent has remained unchanged. At the end of World War II, two new weapons, the atomic bomb and the guided missile, complicated this mission, presenting the Army with unprecedented technological challenges. These two new factors in the national defense equation led the Army to continue its investigations and experiments with missiles and space-based communications and sensor technologies to field new weapons systems.

By the mid-1950s, under the threat of Soviet nuclear attack, the Army and one of its major missile contractors concluded that ballistic missile defense was both technologically feasible and affordable. The Army's early efforts in developing guided missile technology resulted in the first successful American missile flights, earth-orbiting satellites and the first experimental communications, meteorological and reconnaissance satellites. The synergy of the Army's space and missile defense efforts was broken when the Eisenhower Administration established NASA, civilianizing space exploration and reallocating military space missions among the armed services. While retaining proponency for ballistic missile defense, the Army was stripped of its space assets and responsibilities.

Forced out of space, the Army continued to make technological progress in pursuit of ballistic missile defense, culminating in the deployment of the only BMD system in the western world. As it developed the various missile defense systems (NIKE-ZEUS, NIKE-X, SAFEGUARD, SENTRY, *et al*) the Army missile defense organizations evolved into entities that combined research and development, testing, evaluation, and acquisition functions in one place. Slowly and with great difficulty, Army missile defense organizations became functions-based organizations. While this was not the result of an overt design, it did promote collaboration and worked to eliminate duplication of effort and pointless competition for scarce resources.

As missile defense technology became more sophisticated with the advances in computers and sensors, the Army began to experiment with lasers, and particle beams as well as kinetic methods to destroy enemy satellites and guided missile warheads. President Reagan's 1982 National Space Policy, his 1983 Strategic Defense Initiative as well as the invasion of Grenada refocused the Army's attention on space and missile defense. When this renewed attention was combined with internal debates about doctrine and the publication of AirLand Battle Doctrine in 1982, the Army readied itself to reclaim its role in space and to link it to its missile defense mission.

It was not until that moment that the Army determined the ground commander's needs required it to return to space. As AirLand Battle doctrine developed, the entire conception of the battlefield expanded. The Army now concerned itself with the Deep Battle (with a need to see and strike deeply) and with the Rear Battle (with its own needs for expanded command and control). Space-related activities offered the ground commander unique platforms for observation, positioning and communications over a greatly expanded area of concern; that of the operational level battlefield. Missile defense offered a new method of force protection against a new and growing threat.

This doctrine meshed with the Army's long-standing interest in manned space flight. Initially excluded from the military astronaut program by the requirements for experience as a jet pilot and for advanced degrees in the natural sciences, medicine or engineering, or equivalent experience, the Army's first astronaut candidate was chosen in 1978 as a Space Shuttle astronaut. As part of the Military Man in Space program the Army would perform experiments that enhanced its war fighting capabilities. As participation increased, the Army Astronaut Detachment was formed at the Johnson Space Center.

In a deliberate way, the Army began to centralize its space programs and assets and over eight years brought them together in a unified command. In 1992, the Army Strategic Defense Command became the Army Space and Strategic Defense Command; one of its missions was to provide an Army focal point for space and missile defense matters. With the formation of this command, the Army entered the concluding phase of centralizing research and development of space and missile defense assets to benefit the warfighting soldier in the field. In 1993, the Army Space Technology Research Office was transferred from the Communications-Electronics Command to the Space and Strategic Defense Command. This office managed near and possible far-term space research and development programs and provided a developer focus within the Army and with outside agencies and was renamed the Space Applications Technology Program. In 1994, the Army Space Program Office was transferred to the Space and Strategic Defense Command. This office, formed in 1973, has responsibility for the Army's share of the Tactical Exploitation of National Capabilities Program. Thus by 1994, the Army had place its space and missile defense assets in one command.

As the Army Space and Strategic Defense Command moved to centralize support of space and missile defense programs, the reorganizations in the wake of the Cold War and the Gulf War threatened to break apart the command and distribute its functions throughout the Army. In a 1996 memorandum, the Vice Chief of Staff of the Army General Ronald H. Griffith overruled the finding of the "HQDA Redesign Functional Area Assessment." The memo noted "that although the Redesign Functional Area Assessment recommended realignment...with the Training and Doctrine Command (TRADOC)....a number of factors have caused us to reconsider this recommendation."[1]

General Griffith noted that the command carried "out responsibilities in scope and magnitude unlike other Army organizations." It had "a significant operational mission in support of the Warfighting CINCs" because it was the Army component of the U.S. Space Command. In its role as "an executing agent for the Ballistic Missile Defense Organization" it has "a complex array of funding and tasking responsibilities," and "is directly responsible to the Army Acquisition Executive" regarding acquisition matters. In the course of "accomplishing these missions, the command works with numerous non-departmental agencies, the OSD staff and other military services."

It was clear to the Vice Chief of Staff that "these functions do not integrate well into any of the current major commands." Additionally, he acknowledged "a growing need exists for a 'proponent-like' Army facilitator to integrate space and missile defense solutions within the Army and act as the Army advocate in Joint Warfighting forums." Therefore, he noted that "TRADOC, the architect of the future Army" and Space and Strategic Defense Command, "the technical, experimental and operational expert of space and missile defense" will join together to "leverage each other's capabilities to guide the development of Army/Joint space and missile defense capabilities to best meet our Army and Force XXI goals." Following the guidance in the memo, the command and TRADOC developed a memorandum of agreement documenting the "relationship between the two commands" and addressed "the requirements linkage between the two activities." This led directly to the memorandum of agreement that established the Space and Missile Defense Battle Lab—the only battle lab outside TRADOC.

The Army's efforts in space and missile defense ran on parallel paths for many years. As they became intertwined, the Army always viewed space as a medium of operations; operations can occur to, from and in space. In the process of forming its newest major command, the Army moved to centralize its space and missile defense activities. Over the years, as the Space and Missile Defense Command evolved, it retained its emphasis on a functional organization structure that watched over the development of various systems from their earliest stages to fielding. Its functional structure may serve as a model in the Department of Defense, for changes in the way the armed forces are organized are rapidly taking shape. The drive to reform and rationalize the armed forces is being driven by the national emergency engendered by the terrorist attacks against the World Trade Center and the Pentagon on 11 September 2001.

In 2002, U.S. Space Command and U.S. Strategic Command were merged to form a new U.S. Strategic Command to eliminate redundancies in the command process and streamline the decision making process. According to Secretary of Defense Donald Rumsfeld, the change was made because the missions of both commands had converged to the extent that their merger became practical. The new merged command would be responsible for early warning of and defense against both missile attack and long range conventional attack. In addition, the new command controls American nuclear forces, military space operations, computer network operations, strategic warning and global planning. The USASMDC is the Army Service Component Command of the new unified command. In 2003, the president signed Unified Command Plan Change 2, which assigned global strike, information operations, space C^4ISR and integrated missile defense responsibilities to the U.S. Strategic Command.

Since its inception, USASMDC and its predecessor organizations have dealt with space and missile defense technology. Its place in the U.S. Strategic Command presents it and the Army with a new set of technological and organizational challenges. Throughout its history, the USASMDC evolved to meet the needs of our nation, warfighters and allies. The process continues today and as Lieutenant General Joseph Cosumano, USASMDC Commanding General, recently observed, "There is a lot of change out there on the horizon."[2]

End Notes

[1]Memorandum for Space and Strategic Defense Command, Subject: Realignment of the Space and Strategic Defense Command, 12 July 1996. All direct quotations from this document.
[2]General Cosumano speaking at a U.S. Army Space and Missile Defense Association luncheon in Huntsville on 24 January 2003, quoted in Debra Valine, "Space and Missile Defense Remains a Command," *The Eagle* February 2003: 1.

Bibliography

Primary Sources

Unpublished Material

U. S. Army Space and Missile Defense Command Archives: This study rests on the material held in these archives. The collections used include the current command and its predecessor organizations as enumerated in the notes.

Center for Army Lessons Learned Restricted Database. The material in this archive is available to registered users. All of the material used in this study is unclassified.
Central Command, Somalia, Operation Restore Hope (1992), Operation Continue Hope (1993).
10th Mountain Division (LI). "Somalia After Action Report for the Secretary of Defense, Complete Lessons Learned."
10th Mountain Division (LI). "After Action Report: Executive Summary."

Combined Arms and Fort Leavenworth Archives. Held in the Special Collections Department, Combined Army Research Library, Fort Leavenworth, Kansas. The documents used are from the Combined Arms Combat Developments Agency (CACDA) collection.
"Briefing: WRASSE."
The Army Position and Navigation Master Plan (May 1986).

Memorandum from Jacques Gansler to Secretary of the Army and Director, BMDO, 15 November 1999, Subject: Designation of Land-Based National Missile Defense System Land Service.

Memorandum from Kenneth J. Oscar to Commanders U.S. Army Space and Missile Defense Command, Acting Program Executive Officer Air and Missile Defense and U.S. Army Aviation and Missile Command, Subject: Program Realignments, 6 September 2001.

Memorandum from Secretary of Defense Donald Rumsfeld, Subject: Missile Defense Program Direction, 2 January 2002 with attachment "Missile Defense Program Direction."

Memorandum of Agreement between the United States Army Training and Doctrine Command and the United States Army Space and Strategic Defense Command signed 11 May 1995.

Memorandum of Agreement between the United States Army Training and Doctrine Command and the United States Army Space and Strategic Defense Command signed 18 February 1997.

Published Material

Gates, Robert. Intelligence Analysis on the Long-Range Missile Threat to the United States, 4 December 1996, Senate Select Committee on Intelligence. http://www.fas.org/irp/congress/1996_hr/s961204p.htm

Harvey, Jan V., Colonel and Colonel (ret) Alwyn H. King. "Space: The Army's New High Ground," *Military Review* 65.7 (July 1985): 38-51.

Memorandum from Henry Kissinger to President Nixon, 5 March 1969, Subject: Modified Sentinel System with attachment "Issues Concerning ABM Deployment." Declassified on 7 July 1998 by the National Archives. http://www.gwu.edu/~nsarchiv/NSAEBB/NSAEBB36/18-04.htm.

National Air Intelligence Center, *Ballistic and Cruise Missile Threat*. Wright Patterson Air Force Base, September 2000.

National Security Archive. *National Security Archive Electronic Briefing Book No. 26*: "The Chinese Nuclear Weapons Program: Problems of Intelligence Collection and Analysis, 1964-1972," edited by William Burr (31 March 2000), http://www.gwu.edu/~nsarchiv/NSAEBB/NSAEBB26/index.html.

National Defense Authorization Act for Fiscal Year 1996, Subtitle C – Ballistic Missile Defense Act of 1995, H.R. 1530. http://www.fas.org/spp/starwars/congress/1995/h950614i.htm

National Missile Defense Independent Review Team. "Executive Summary," 13 June 2000.

Office of Technology Assessment. *MX Missile Basing*. Washington, D.C.: Government Printing Office, 1981.

Ovitt, Spalding W. and L.G. Bowers, editors. *The Balloon Section of the American Expeditionary Forces*. New Haven: Tuttle, Morehouse and Taylor, 1919.

President Ronald Reagan, 13 July 1985. "Radio Address to the Nation on the Strategic Defense Initiative." http://www.reagan.utexas.edu/resource/speeches/1985/71385a.htm.

Public Law 104-106, 10 February 1996. National Defense Authorization Act for Fiscal Year 1996. http://lcweb2.loc.gov/law/usa/us040106.pdf.

Public Law 106-38. http://thomas.loc.gov

Remarks by the President to Students and Faculty at National Defense University, Washington, D.C., 1 May 2001. http://www.whitehouse.gov/news/releases/2001/05/20010501-10.html.

Remarks by the President on National Missile Defense, 13 December 2001. http://www.whitehouse.gov/news/releases/2001/12/20011213-4.html.

Remarks by the President at 2002 Graduation Exercise of the United States Military Academy, 1 June 2002. http://www.whitehouse.gov/news/releases/2002/06/20020601-3.html.

Republican Contract with America, 1995. http://www.house.gov/Contract/CONTRACT.html

Report of the Panel on Reducing Risk in Ballistic Missile Defense Flight Test Programs, 21 February 1998. http://www.fas.org/spp/starwars/program/welch/welch-2.htm.

Roe, Linas A. Major and Major Douglas H. Wise, "Space Power is Land Power: The Army's Role in Space," *Military Review* 66.1 (January 1986): 4-17.

Speech by President William J. Clinton, at Georgetown University, on National Missile Defense Deployment, 1 September 2000. http://www.ceip.org/files/projects/npp/resources/Presidentdelaysnmd.htm

State of the Union Address by President George H.W. Bush, 29 January 1991. http://www.cspan.org/executive/transcript.asp?cat=current_event&code=bush_admin&year=1991.

Statement by the President, "President Announces Progress in Missile Defense Capabilities." White House Press Office, 17 December 2002.

"Text of Reagan and Pentagon Statements on MX Missile Proposal," *New York Times*, 23 November 1982, A-14.

Transcript of the Joint Media Availability – Secretary of Defense Rumsfeld and Secretary-General Robertson, 8 March 2001. http://www.defenselink.mil/news/Mar2001/t03082001_t308sd2a.html.

United States. Department of Defense. Memorandum for Correspondents. Memorandum No 049-M, 3 April 1997.

United States Congress. House of Representatives. Armed Services Committee. Executive Summary of the Report of the Commission to Assess the Ballistic Missile Threat to the United States, July 15, 1998. http://www.house.gov/hasc/testimony/105thcongress/BMThreat.htm.

United States Congress. House of Representatives. Armed Services Committee. Report of the Commission to Assess the Ballistic Missile Threat to the United States. Executive Summary. July 15, 1998. http://www.house.gov/hasc/testimony/105thcongress/BMThreat.htm

United States Congress. House of Representatives. National Security Committee. Statement for the Record by Richard N. Cooper, Chairman, National Intelligence Council, *Emerging Missile Threats to North America during the Next 15 Years*, for 28 February 1996 Hearings. (Cooper's testimony, which was prepared, but not delivered.) http://www.ceip.org/files/Publications/NIE1995Cooper.asp?p=8&from=pubtype

United States. Department of State. *Foreign Relations of the United States 1964-1968.* Volume XIV: *Soviet Union*, edited by David C. Humphrey and Charles S. Sampson General Editor: David S. Patterson, Department of State Publication 10779. http://www.state.gov/r/pa/ho/frus/johnsonlb/xiv/1400.htm.

United States Army. *Army Space Master Plan.* Washington, D.C.: Department of the Army, 2000.

United States Army. Army Ballistic Missile Agency, Redstone Arsenal, Alabama. *Development Proposal for Project Adam.* ABMA Report No. D-TR-1-58, 17 April 1958.

United States Army. Space and Missile Defense Command. *Directed Energy Master Plan.* Huntsville: Space and Missile Defense Command, January 2000.

United States Army. Training and Doctrine Command. *Concept for Space Operations in Support of the Objective Force.* Final Coordination Draft. Fort Monroe: U.S. Army Training and Doctrine Command, 22 February 2003.

Welch Report Briefing, 1999. http://www.acq.osd.mil/bmdo/bmdolink/pdf/welch.pf.

White House Fact Sheet. "White House on National Missile Defense," 1 September 2000.

White House Press Release. Statement by the President, 13 June 2002, http://www.whitehouse.gov/news/releases/2002/06/20020613-9.html.

Secondary Sources

Published Books and Articles

"Airborne surveillance testbed successfully tracks targets," *The Redstone Rocket*, 26 June 1992: 4.

Aitkin, Hugh G.J. *Syntony and Spark—The Origins of Radio*. Princeton: Princeton University Press, 1985.

Allison, David K. *New Eye for the Navy: The Origin of Radar at the Naval Research Laboratory*. Washington, D.C.: Naval Research Laboratory, 1981.

Anderson, Edward G., III, Lieutenant General. "The New High Ground," *Armed Forces Journal International*, October 1997: 66.

"Appropriations Slice $30 Million Intended for FY-95 THAAD Testing," *Inside the Army,* 8 August 1994.

"Army Happy With Use of Cruise Missile Defense Aerostat in Roving Sands," *Inside the Army*, 1 July 1996.

"Army, OSD Confident THAAD EMD Design will Fix Missile's Problems," *Inside Missile Defense*, 20 October 1999.

"Army recommends ERINT to DAB for Patriot Improvement," *BMD Monitor*, 25 February 1994.

"Army Weaponry and Equipment," *Army* 44 (October 1984): 296-297.

Asker, James. "Rockwell Selected as Sole Contractor for $100-Million ASAT Design Effort," *Aviation Week and Space Technology*, 23 July 1990.

Associated Press. "Missile Defense Plan Draws Some Fire," *The Guardian*, 18 December 2002.

Atkinson, Joseph D. and Jay M. Shafritz. *The Real Stuff: A History of NASA's Astronaut Recruitment Policy*. New York: Praeger, 1985.

Baker, Caleb. "SDIO, Army Ends Dispute Over Theater Defense," *Defense News*, 25 November 1991: 4, 29.

Ball Aerospace. "Optical Aircraft Measurement Program." http://www.ball.com/aerospace/oamp.html.

Ballistic Missile Defense Organization. *Ballistic Missile Defense Organization Annual Historical Review Fiscal Year 1978, 1 October 1977 – 30 September 1978.* Huntsville: Ballistic Defense Organization, n.d.

_____. *Ballistic Missile Defense Organization Annual Historical Summary, FY81.* Huntsville: Ballistic Missile Defense Organization, n.d.

_____. *Ballistic Missile Defense Organization Annual Historical Review Fiscal Year 1986.* Huntsville: Ballistic Missile Defense Organization, n.d.

_____. "National Missile Defense: An Overview (1993-2000)." http://www.acq.osd.mil/bmdolink/html/nmdhist.html.

Bates, Charles C. and John F. Fuller, *America's Weather Warriors, 1814-1985.* College Station: Texas A&M University Press, 1986.

Baucom, Donald R. *The Origins of SDI, 1944-1983.* Lawrence: University Press of Kansas, 1992.

_____. "Missile Defense Milestones 1944-2000," http://www.acq.osd.mil/bmdo/bmdolink/html/milstone.html.

Bell Laboratories. *ABM Research and Development at Bell Laboratories: Kwajalein Field Station.* Whippany, NJ: Bell Laboratories for U.S. Army Ballistic Missile Defense Systems Command, 1975.

_____. *ABM Research and Development at Bell Laboratories: Project History.* Whippany, NJ: Bell Labs for the U.S. Army Ballistic Missile Defense Systems Command, 1975.

Bilstein, Roger E. *Orders of Magnitude: A History of NACA and NASA, 1915-1990.* Washington, D.C.: National Aeronautics and Space Administration, Office of Management, Scientific and Technical Division, 1989.

"Blair, William S." *Dictionary of American Biography* (New York: Scribner, 1943-) supplement VII.

Boehm, Joshua with Craig Baker, Stanley Chan and Mel Sakazaki. *A History of United States National Security Space Management and Organization.* www.fas.org/spp/eprint/article03.html accessed on 3 July 2002.

Boffey, Philip. *Claiming the Heavens: The New York Times Complete Guide to the Star Wars Debate.* New York: Times Books, 1988.

Bosma, John and Richard Whelan. *Guide to the Strategic Defense Initiative.* Arlington: Military Space, 1985.

Botting, Douglas. *The Epic of Flight: The Giant Airships.* Alexandria: Time-Life Books, 1981.

Brewer, Cynthia. "TCMP born in Desert Storm," *The Eagle*, September/October 1996: 13.

Broad, William J. "White Sands to test laser," *El Paso Herald-Post*, 3 October 1997: A-1, A-2.

Browne, Malcolm W. "Weapon that Fights Missiles Could Alter World Defense Focus" *New York Times*, 4 December 1978: 1, 3.

"Budget Constraints put ERINT and PATRIOT Interceptors on Collision Course," *Inside the Pentagon*, 26 November 1992.

Buderi, Robert. *The Invention that Changed the World: How a Small Group of Radar Pioneers Won the Second World War and Launched a Technological Revolution.* New York: Simon and Schuster, 1996.

Burkey, Martin. "Defense Official Urges Planning for Space Warfare," *Huntsville Times*, 9 February 1998: D-2.

Burkey, Martin. "ERINT missile test rates a success," *The Huntsville Times*, 16 February 1994: B-2.

Burrows, William E. *Deep Black: Space Espionage and National Security.* New York: Random House, 1986.

Burrows, William E. *This New Ocean: The Story of the Space Age.* New York: Random House, 1998.

Butrica, Andrew J., editor. *Beyond The Ionosphere: Fifty Years of Satellite Communication.* Washington, D.C.: National Aeronautics and Space Administration, NASA History Office, 1997.

Cannon, Lou. "Reagan's Big Idea. (Ronald Reagan's Strategic Defense Initiative)" *National Review*, 22 February 1999. http://www.findarticles.com/cf_natrvw/m!282/199_Feb_22/53703734/print.jhtml

Capaccio, Tony. "Lockheed Defensive Missile Said Ready As U.S. Eyes Iraq Options," *Bloomberg.com*, 14 August 2002.

Carlson, Christina M. and Robert Lyon. *Last Line of Defense: NIKE Missile Sites in Illinois.* Denver: U. S. National Park Service, 1996.

Centre for Defense and International Security Studies. *Ballistic Missiles: The German V-2 Campaign, 1944-1945.* http://www.cdiss.org/v2.htm.

_____. *Countering the V-1 & V-2 in WWII.* http://www.cdiss.org/scdnt2.htm.

_____. *Cruise Missiles: A Brief History: 1900-1945.* http://www.cdiss.org/cmhist.htm.

Chandler, Charles DeForest and Frank P. Lahm. *How Our Army Grew Wings: Airmen and Aircraft before 1914.* New York: The Ronald Press Company, 1943.

Cirincione, Joseph. *The Persistence of the Missile Defense Illusion*, http://www.ceip.org/Programs/npp/bmdhistoyr.htm.

Clark, Ronald. *The Man Who Broke Purple: The Life of Colonel William F. Friedman, Who Deciphered the Japanese Code in World War II.* Boston: Little, Brown and Co., 1977.

Coines, Greg. "Senior NSC Official Blasts Dole-Gingrich "Defend America" Bill," *Defense Daily*, 9 May 1996.

Collier, Joe Guy. "THAAD 'smokes split-off warhead," *The Huntsville Times*, 2 August 1999: A-1.

Collins, John M. *U.S.-Soviet Military Balance, 1980-1985.* Washington, D.C.: Pergamon-Brassey's, 1985.

Compton, David. *Where No Man Has Gone Before: A History of Apollo Lunar Exploration Missions.* Washington, D.C.: National Aeronautics and Space Administration, Office of Management, Scientific and Technical Information Division, 1989.

Compton, Jeff. "NMD Target vehicle flies demonstration," *The Eagle*, October 2000: 22.

Costello, John, Lieutenant General. "Space and Missile Defense," *Army*, December 1998:12.

Covault, Craig. "USAF Missile Warning Satellites Providing 90 Sec. Scud Attack Alert," *Aviation Week and Space Technology,* 134.3 (January 21, 1991): 60.

Currie-McDaniel, Ruth and Claus Martel. *The U.S. Army Strategic Defense Command, Its History and Role in the Strategic Defense Initiative.* 3rd edition. Huntsville: U. S. Army Strategic Defense Command, 1989.

"DAB Selects Ground Based Mode for Initial KEW ASAT," *Defense Daily*, 15 December 1989, 405.

"Data 'Drop-Out,' Mapping Errors are Possible Causes of ERINT Failure," *Inside the Army*, 28 June 1993.

Davis, Brett. "Missile System needs more tests, report says," *The Huntsville Times*, 10 July 1996: B-2.

_____ and Martin Burkey. "THAAD thud may be $15M penalty," *The Huntsville Times*, 10 July 1998: A-11.

Davis, Connie M. "Aerostat Rises to the top at Roving Sands '96," *The Eagle*, July 1996: 11.

Donnelly, John. "Perry Puts U.S.-Israeli Tactical Laser on Fast Track," *Defense Week*, 20 May 1996: 1.

Dupont, Daniel G. "Army to Propose ACTD for Low-Cost, Tactical High Energy Laser System," *Inside the Army*, December 1995.

_____. "Army Space Command Sent to Train Ft. Drum Soldiers: Somalia Operation May Be Army's Largest Use of Space Technology to Date," *Inside the Army*, 14 December 1992: 3.

_____. "Pentagon Decides to Deploy Patriot PAC-3 Missiles to Persian Gulf Area," *Inside the Army,* 16 February 1998: 1.

_____. "SMDC Eyes Tactical Laser for Korea, Touts Directed Energy Development," *Inside the Army*, 3 May 1999: 1

_____. "SSDC Working to Create Battle Lab for Theater Missile Defense, Space," *Inside the Army*, 24 October 1994: 1.

_____ and Jeffrey Moag. "Lockheed Exceeds Its THAAD Cost Estimates by $83 Million," *Inside the Army* 15 August 1994.

_____. "While U.S., Israel Discus Tactical Laser Prototype, Army Looks to the Future," *Inside Missile Defense*, 16 May 1996.

Eisman, Dale. "Bill Introduced in Congress to Make Anti-Missile Shield A National Policy." *The Virginian-Pilot* 6 August 1998.

Engelhardt, Joseph P., compiler. *Desert Shield and Desert Storm: A Chronology and Troop List for the 1990-1991 Persian Gulf Crisis*. Carlisle Barracks: Strategic Studies Institute, U.S. Army War College, 25 March 1991.

"ERINT Hits Air-Breathing Target Drone," *Aviation Week and Space Technology*, 13 June 1994: 26.

"ERINT Scores successful intercept of simulated TBM," *Inside the Pentagon*, 2 December 1993.

Evans, Donald W. *U.S. Army Use of Space-Based Systems.* www.globalsecurity.org/space/library/report/1990/EDW.htm, accessed on 4 February 2003.

"Extended range interceptor wins approval for Patriot," *Redstone Rocket*, 23 February 1994: 3.

Fagan, M.D., editor. *A History of Engineering and Science in the Bell System: National Service in War and Peace (1925-1975).* Whippany: Bell Telephone Laboratories, 1978.

Fantasia, LuAnne. "Carr's Crew Ready for Roving Sands '99," *The Eagle*, May 1999: 1.

_____. "Take THAAD!!," *The Eagle*, Special Edition 1999.

Federation of American Scientists. "Anti-Ballistic Missile Treaty Chronology." http://www.fas.org/nuke/control/abmt/chron.htm.

Friedman, Norman. *Desert Victory - The War for Kuwait.* Annapolis: Naval Institute Press, 1991 and http://www.pbs.org/wgbh/pages/frontline/gulf/weapons/scud.html.

Gaffney, Frank. "Shifted trajectory on anti-missile defense," *Washington Times*, 20 June 1995: 16.

Garamone, Jim. "SECDEF outlines need for national missile defense," *Kwajalein Hourglass*, 21 August 2001: 5.

Gerhardt, Igor D. "Space the Air Land Battle, *Army*, 40.6 (June 1990): 43-47.

Gertz, Bill. "Perry: Missile Defense Unnecessary," *Washington Times*, 26 April 1996: 6.

_____. "Shared satellite laser test weighed," *Washington Times*, 2 January 1998: 1.

_____. "Single-site missile defense leaves Alaska, Hawaii naked," *Washington Times*, 9 May 1997.

Glashow, Jason. "Hughes, Martin, Rockwell now in race: Contractors are Concerned GBI Downselect Could Yield Two Loses," *Inside the Army*, 3 January 1994.

Gormley, Dennis M. *Double Zero and Soviet Military Strategy: Implications for Western Security.* London: Jane's Publishing Company, 1988.

Graham, Bradley. "Missile Defense Program Changes Course," *The Washington Post*, 5 August 2002: A-6.

Green, Constance McLaughlin, Harry C. Thomson and Peter C. Roots. *The Ordnance Department: Planning Munitions for War* (*United States Army in World War II: The*

Technical Services). Washington, D.C.: U.S. Army Center of Military History, 1955, reprinted 1990.

"Ground-based Midcourse Defense Segment Exoatmospheric Kill Vehicle" Fact Sheet. Released by Raytheon, 2001.

Hall, R. Cargill. "The Origins of U.S. Space Policy: Eisenhower, open Skies and Freedom of Space," *Colloquy*, 14 (December 1993): 5-6, 19-24.

Hasenauer, Heike. "Army Takes the Lead in ASAT," *Soldiers,* August 1989.

_____. "The Army Astronaut Program," *Soldiers*, 45.2 (February 1990): 18.

Holzer, Robert. "Navy Panel May Speed Missile Tests" *Defense News*, 31 July – 6 August 1995: 38.

Hsu, Emily. "JLENS Officials Awaiting Army Decision on Homeland Defense Role," *Inside Missile Defense*, 26 June 2002: 1, 5, 6.

"Independent Study Uncovers Management Problems with THAAD Program," *Inside the Army* 26 December 1994.

Institute for Foreign Policy Analysis. *National Missile Defense: Policy Issues and Technological Capabilities*, July 2000. www.ifpa.org.

"Iraqis Still Receive Weather Data from US Satellites," *Aviation Week and Space Technology*, 134.4 (January 21, 1991): 26.

"Is Space the Pentagon's Final Battleground?" *Business Week*, 15 June 1998.

Jackson, Robert. *Airships* Garden City, N.Y.: Doubleday and Co., 1973.

Jarrell, Ruth and Mary T. Cagle. *History of the Plato Antimissile Missile System: 1952-1960.* Redstone Arsenal: U.S. Army Ordnance Missile Command, 1961.

Kagan, Frederic W. "Star Wars in Real Life: Political Limitations on Space Warfare," *Parameters*, Autumn 1998:112-130.

Kahn, David. *The Codebreakers: The Story of Secret Writing.* New York: Macmillan Company, 1967.

Kennett, Lee. *The History of Strategic Bombing.* New York: Scribners, 1982

Kiernan, Vincent. "Defense Panel Gives Green Light to SDI Ground-Based Radar," *Space News*, 30 July –5 August 1990: 17.

_____. "Lengthy Delay Hits Laser ASAT Work" *Space News*, 20 August 1990: 1.

_____. "SDI: Setbacks Expected – Critics Blast Program After Last Month's LEAP Failure" *Space News*, 13 July 1992: 3, 29.

Kitchens, James, III. *A History of the Huntsville Division – 15 October 1967 – 31 December 1976*. Huntsville: U. S. Army Corps of Engineers, 1978.

Langreth, Robert. "Sons of the Patriot," *Popular Science*, June 1993: 1005.

Larson, Melinda. "Kauai to Kwajalein fight support by KMR: STARS mission a stellar success," *Kwajalein Hourglass*, 28 August 1993: 1, 3.

Lebow, Eileen. *A Grandstand Seat: The American Balloon Service in World War I*. Westport: Praeger, 1998.

Lewin, Ronald. *ULTRA Goes to War*. New York: McGraw-Hill Book Co., 1978.

Lewis, Emanuel Raymond. *Seacoast Fortifications of the United States: An Introductory History*. Washington, D.C.: Smithsonian Institution Press, 1970.

Lewis, Tom. *Empire of the Air: The Men Who Made Radio*. New York: Edwin Burlingame Books, 1991.

Lockheed Missiles and Space. "Homing Overlay Experiment." http://lmms.external.lmco.com/newsbureau/photos/hoe1_caption.html.

Lonnquest, John and David Winkler. *To Defend and Deter: The Legacy of the United States Cold War Missile Program*. Rock Island: U.S. Army Construction Engineering Research Laboratories, 1996.

McDonnell Douglas Astronautics Company. *Ballistic Missile Defense: A History of Achievement*. Huntington Beach: McDonnell Douglas,1982.

McDougall, Walter A. *The Heavens and the Earth: A Political History of the Space Age*. New York: Basic Books, 1985.

McMahon, K. Scott. *Pursuit of the Shield*. New York: University Press of America, 1997.

Mann, Paul. "Next President Faces Missile Defense Knot," *Aviation Week and Space Technology*, 18 September 2000: 27.

"Missile Defense Plan Draws Some Fire," *The Guardian*, 18 December 2002.

Moeller, Stephen. "Vigilant and Invincible: United States Army Air Defense Command," *Air Defense Artillery*, May-June 1995.

"Missile target vehicle has successful test at White Sands Missile Range, N.M.," *Redstone Rocket*, 12 November 1997: 20.

Moore, Jack. "HELSTF puts Tularosa Basin into laser weapons age," *Alamogordo Daily News*, 8 March 1992: 2-E.

Mulholland, David. "Army Ties Balloon Radar to Missile Defense," *Defense News*, 17 May 1999: 1.

Nartker, Mike. "U.S. Plans: Activist Group Provides More Details on Decoys Used in Intercept Test," *Global Security Newswire*.

Naylor, Sean. "The Lessons of Anaconda," *New York Times on the Web*, 2 March 2003. http://www.nytimes.com/2003/03/02/opinion/02NAYL.html?pagewanted=all&position=top, accessed on 2 March 2003.

New Technology, Inc., *History of Ballistic Missile Defense Developments: A Synopsis*. Huntsville: BMDATC, 1983

Newell, Clayton R., Lieutenant Colonel. "The Army and Space," *Army*, 37.9 (September 1987): 59-61.

Nimmen, Jane van, Leonard C. Bruno, Linda N. Ezell, *NASA Historical Data Book*. 4 volumes. Washington, D.C.: Scientific and Technical Information Division, National Aeronautics and Space Administration, 1988-1994.

Opall, Barbara. "Arrow to Face Chemical Target Test," *Defense News*, 31 May – 6 June 1993: 30.

_____. "Missile funds shift to small, mobile systems," *Army Times*, 6 December 1993.

_____. "New THAAD Puts on Weight – Redesign of Larger Missile May Add Cost, Limit Availability," *Defense News*, 22-28 March 1993: 3, 28.

_____. "Strategic Accord Inhibits Advances in TMD Programs," *Defense News*, 4-10 October 1993.

_____ and Sharone Parnes. "Test Halts Cast Shadow on Arrow," *Defense News*, 19-25 July 1993: 4, 50.

Ordway, Frederick I., Mitchell R. Sharpe and Ronald C. Wakeford. "Project Horizon: An Early Study of a Lunar Outpost," *Acta Astronautica*, 17. 10:1105-1121.

Papp, Daniel S. "From Project Thumper to SDI: The Role of Ballistic Missile Defense in US Security Policy" *Aerospace Power Journal*, Winter 1987-1988. http://www.airpower.maxwell.af.mil/airchronicles/apj/apj87/papp.html.

Parker, Richard. "Lockheed told to overhaul THAAD," *The Huntsville Times*, 9 July 1998: A-1, A-9.

Payne, Keith. *Strategic Defense: "Star Wars" in Perspective.* Langham, MD: Hamilton Press, 1986.

Pearlstein, Steven. "The Missile Makers' Next Big Target," *Washington Post*, 5 August 1992: 81.

Perry, Walter, Bruce Pirnie, John Gordon, IV. *The Future of Warfare: Issues from the 1999 Cycle Army After Next Study Cycle*. Santa Monica: RAND Corporation, 2001, http://www.rand.org/publications/MR/MR1183/ accessed on 13 March 2002.

"Plan may foment administration battle with Senate: Russia Considers U.S. Plan to Clear ABM Treaty for THAAD," *Inside the Pentagon*, 9 December 1993.

Pogue, Forrest. *The Supreme Command.* (*United States Army in World War II: European Theater of Operations*). Washington D.C.: U.S. Army Center of Military History, 1954, 1989.

Public Affairs Office, White Sands Missile Range. "The Bumper Project." http://www.wsmr.army.mil/paopage/Pages/bump.htm, accessed on 30 January 2003.

_____. "The Corporal," http://www.wsmr.army.mil/paopage/Pages/Corppr.htm, accessed 31 January 2003.

_____. V-2 in America," http://www.wsmr.army.mil/paopage/Pages/V-2.htm, accessed on 30 January 2003.

"Race car technology puts THAAD team in winner's circle," *The Redstone Rocket*, 17 July 2002: 7.

Raines, Rebecca Robbins. *Getting the Message Through: A Branch History of the U.S. Army Signal Corps*. Washington, D.C.: U.S. Army Center of Military History, 1996.

"Raytheon Tracks Satellite with Prototype Radar," *Space Daily*, 18 September 1998, http://www.spacedaily.com/news/radar-98a.html.

Riebelling, Sandy. "THAAD System," *Redstone Rocket*, 5 May 2002.

Robinson, Clarence A., Jr. "Amy Pushes New Weapons Effort," *Aviation Week and Space Technology*, 16 October 1978.

_____. "Missile Engineering, Prototype Site Defense Construction Set," *Aviation Week & Space Technology*, 29 April 1974.

_____. "Soviets Grasping Strategic Lead," *Aviation Week & Space Technology*, 30 August 1976: 14-18.

_____. "Strategic Shifts – 3: Missile Defense Radar System Tests Set," *Aviation Week and Space Technology*, 20 September 1976.

Robblee, Paul A., Jr. "The Army's Stake in Emerging Space Technologies," *Parameters* 18.4 (December 1988): 113-119.

"Rockwell, Hughes get SSDC contract go-ahead," *Redstone Rocket*, 1 June 1994: 8.

Roosevelt, Ann. "Soldiers Help Save Money on THAAD Program," *Defense Week*, 26 June 2000: 3.

Rosenberg, Eric. "Reagan's 'Star Wars' Vision Finds Life Under Clinton," *Legislative Focus*, 13 May 1999: 26.

Rosenfeld, Stephanie G. "SMDC Requests Approval to Make JLENS An Official Acquisition Program," *Inside Missile Defense* 10 March 1999.

SASC-funded Nautilus laser seen winner in conference," *Aerospace Daily*, 10 May 1996.

Scott, William B. "Wargames Underscore Value of Space Assets for Military Operations," *Aviation Week and Space Technology*, 28 April 1997:60-61.

"Scuds blow up missile design," *Army Times*, 5 April 1993.

"SDC awards contract to McDonnell Douglas," *The Redstone Rocket*, 5 October 1988: 2.

Shepherd, Marquis. "Army Unit to Bring Technology of Space Down to Earth for Troops," *Kansas City Times*, 13 January 1988: B-3.

Sherrill, Gerda. "JLENS hits over-the-horizon cruise missile," *The Eagle*, October 2000: 19.

_____. "New theater missile defense target demonstrated," *Redstone Rocket*, 22 December 1993: 15.

_____. "TCMP Successful Again," *The Eagle*, September/October 1996: 12.

_____. "SSDC strapdown test a success," *The Redstone Rocket*, 21 September 1994: 6.

Sieff, Martin. "Space Dependence: A Vulnerability," *Washington Times*, 2 February 1998:12

Siegel, Steven. "Army Space Institute," *Army Trainer*, Summer 1987:20-21.

Sirak, Michael C. "Participation in CEC-Patriot Link Set for Next Month; Administration Requests $25 Million for JLENS in Fiscal Year 2000 Plan," *Inside Missile Defense*, 10 February 1999.

_____. "U.S. Wrestles With Location, Number of NMD Sites," *Inside Missile Defense*, 7 April 1999.

"Space and Missile Defense Command Seeking Information on Mobile THEL," *Inside the Army,* 9 August 1999.

"Spacecraft Played A Vital Role in Gulf War Victory," *Aviation Week and Space Technology*, 134.16 April 22, 1991: 91.

Springer, Anthony M. "Project Adam: the Army's Man in Space Program," *Quest* Summer/Fall 1994:46-47.

_____. "Securing the High Ground: The Army's Quest for the Moon," *Quest,* 7.2: 32-38.

Stares, Paul B. *The Militarization of Space: U.S. Policy, 1945-1984*. Ithaca: Cornell University Press, 1985.

"STARS missile has successful flight," Redstone *Rocket*, 27 July 1994: 2.

"Stars Booster Deploys Objects for MSX Test," *Aviation Week and Space Technology*, 16 September 1996.

Stein, Kenneth J. "New Missile Defense Systems Studied," *Aviation Week and Space Technology*, 11 October 1976: 46-47.

"Strategic Defense test flight 'an unqualified success,'" *The Redstone Rocket*, 6 February 1991: 3.

Sullivan, Gordon R. and Michael V. Harper. *Hope Is Not A Method: What Business Leaders Can Learn From America's Army*. New York: Times Books/Random House, 1996.

Swenson, Loyd S., Jr., James M. Grimwood, Charles C. Alexander. *This New Ocean: A History of Project Mercury*. Washington, D.C.: National Aeronautics and Space Administration, NASA History Office, Office of Policy and Plans, 1998.

System Planning Corporation at the direction of the Strategic Defense Initiative Organization. *Ballistic Missile Proliferation: An Emerging Threat 1992*. Arlington, VA: System Planning Corporation, 1992.

"TAA validates requirement for NMD – Total Army Analysis 2001 Adds THAAD Battalion to Corps Force," *Inside the Army*, 5 July 1993.

"Tactical High Energy Laser" *ADA Magazine Homepage* Summer 1998. http://147.71.210.21/summer98/newpage16.htm.

Terrett, Dulany. *The Signal Corps: The Emergency (to December 1941) (United States Army in World War II, The Technical Services*). Washington, D.C.: U.S. Army Center of Military History, 1956, reprint ed., 1986.

"Third LEAP test misses intercept," *BMD Monitor*, 2 July 1993;

"Top Republicans offer Defend America Act of 1996," *BMD Monitor*, 5 April 1996.

Tronolone, Michael, Jr. (CPT). "More Than 50 Years of Nuclear Terror: A History of the Ballistic Missile Threat," *ADA Magazine*, http://147.71.210.21/adamag/August%202000/TBMHidst.html.

United States. Department of Defense. *Annual Report to the President and Congress 1996*. Washington, D.C.: Government Printing Office, 1996.

United States. General Accounting Office. *Ballistic Missile Defense Information on Directed Energy Programs for Fiscal Years 1985 Through 1993*. GAO/NSIAD-93-182, June 1993.

_____. Ballistic Missile Defense: Records Indicate Deception Program Did Not Affect 1984 Results. July 1994.

_____. *Missile Defense: Status of the National Missile Defense Program*. GAO/NSIAD-00-131, May 2000.

_____. *National Space Issues: Observations of Defense Space Program and Activities*, 16 August 1994. GAO/NSIAD-94-253, 10.

"U.S. Army's HEDI Concept Evolves Into More Capable Two-Stage Interceptor," *Aviation Week & Space Technology*, 7 January 1991: 62.

U. S. Army Center for Army Lessons Learned. Center for Army Lessons Learned (CALL) Newsletter No. 91-3, *The Ultimate High Ground!: Space Support to the Army; Lessons from Operations DESERT SHIELD and DESERT STORM*, http://call.army.mil/products/newsltrs/91-3/ accessed on 10 January 2003.

U. S. Army Combined Arms Center History Office. *U. S. Army Combined Arms Center Annual Historical Review 1986*. Fort Leavenworth: Combined Arms Center History Office, 1987.

_____. *U. S. Army Combined Arms Center Annual Historical Review 1987*. Fort Leavenworth: Combined Arms Center History Office, 1988

_____. *U. S. Army Combined Arms Center Annual Historical Review 1988*. Fort Leavenworth: Combined Arms Center History Office, 1989.

_____. *U. S. Army Combined Arms Center Annual Historical Review 1991*. Fort Leavenworth: Combined Arms Center History Office 1992.

_____. *U. S. Army Combined Arms Center Annual Historical Review 1994*. Fort Leavenworth: Combined Arms Center History Office, 1996.

U.S. Army SAFEGUARD System Command. *Annual Historical Summary 1 July 1969 to 30 June 1970*, Volume II: *Supporting Documents* Huntsville: SAFSCOM, n.d.

_____. *Annual Historical Summary 1 July 1970 to 30 June 1971*. Volume II: *Supporting Documents* Huntsville: SAFSCOM, n.d.

U. S. Army Space and Missile Defense Command Historical Office and Science Applications International Corporation. *DCE: Army Leadership at the Cutting Edge of Space Control*. Huntsville: U. S. Army Space and Missile Defense Command, 2003.

_____. *The Joint Tactical Ground Station: Fielding and Operational Lessons Learned*. Huntsville: U.S. Army Space and Missile Defense Command, 2003.

U. S. Army Space and Missile Defense Command Public Affairs Office Fact Sheet. JLENS Over-the-Horizon Surveillance and Tracking Fact Sheet, dated November 1997.

_____. Low Cost Cruise Missile Defense Initiative Low Cost Interceptor Fact Sheet, dated April 2002.

U.S. Army Space and Missile Defense Battle Lab, "Army Space Exploitation Demonstration Program," 21 August 1998.

U. S. Army Space and Strategic Defense Command Public Affairs Office News Release. "Midcourse Space Experiment goes into orbit," undated [April 1996].

U. S. Army Space and Strategic Defense Command Public Affairs Office News Release. "KE ASAT hover test is highly successful," 12 August 1997.

U. S. Army Strategic Defense Command Historical Office. *U. S. Army Strategic Defense Command – Annual Historical Review Fiscal year 1986.* Huntsville: U. S. Army Strategic Defense Command, n.d.

_____. *U.S. Army Strategic Defense Command Annual Historical Review Fiscal Year 1987.* Huntsville: U. S. Army Strategic Defense Command, 1990.

U.S. Army Training and Doctrine Command. *Knowledge and Speed: Battle Force and the U.S. Army of 2025: The 1998 Annual Report on The Army After Next Project to the Chief of Staff of the Army.* Fort Monroe: TRADOC, 1998.

_____. Deputy Chief of Staff for Doctrine. "Army Space Reference Text, Chapter 2: Army Space History," http://www.fas.org/spp/military/docops/army/ref_text/, accessed on 4 February 2003.

_____. Military History Office. *Annual Historical Review 1986.* Fort Monroe: Training and Doctrine Command Military History Office, 1987.

U. S. Department of Commerce. Office of Technology Assessment. *MX Missile Basing.* Washington, D.C.: 1981.

U.S. Department of Defense, "Missile Defense Operations Announcement," 17 December 2002.

U. S. National Park Service. *Last Line of Defense: NIKE Missile Sites in Illinois.* Denver: U.S. National Park Service, 1996.

"Update: SMDC Commander on Space," *BMD Monitor*, 30 April 1999:2.

Valine, Debra. "Low Cost Interceptor seeking ways to balance cost per kill," *The Eagle*, October 2000: 6.

_____. "Space and Missile Defense remains a Command," *The Eagle*, February 2003: 1.

Vaughn, Skip. "National missile defense establishes radar project," *Redstone Rocket*, 12 July 1999: 16.

_____. "Radar product office joins THAAD missile team," *Redstone Rocket*, 5 July 1995: 1.

Ventry, Lord and Eugene M. Kolesnik. *Jane's Pocket Book of Airships.* New York: Collier Books, 1976.

Walker, James A., Frances Martin, and Sharon S. Watkins. *Strategic Defense: Four Decades of Progress.* Huntsville: U.S. Army Space and Strategic Defense Command, 1995.

Weinberger, Caspar. "Testing the MIRACL" *Forbes* 6 October 1997: 37.

Werrell, Kenneth P. *Archie, Flak, AAA, and SAM: A Short Operational History of Ground-based Air Defense*. Maxwell Air Force Base: Air University Press, 1988, http://www.strandlab.com/buzzbombs/index.html.

Winograd, Erin Q. "Shelton Defends Stance On NMD, Labeling Technology 'Just Not There' Yet," *Inside the Army*, 14 September 1998: 1.

Winter, Thomas C., Jr., Major. "The Army's Role in Space," *Military Review* 48.7 (July 1968): 82-86.

Yaranella, Ernest. *The Missile Defense Controversy – Strategy, Technology, and Politics, 1955-1972*. Lexington: University Press of Kentucky, 1977.

Unpublished Studies

Kelly, Ricky B. Centralized Control of Space: The Use of Space Forces by a Joint Force Commander. Masters Thesis, School of Advanced Airpower Studies, Air University, Maxwell Air Force Base, Alabama, September 22, 1994.

Killegrew, John W. The Impact of the Great Depression on the Army, 1929-1936. Ph.D. dissertation, Indiana University, 1960.

Miller, Robert A. The United States Army during the 1930s. Ph.D. dissertation, Princeton University, 1973.

Nichols, Matthew. "Early Concepts for Space-Based Ballistic Missile Defense," unpublished paper presented at the Conference of Army Historians, August 2002.

Parkinson, Russell J. Politics, Patents and Planes: Military Aeronautics in the United States, 1863-1907. Ph.D. dissertation, Duke University, 1963.

U.S. Army Space and Missile Defense Command Historical Office and Science Applications International Corporation. *Space Warriors: The Army Space Support Team*, 1999. Unpublished ms held at USASMDC Historical Office.

U. S. Army Strategic Defense Command Historical Office. U.S. Army Strategic Defense Command Annual Historical Review, Fiscal Year 1989.

Appendix A
Commanders and Directors

U.S. Army Space and Missile Defense Command (USASMDC)

Commanders

Lieutenant General Joseph M. Cosumano, Jr.: April 2001 – present

Brigadier General John M. Urias: March 2001 - April 2001, Commander

Lieutenant General John Costello: October 1998 - March 2001

Colonel (P) Steven W. Flohr: August - October 1998, Interim Commander

Lieutenant General Edward G. Anderson III: October 1997 - August 1998
Commander, USASSDC, October 1996 - September 1997

Deputy Commanders for Research, Development and Acquisition (Huntsville)

Major General John M. Urias: February 2001 - present
Effective December 2001, Brigadier General Urias was dual-hatted as the PEO for the Air and Missile Defense. Promoted to Major General on 1 July 2002.

Deputy Commanders Operations/ARSPACE (Colorado Springs)

Brigadier General Richard V. Geraci: August 2000 - present
Promoted to Brigadier General on 1 October 2001. Effective 16 October 2001, the Deputy Commanding General for Space is dual hatted as the Chief of Space Information Operations Element (Reach-back Element) (SIOE(RE)) of the U.S. Space Command. Following the 2002 merger of U.S. Space Command, U.S. Strategic Command, the SMDC DCG/Space works for the U.S. Strategic Command's Director of Space Operations.

Deputy Commanders

Brigadier General Steven W. Flohr: October 1997 - November 1999
Deputy Commander, USASSDC, September 1997. Promoted to Brigadier General on 1 November 1998.

Space and Missile Defense Technical Center
(formerly the Missile Defense and Space Technology Center)

Directors

Mr. Jess F. Granone: May 1999 - present

Mr. William C. Reeves, Jr.: April - May 1999, Acting Director

Mr. Jess F. Granone: January 1999 - March 1999, Acting Director

Dr. James R. Fisher: February 1995 - December 1998

Space and Missile Defense Battle Lab
(formerly the Missile Defense Battle Integration Center)

Directors

Mr. Larry H. Burger: February 1997 - present

Mr. Larry H. Burger: November 1996 - February 1997, Acting Director - BIC

Dr. Robin Buckelew: January 1995 - November 1996

Test and Evaluation Center
(formerly the Space and Missile Defense Acquisition Center)

Directors

Brigadier General John M. Urias: February 2001 - Present

Dr. Linda Gentle: November 1999 - February 2001, Acting Director

Brigadier General Steven W. Flohr: November 1998 - November 1999

National Missile Defense TRADOC Systems Manager

Directors

Colonel Jeffrey C. Horne: June 2000 - present

Colonel Ronald E. Ouellette: November 1999 - June 2000

Colonel Robert K. Billings: August 1999 - November 1999

Office of Technical Integration and Interoperability (OTII)

Directors

Mr. William C. Reeves, Jr.: July 2000 - present

Force Development and Integration Center (FDIC)

Directors

Mr. Terry Nelson: 17 March 2003 - Present - Acting Director

Colonel Glen C. Collins, Jr.: August 2000 - Deployed 12 March 2003

Colonel Robert Gregg: August 1999 - August 2000

Colonel Robert K. Billings: 1997 - August 1999

U.S. Army Space Command

Commanders

Brigadier General Richard V. Geraci: August 2000 - present
Promoted to Brigadier General 1 October 2001.

Colonel John V. Klemencic: May 2000 - August 2000

Colonel Michael W. McKeeman: July 1998 - May 2000

Colonel Steven A. Bowman: May 1998 - July 1998, Interim Commander

Colonel Otis B. Ferguson, Jr.: September 1996 - May 1998

Colonel William Hoyman: January 1996 - September 1996

Colonel E. Paul Semmens: July 1994 - December 1995

Colonel Terry L. Burns: March 1994 - June 1994

Brigadier General Gregory A. Rountree: July 1993 - February 1994
Promoted to Brigadier General on 18 February 1994.

Colonel Terry L. Burns: May 1993 - July 1993, Acting Deputy to Commander

Colonel Michael W. Keaveney: April 1991 - May 1993

Colonel Ronan I. Ellis: May 1989 - March 1991

Colonel Joe B. Thurston, Jr.: April 1988 - May 1989, Commander
(Army Space Agency, reorganized as the Army Space Command on 7 April 1988.)

U.S. Army Space Forces

Commanders

Colonel David W. Shaffer: August 2002 - present

Colonel William J. Partridge: March 2001 - August 2002

Colonel John V. Klemencic: August 2000 - March 2001

193rd Space Battalion - Colorado Army National Guard

(Activated 28 September 2001)

Lieutenant Colonel Michael Yowell: September 2001 - present

1st Space Battalion

(Activated 15 December 1999)

Lieutenant Colonel Scott F. Netherland: November 2001 - present

Lieutenant Colonel Timothy R. Coffin: December 1999 - November 2001

1st Satellite Control (SatCon) Battalion

(Activated 1 November 1995)

Lieutenant Colonel Mearen Bethea: June 2002 - present

Lieutenant Colonel Winston L. Davis: June 2000 - 28 June 2002

Major Patrick H. Rayermann: December 1995 - May 1996

Lieutenant Colonel Lynn E. Weber: November 1995 - December 1995

U.S. Army Kwajalein Atoll

Commanders

Colonel Jerry Brown: July 2002 - present

Colonel Curtis L. Wrenn, Jr.: July 2000 - July 2002

Colonel Gary K. McMillen: July 1998 - July 2000

Colonel Scott B. Cottrell: July 1996 - July 1998

Colonel David E. Spaulding: August 1994 - July 1996

Colonel Crosby E. Hazel: August 1992 - August 1994

Colonel John J. MacNeill: 1990-1992

Colonel Philip R. Harris: 1988 - 1990

Colonel Richard G. Chapman, Jr: August 1986 - August 1988

Colonel James R. Allred: July 1986 - August 1986

Colonel William A. Spin: July 1984 - July 1986

Colonel John W. Banks, Jr.: June 1982 - July 1984

Colonel Peter F. Witteried: 1980 - 1982

Colonel John H. Reeve: 1978 - 1980

Colonel Ernest A. Van Netta: 1976 - 1978

Colonel Robert L. Russell: 1973 - 1976

Colonel Jesse L. Fishback: 1971-1973

Colonel Donald B. Millar: 196 - 1971

Colonel Frank C. Healy: 1967 - 1968

Colonel Melvin D. Clark: 1965 - 1967

Colonel Glen H. Crane: 1964 - 1965
(Assumed control of Kwajalein Pacific Missile Range Facility from the Navy in July 1964.)

Captain H.D. Allen, (USN): 1964 - 1964

Commander H.R. Gordinier (USN): 1964 - 1964

Captain P. Holmberg (USN): 1961 - 1963

Captain G. Smith (USN): 1959 - 1961

High Energy Laser Systems Test Facility

Directors

Thomas Hodge: August 2002 – present

Lieutenant Colonel Lyn Tronti: July 2000 - August 2002

Colonel Ronald J. Nelson: July 1998 - May 2000

Colonel Larry D. Anderson: June 1995 - July 1998

Colonel George P. Lasche: December 1993 - May 1995

Major Vernon C. Bice: November 1993 - December 1993, Acting Director

Colonel Henry W. Meyer, Jr.: November 1992 - October 1993

Colonel Richard L. Knox: November 1991 - October 1992

Colonel James E. Green: October 1990 - July 1991 Acting Director

Army Space Program Office

Directors

Colonel Steven Fox: July 2001 - present

Colonel Darrell Lance: June 1999 - July 2001

Colonel Melvin L. Heritage: December 1997 - June 1999

Colonel Arthur R. Marshall, Jr.: June 1994 - December 1997

Colonel Sherwood C. "Woody" Spring: July 1989 - June 1994

Colonel Charles J. Sollohub: May 1984 - July 1989

Colonel Robert A. Schow, Jr.: June 1981 - May 1984

Colonel Hugh H. Trumbull: August 1978 - June 1981

Colonel Ronald Lemanski: November 1975 - August 1978

Colonel Robert A. Ready: July 1973 - November 1975

Garrison Commander Fort Greely, Alaska

Commanders

Major Marie Grimmer: October 2002 - present
(The USASMDC assumed responsibility for the Fort Greely Garrison on 1 October 2002)

U.S. Army Space and Strategic Defense Command (USASSDC)

Commanders

Lieutenant General Edward G. Anderson III: October 1996 - October 1997

Lieutenant General Jay M. Garner: September 1994 - October 1996

Lieutenant General Donald M. Lionetti: August 1992 - September 1994

Deputy Commanders

Colonel (P) Steven W. Flohr: September 1997 - October 1997

Brigadier General Robert E. Armbruster: November 1996 - April 1997
Promoted to Brigadier General on 12 November 1996

Brigadier General Willie B. Nance, Jr.: November 1995 - October 1996
Promoted to Brigadier General on 3 April 1996. Assumed duties as PEO for Tactical Missiles on 15 July 1996. Under an agreement with General Hite, Brigadier General Nance continued to serve as the DCG until October.

Major General Jan A. Van Prooyen: August 1992 - June 1995
Promoted to Major General on 12 April 1995.

U.S. Army Strategic Defense Command (USASDC)

Commanders

Brigadier General William J. Schumacher: July 1992 - July 1992, Acting Commander

Lieutenant General Robert D. Hammond: July 1988 - June 1992
Appointed Program Executive Officer for Strategic Defense in October 1988.

Brigadier General Robert L. Stewart: May 1988 - July 1988, Acting Commander

Lieutenant General John F. Wall: July 1985 - May 1988

Deputy Commanders

Brigadier General William J. Schumacher: February 1992 - July 1992

Colonel Robert S. Troth: July 1991 - February 1992, Acting Deputy Commander

Major General John S. Peppers: November 1989 - July 1991

Brigadier General Robert L. Stewart: July 1987 - October 1989

Major General Eugene Fox: July 1985 - July 1987

Ballistic Missile Defense Organization (BMDO)

Program Managers

Major General Eugene Fox: September 1984 - July 1987

Major General Elvin R. Heiberg, III: May 1983 - September 1984

Major General Grayson D. Tate, Jr.: June 1979 - May 1983
Ballistic Missile Defense Program Manager, (June 1979 - September 1982) Commander Ballistic Missile Defense Systems Command (When the PM Office transferred to Washington, in September 1982, Major General Tate assigned the BMDSCOM duties to Colonel Robert J. Feist.)

Major General Stewart C. Meyer: November 1977 - June 1979
BMD Program Manager, Commander, BMDSCOM and Commander, BMDATC

Brigadier General John G. Jones: August 1976 - October 1977
BMD Program Manager and September 1975 - October 1977 Commander BMDSCOM

Major General Robert Creel Marshall: August 1974 - August 1976
BMD Program Manager and July 1969 - April 1973 Commander, BMDSCOM

Lieutenant General Walter P. Leber: May 1974 - August 1974, BMD Program Manager

Deputy Program Mangers

Brigadier General William J. Fiorentino: October 1984 - June 1985

Mr. Jack H. Kalish: August 1983 - -----
Deputy BMDPM - Washington

Major General Eugene Fox: August 1983 - September 1984
Deputy BMDPM - Huntsville

Ballistic Missile Defense Systems Command (BMDSCOM)

Commanders

Brigadier General Eugene Fox: July 1983 - January 1986

Colonel Robert J. Feist: September 1981 - July 1983

Major General Grayson D. Tate, Jr.: June 1979 - September 1982

Major General Stewart C. Meyer: November 1977 - June 1979

Colonel Roger D. Powell: October 1977 - November 1977, Acting Commander

Brigadier General John G. Jones: September 1975 - October 1977

Major General Bates C. Burnell: May 1974 - August 1975

Major General Robert Creel Marshall: July 1969 - April 1973

SAFEGUARD

Systems Manager

Lieutenant General Walter P. Leber: April 1971 - August 1971

Lieutenant General Alfred D. Starbird: March 1969 - March 1971

Commanders

Brigadier General Bates C. Burnell: April 1973 - May 1974

Brigadier General Robert Creel Marshall: October 1969 - April 1973,
July 1969 - October 1969, Acting Commander

SENTINEL

Systems Manager

Lieutenant General Alfred D. Starbird: November 1967 - March 1969

Commanders

Brigadier General Ivey O. Drewry: November 1967 - July 1969

NIKE-X Project Office

Lieutenant General Austin W. Betts: September 1966 - November 1967
Acting NIKE-X System Manager in addition to principal assignment as Army Chief of R&D. Remained Chief of R&D after November 1967.

Brigadier General George Mayo, Jr.: December 1966 - November 1967
Deputy NIKE-X System Manager (Plans). Received his promotion to Brigadier General in February 1967.

Brigadier General Ivey O. Drewry: 1962 - 1967
NIKE-ZEUS and subsequently NIKE-X Project Manager

Program Executive Officer for Air and Missile Defense

(formerly PEO Missile Defense and PEO Global Protection Against Limited Strikes (GPALS))

Brigadier General John M. Urias: December 2001 - present
Dual Hatted as the PEO and the DCG, RDA for USASMDC.

Dr. Shelba Proffitt: February 2001 - December 2001, Acting PEO-AMD.

Brigadier General John M. Urias: September 1999 - January 2001
Promoted to Brigadier General on 31 January 2000.

Brigadier General Daniel P. Montgomery: March 1996 - September 1999
Promoted to Brigadier General on 8 November 1996.

Brigadier General Richard A. Black: January 1994 - March 1996

Mr. Alan D. Sherer: September 1993 - January 1994, Acting PEO-MD.

Major General William S. Chen: July 1992 - September 1993

Missile Defense Agency

(formerly Ballistic Missile Defense Organization and Strategic Defense Initiative Organization)

Directors

Lieutenant General Ronald T. Kadish (USAF): January 2002 – Present
Missile Defense Agency was established 2 January 2002. June 1999 - January 2002 Director, Ballistic Missile Defense Organization (BMDO)

Lieutenant General Lester L. Lyles (USAF): August 1996 - May 1999

Rear Admiral Richard West (USN): June 1996 - July 1996, Acting Director

Lieutenant General Malcolm R. O'Neill (USA): November 1993 - May 1996
Director BMDO; August 1993 - November 1993, Acting Director BMDO; January 1993 - August 1993, Acting Director SDIO when renamed 13 May 1993. Major General

O'Neill was confirmed on 19 November and promoted to Lieutenant General on 22 November 1993.

Ambassador Henry F. Cooper: July 1990 - January 1993

Lieutenant General George S. Monahan, Jr. (USAF): February 1989 - June 1990

Lieutenant General James A. Abrahamson (USAF): March 1984 - January 1989

Appendix B
Army Space and Missile Defense Chronology

1937

May 1937 Colonel William Blair, Director, Army Signal Corps Laboratory, Fort Monmouth, granted patent for first Army/military radar.

1942

3 October 1942 The first A-4 rocket was successfully launched from Peenemünde. The A-4 was known by the allies as the V-2.

1944

31 January - 4 February 1944 Elements of the 7th Infantry Division and the 4th Marine Division began an assault on Enubuj, Mellu, and Ennuebing Islands in the Marshall Islands. On 1 February, the 7th Infantry Division led an assault Kwajalein, while the 4th Marine Division landed on Roi-Namur. The Marines secured Roi-Namur on 2 February 1944. The American flag was raised over Kwajalein on 4 February.

12-13 June 1944 The German Army launched the first ten V-1 rockets against the city of London. The V-1 was a pilot-less aircraft that flew at a speed of 400 mph.

8 September 1944 The German Army launched the first V-2 missiles in an attack on London. The V-2 was a supersonic ballistic missile with a top speed of 3300 mph.

1945

January 1945 At the request of the Army Ordnance Department, Bell Telephone Laboratories began work on an anti-aircraft missile that later became the NIKE.

4 July 1945 A delegation of American officers, sent to Europe to investigate the use of ballistic missiles during World War II, recommended that the U.S. undertake a research and development program to develop defenses against these new weapons.

1946

January 1946 Wernher von Braun and 127 German missile experts are brought to the United States (Fort Bliss, Texas) under Operation PAPERCLIP.

January 1946 The Army Signal Corps bounced radio signals off the moon and received the reflected signal back on earth. This did not provide an effective communications link, but proved it was possible to send radio signals through space and back to earth with moderate power.

4 March 1946 The Army Air Force awarded two contracts to study antimissile missiles.

4 April 1946 Stalin told U.S. Ambassador Walter Bedell Smith that the "United States [had] definitely aligned itself against the U.S.S.R.."

16 April 1946 The Army launched the first reconstructed V-2 from White Sands Proving Ground, New Mexico.

29 May 1946 The Stilwell Board Report, convened to determine what equipment U.S. ground forces would require following World War II, recommended the development of defenses against ballistic missiles.

1 July 1946 The beginning of "Operation Crossroads," the atomic tests at Bikini. A B-29 called "Dave's Dream", flew from Kwajalein for the Island of Bikini.

December 1946 The Department of War accepted Dr. Vannevar Bush's judgment, dismissing most missile and space research and development.

1947

24 May 1947 The first full-scale flight test of a U.S. ballistic missile, the CORPORAL.

18 July 1947 With the agreement of the U.S. the Trust Territory of the Pacific Islands (TTPI) was placed under the trusteeship system established under the Charter of the U.N. The agreement went into effect upon approval by the Security Council (2 April 1947) and the U.S. government. Executive Order No. 9875, delegated interim authority and responsibility for the civil administration of the territory to the Secretary of the Navy.

1949

24 February 1949 A two-stage BUMPER missile consisting of a German V-2 with a WAC (Without Altitude Control) CORPORAL as the second stage was fired at White Sands Missile Range (WSMR). The launch, part of the Army missile program headed by Dr. von Braun and his team, achieved an altitude of 250 miles, a world record at that time. This was the first penetration of outer space. (The rocket did not remain in orbit because it lacked sufficient velocity.)

June 1949 The Army transferred von Braun and the Ordnance Research and Development Division Sub-Office (Rocket) from Fort Bliss, Texas, to Redstone Arsenal, Huntsville, Alabama. The move was completed in 1950.

August 1949 The Soviet Union exploded an atomic bomb. This provided the impetus for the U.S. to develop a hydrogen bomb, the Army to build antiaircraft emplacements around strategic locations, and led to a reappraisal of U.S. national security policy.

1950

April 1950 NSC 68 (National Security Council Memorandum 68) established containment policy against Communist aggression as U.S. national policy. This memorandum was signed in September 1950 and remained classified for twenty five years.

July 1950 The Army formed the Anti-Aircraft Command later known as the U.S. Army Air Defense Command (ARADCOM). Headquartered at Ent AFB, Colorado Springs, CO, this organization was "to provide Commander in Chief, North American Defense Command with combat ready Army Air Defense Forces to defend critical localities in CONUS." From July 1961, the ARADCOM participated in the BMD program. As of 1972, they were "to provide specified support in user deployment planning activities and to assume command of the tactical SAFEGUARD forces upon deployment."

30 September 1950 President Truman approved National Security Council 68, the first comprehensive statement of American national strategy from the NSC.

1951

29 June 1951 President Harry S Truman signed Executive Order 10265, revoking Executive Order No. 9875 and transferring administration of the TTPI from the Secretary of the Navy to the Secretary of the Interior.

1953

20 August 1953 First launch of the Redstone rocket at Cape Canaveral, FL. Beginnings of the Army's space effort, part of the program at Redstone Arsenal, AL. The Redstone served as a space launcher and, in 1958, as a tactical ballistic missile stationed in Germany.

1954

February 1954 Operational hydrogen bomb detonated by the United States. The U.S.S.R. detonated its first thermonuclear weapon in August 1953.

1 March 1954 Congress approved U.S. participation in International Geophysical Year 1957-1958 (IGY) program.

5 May 1954 The *New York Times* reported that the Soviet Union might be besting America in rocket and missile development, to include new "supersonic" missiles capable of intercontinental nuclear strikes. The press dubbed these ICBMs the "ultimate weapons" for which there was no defense.

6-7 December 1954 Joint Army-Navy conference on Project ORBIT. As a result of this conference the Project ORBIT proposal was pressed in December 1955.

1955

February 1955 The Army Ballistic Missile Agency (ABMA), in Huntsville, AL, contracted with Western Electric Company and Bell Telephone Laboratories, for a study to explore the feasibility of a defense against strategic ballistic missiles, in particular ICBMs.

29 July 1955 President Eisenhower announced the United States' intention to launch a satellite during the IGY (1957-1958).

1 December 1955 President Dwight D. Eisenhower approved NSC recommendation (NSC 1484) to assign the ICBM and the IRBM "joint" highest national priority. Some sources trace this decision to a State Department study which concluded that, in the missile race, if the Soviet Union produced a long range ballistic missile first it would greatly reduce confidence in American technological superiority.

1956

1 February 1956 The ABMA established as a Class II activity on Redstone Arsenal (General Orders 68, 22 December 1955).

May 1956 "The Special Assistant for Guided Missiles to the Secretary of Defense disapproved the Army's request that the Jupiter-C be designated the backup to the Vanguard." Officials decided that "the need for ballistic missiles for retaliatory strikes was a national priority and it was feared that trying to meet two or more projects simultaneously would dilute the Army's tactical and intermediate ballistic missile work."

September 1956 The JUPITER-C achieved the first deep penetration of space with an altitude of more than 682 miles and a range of 3,355 miles.

2 October 1956 "The Special Assistant for Guided Missiles to the Secretary of Defense informed the Secretaries of the Army and the Air Force that in the antimissile field, the Air Force would have responsibility for developing the early warning system and the Army would have responsibility for active defense system. The assignment to the Army was justified on the grounds that major targets were already defended by NIKE sites, NIKE-II appeared to be the only project beyond the study stage capable of accomplishing the mission, and there was "a basic similarity between the anti-ICBM problem and the anti-missile missile for field army use."

1 November 1956 The Department of the Army (Army Chief of Research and Development) authorized the Chief of Ordnance to begin phased development of the NIKE II System.

15 November 1956 NIKE II, the product of an 18-month study by Bell Labs and Western Electric, was officially named NIKE-ZEUS.

26 November 1956 Secretary of Defense Charles Wilson issued a directive settling the Army-Air Force dispute over defense responsibilities for the Continental United States. The Army was responsible for land-based surface-to-air missile systems protecting specific cities or vital installations that required radars near the launching sites to give the missiles guidance information. Intercepts would occur within 100 nautical miles of targets in the atmosphere. The Air Force was responsible for land-based surface-to-air missile systems for area defense, without reference to specific sites. A network of radars far from the launch sites gave the missiles guidance information; intercepts would occur outside the atmosphere. The directive did not distinguish between ballistic missile defense and antiaircraft defense and failed to specify which service would control deployed systems.

1957

26 February 1957 Secretary of Defense Charles E. Wilson announced at a press conference that the JUPITER missile program was "in effect" canceled. One day later Major General John B. Medaris announced that the current development work on the JUPITER IRBM would continue unabated. The JUPITER was the primary project of the ABMA.

9 July 1957 The ABMA sent a letter to Chief of Ordnance outlining its potential satellite launching capability.

26 August 1957 The Soviet Union announced their first successful ICBM test flight, the SS-6, "a single stage missile with clustered engines that developed twice the power of the American Atlas or Titan ICBMs."

September 1957 A joint Atomic Energy Commission and DoD study concluded that it was feasible to develop a warhead for NIKE-ZEUS.

4 October 1957 The U.S.S.R. launched SPUTNIK the world's first artificial satellite into orbit, using an SS-6 Sapwood ICBM. The payload weighed 184 pounds. The satellite carried only a radio beeper which transmitted for 21 days before its batteries wore out.

5 October 1957 Dr. Wernher von Braun briefed Secretary of Defense McElroy on the Jupiter-C/Redstone system for immediate satellite launch. He promised the first U.S. satellite in 60 days. Major General Medaris asked for 90 days to complete the mission.

3 November 1957 The Soviet Union launched the world's second satellite, Sputnik II, into low earth orbit. Onboard the 1,119 pound satellite was a live dog named Laika. By launching such a heavy satellite, the Soviets demonstrated that they could also launch a nuclear weapon(s) and possibly de-orbit them on command. It remained in orbit until 13 April 1958.

7 November 1957 In a major televised address on science and security, President Eisenhower announced that Army scientists had successfully solved the problem of ballistic missile reentry.

8 November 1957 The President directed the Army to orbit a satellite by March 1958. The Secretary of Defense ordered the ABAMA to prepare a Jupiter-C missile to launch the Explorer I satellite, as part of the IGY program. Mission was completed 31 January 1958.

December 1957 The Gaither report credited the Soviet Union with a substantial lead in long-range ballistic missiles and gave rise to the so-called "missile gap."

1958

16 January 1958 Secretary of Defense Neil McElroy issued guidance to the Air Force and the Army on the BMD issue. The Army was directed to continue with the ZEUS system components (missile, launch site, radars, and computer components). The Air Force was to continue developing early warning radars, tracking and acquisition radars and communications links, ensuring that they were compatible with the NIKE-ZEUS system. Missile work on the Air Force's WIZARD system, however, was terminated. This decision also assigned "the direction of this development effort [BMD] to the Advanced Research Projects Agency (ARPA), to make [the] most effective use of our overall national capability." President Eisenhower announced the creation of ARPA, established in response to Sputnik, in his 1958 State of the Union Address.

The memoranda only addressed the development of the system. There was no mention of organizational control over a deployed system.

22 January 1958 The NSC assigned the highest national priority ("S-Priority") to the NIKE-ZEUS antimissile missile development program.

31 January 1958 The United States Army launched Explorer I, the first U.S. satellite to orbit around the earth, using a JUPITER-C rocket. The launch was completed 84 days after the ABMA was given the mission. The 18.13-pound satellite measured high altitude radiation belts and discovered the cosmic radiation belt identified by Dr. James A. Van Allen. In his State of the Union Address, President Eisenhower had given top priority to the development of satellite and missile defense systems, making them comparable to ICBM and IRBM research efforts. Explorer I returned to earth's atmosphere on 31 March 1970.

7 February 1958 The Department of Defense (DoD) established the ARPA, which was responsible for the nation's outer space program.

17 March 1958 Vanguard I successfully launched. The Army Signal Corps designed and built Vanguard I cloud cover satellite solar converters for NASA and proved the practicability of solar converters.

20 March 1958 The DA established the U.S. Army Ordnance and Missile Command (AOMC) at Redstone Arsenal. The AOMC was composed of the ABMA; the Jet Propulsion Lab, Pasadena, CA; WSMR; Redstone Arsenal; and the newly created U.S. Army Rocket and Guided Missile Agency (ARGMA) Effective 31 March 1958

26 March 1958 Explorer III placed in orbit. It was the first American satellite to store information on tape and play it back when it received a command from a ground station.

1 April 1958 NIKE-ZEUS project is under the ARGMA, which supplanted the RAMMSO. The ARGMA was organized as a subordinate element of the AOMC, at Redstone Arsenal (AOMC General Orders 6). The stated mission of the ARGMA was the development, procurement, production, industrial engineering, industrial mobilization, maintenance and repair part supply, and stock control of ordnance rockets and guided missiles. This charge encompassed all Army surface-to-surface and surface-to-air missiles except the ballistic missiles for which the ABMA was responsible.

2 April 1958 The President recommended to Congress that a civilian agency be established to direct non-military space activities.

17 April 1958 Project "Man Very High" redesignated Project ADAM. Formal proposals were submitted to OCRD and ARPA in May and June. Director of ARPA advised the Secretary of the Army, in an 11 July 1958 memo that Project ADAM "was not considered necessary to 'Man-in-Space' program and therefore would not be funded by ARPA." They added that should the project proceed with another organization, they "would be pleased to receive information on applicable data."

June 1958 The National Aeronautics and Space Act was signed. This act created the National Aeronautics and Space Administration (NASA) as of 1 October 1958. NASA was given a broad charter for aeronautical and space research. The core of NASA's facilities came from the disbanded National Advisory Committee for Aeronautics. The Air Force would continue development of ICBMs and the Navy could continue the development of sea-launched rockets although the Navy did transfer Project Vanguard and part of the Naval Research Lab to NASA in November 1958. The Army could continue to develop IRBMs but would transfer much of its rocket program to NASA. Most NASA facilities, launch sites and test ranges have been and continue to be, built under the supervision of the Army Corps of Engineers.

16 July 1958 Army proposal for Space Payloads forwarded to Director ARPA by Secretary of Army with recommendation that it be approved in conjunction with Plan B 12-vehicle Juno IV program.

25 July 1958 ARPA Order 10-59 issued to AOMC, approving meteorological payload. The order made no mention of launching vehicle.

26 July 1958 The Army launched a Juno I rocket which placed the Explorer IV satellite into an elliptical, inclined orbit. The satellite measured the results of a high altitude nuclear explosion and took measurements of the sun for three months. Its orbit decayed in October 1959.

29 July 1958 President Eisenhower created NASA to avoid the militarization of space. On 21 October 1959, the ABMA scientists and engineers transferred to NASA.

15 August 1958 ARPA authorized ABMA to begin work on the Saturn booster.

23 September 1958 After AOMC and ARPA signed a Memorandum of Understanding (MOU) on this date, the SATURN program began at ABMA under Army management.

SATURN design studies were authorized to proceed at Redstone Arsenal for development of a 1.5 million pound thrust, clustered engine first stage.

1 October 1958 NASA officially activated.

November 1958 A NIKE-HERCULES missile destroyed a supersonic target missile traveling faster than 1500 miles per hour at an altitude greater than 60,000 feet. This was the first intercept of a very high altitude supersonic target missile.

26 November 1958 The U.S. manned satellite space program using the REDSTONE as a booster was officially named Project MERCURY.

3 December 1958 Support agreement signed between the Army and NASA. Army rockets/missiles would be used extensively in the early space program, to include the Mercury Redstone manned satellite space program. As part of this agreement the JPL transferred from the AOMC to NASA.

6 December 1958 The Army's Juno II rocket, a modified Jupiter, launched the Pioneer III lunar probe for NASA. It did not reach the moon, but did travel for more than 66,654 miles into outer space and gathered radiation data that indicated the existence of a second radiation belt around the Earth. The launch was also a successful test of the first four-stage JUNO II vehicle.

13 December 1958 JUPITER Missile AM-13 was fired, marking the first successful flight of a JUPITER IRBM incorporating the tactical ballistic shell configuration. The missile also carried a squirrel monkey named Gordo, contributing highly useful data for Army and Navy medical research into space flight. Although Gordo made the flight with no adverse effects, the monkey could not be recovered because the nose cone's flotation device failed.

18 December 1958 The Army Signal Corps placed President Eisenhower's Christmas message to the world on a communications payload satellite – the Signal Communications by Orbiting Relay Equipment (SCORE). The satellite transmitted the President's message on a shortwave frequency to the world below. This was the first time that the human voice was heard from space. The system operated for 12 days responding to 78 transmissions before the batteries failed. The President's message was: "This is the President of the U.S. speaking. Through the marvels of scientific advance, my voice is coming to you from a satellite traveling in outer space. My message is a simple one: Through this unique means I convey to you and all mankind, America's wish for peace on Earth and goodwill toward men everywhere." This was the first time that the human voice was heard from space.

1959

January 1959 NASA published the selection criteria for astronauts. One of the requirements was that all astronauts had to be experienced test pilots. This effectively eliminated Army personnel from consideration as astronaut candidates.

8 January 1959 NASA assigned ABMA the mission to develop eight REDSTONE-type vehicles for use in the MERCURY manned satellite program.

12 February 1959 The DoD's Ballistic Missile Committee approved the test program for NIKE-ZEUS and made Kwajalein the down-range test site.

17 February 1959 Vanguard II satellite launched into low Earth orbit, carrying an Army developed infrared cloud imaging sensor but stability problems precluded imaging efforts.

1 March 1959 The ABMA was assigned responsibility for development of ballistic targets for the NIKE-ZEUS test program.

3 March 1959 The JUNO II launched PIONEER IV, which achieved a velocity greater than the 24,560 miles per hour required to escape the Earth's gravitational force, shot by the Moon at a distance of about 36,000 miles from that body and sped on to become the first U.S. satellite to make an orbit of the Sun.

20 March 1959 Army task force formed under direction of Major General John Medaris of AOMC. The purpose of Project Horizon was to develop a plan to establish a lunar outpost by the quickest practical means.

28 May 1959 Able, a 7 lb rhesus monkey and Miss Baker, a 1 lb squirrel monkey, became the first living creatures to fly in space and return safely. "The two primates were hurled 1600 miles down-range into the Atlantic aboard a Jupiter nose cone."

26 August 1959 First NIKE-ZEUS missile was fired at WSMR. The test was deemed a partial success.

27 August 1959 All national missions related to target missile systems were centralized in ARGMA on this date.

September 1959 The DoD determined that the Air Force would have responsibility for all military space operations, with the exception of the Navy Polaris program.

22 September 1959 The Army and Navy signed an MOA for the use of Kwajalein as a test range for the NIKE-ZEUS program.

13 October 1959 The ABMA launched its last Explorer satellite, Explorer VII, with a Juno II rocket. This satellite studied the X-rays emitted by the sun and their influence on the ionosphere. It also identified the heavy particles constituting cosmic rays and measured the heat emitted by the Earth.

November 1959 The Army transferred its Saturn rocket development program to NASA.

1960

29 February 1960 The ADVENT program was established by ARPA. It would be a single 24-hour, equatorial synchronous, military communications system. The Army would develop the satellite communications equipment and the Air Force would be responsible for the booster and the spacecraft.

15 March 1960 A REDSTONE missile successfully fired from the WSMR lofted a "flying TV station for the first time."

1 April 1960 The Army launched TIROS 1 (Television and Infrared Observation Satellite) into low-earth orbit. TIROS 1 was the first American weather satellite. Both ABMA and Army Signal Corps helped to develop the TIROS 1 and 2 satellites.

3 June 1960 A NIKE-HERCULES antiaircraft guided missile tracked and shot down a CORPORAL ballistic missile at the WSMR marking the first ballistic missile to be killed by a missile.

1 July 1960 The AOMC/ABMA lost all of its space-related missions, along with about 4,000 civilian employees and $100 million worth of buildings and equipment.

10 August 1960 Ninth ZEUS missile (20009) successfully tested at WSMR. This was the first firing of the advanced design ("wingless") missile.

12 August 1960 ECHO I, the first passive relay communications satellite was launched. It demonstrated the feasibility of global communications via satellite.

September 1960 A NIKE-HERCULES missile shot down another NIKE-HERCULES missile at an altitude of 19 miles, the highest known missile kill to date.

4 October 1960 The Army's COURIER 1B satellite was launched into a low-earth orbit. This communications satellite exceeded the storage and transmission capabilities of the earlier SCORE satellite. It was the first communications satellite to be powered by long life solar cells to recharge nickel cadmium storage batteries.

9 December 1960 The Mercury I unmanned capsule was launched on a suborbital flight using an Army Redstone missile.

1961

31 January 1961 A second Mercury test flight carried a chimpanzee named Ham into space. The suborbital mission helped prove the system's operational capabilities in a space environment. Ham survived the flight.

4 March 1961 According to reports, the Soviet V-1000 antimissile completed the first successful missile intercept.

6-28 March 1961 DoD Directive 5160.32 Development of Space Systems assigned responsibilities for satellite development: (1) each service could conduct preliminary research to use satellite technology. (2) Army would continue its ADVENT communications satellite work. (3) Navy would continue its TRANSIT navigation satellite work. (4) Air Force would conduct advanced research and development work and operate all DoD reconnaissance satellites and (5) DoD would review and approve all advanced satellite research and development proposals.

12 April 1961 Yuri Gagarin, aboard a Vostock capsule, became the first man to fly in space and orbit the Earth. He parachuted back to earth after reaching a designated altitude.

5 May 1961 Alan Shepard became the first American to make a sub-orbital flight into space. He flew aboard the Mercury 3 capsule known as the Freedom 7. Launched by a modified Army Redstone rocket, the flight lasted 15 minutes and 22 seconds.

21 July 1961 Virgil Grissom went into space in another suborbital Mercury mission launched by the Army's Redstone Rocket.

22 September 1961 Secretary of Defense approved first two phases of a three-phase plan for the production and deployment of NIKE-ZEUS. The system was to be deployed in defense of twelve metropolitan areas.

11 December 1961 The ARGMA was abolished and its missions and functions were merged with AOMC Headquarters (General Orders 47, 26 December 1961).

14 December 1961 Three ZEUS firings were conducted on this date at three different locations. 1. NIKE-ZEUS ZM-6 was successfully test fired at Point Magu, in the longest and highest test flight made to date. 2. NIKE-ZEUS ZW-34 successfully acquired, tracked and intercepted a high altitude, maneuvering supersonic NIKE-HERCULES target missile over WSMR. This was the first intercept of a HERCULES guided missile, and the first successful integrated system test. – "a major ABM system milepost." 3. NIKE-ZEUS ZK-1 was test fired from Kwajalein against a space point, the first firing from this test facility.

1962

30 March 1962 A Feasibility study was completed on a fast-reaction, surface-to-air missile "which by its rapid acceleration, would maximize the time available to a defense for discriminating between warheads and possible decoys."

27 April 1962 The Secretary of Defense added a new requirement to the NIKE-ZEUS system, to provide the capability for a satellite interception demonstration at Kwajalein, by 1 May 1963. This project was code named MUDFLAP.

12 December 1962 First fully successful NIKE-ZEUS missile intercept of an ICBM. The missile intercept occurred with a miss distance well within acceptable limits. The target was an Atlas D launched from Vandenberg AFB, CA.

17 December 1962 First ZEUS missile, modified for Project MUDFLAP anti-satellite tests was fired at WSMR. It successfully intercepted a designated space point at an altitude of 100 nautical miles.

1963

5 January 1963 At the direction of the Deputy Secretary of Defense Roswell Gilpatrick, NIKE R&D was redirected to a new system which would address the 1970's ICBM threat.

1 February 1963 Office of the NIKE-ZEUS Project Manager established as a Class II activity, assigned to the Headquarters, U.S. Army Materiel Command. Located in Huntsville, the personnel and records of this office were to be transferred from the U.S. Army Missile Command (MICOM) to AMC HQ. The Project Office was thus separated from MICOM, except for administration, training and logistics support.

23 May 1963 NIKE-ZEUS achieved another milestone when it successfully intercepted an AGENA D earth satellite.

4 July 1963 A NIKE-ZEUS fired from Kwajalein successfully intercepted an Atlas "E" launched from Vandenberg AFB CA.

1964

1 February 1964 NIKE-X Project Office replaced the NIKE-ZEUS Project Office. It was organized in the same format as the NIKE-ZEUS Office.

1 July 1964 Kwajalein transferred from the U.S. Navy to U.S. Army control and assigned to the NIKE-X Project Office.

October 1964 The Chinese exploded a nuclear device. At this time, however, they had few bombs and no missile to deliver the device.

14 October 1964 NASA modified their astronaut selection criteria, dropping the jet pilot experience requirement, thereby allowing scientist-astronauts, making Army personnel eligible for astronaut duty.

1965

5 June 1965 The Secretary of the Army approved an organization plan which placed the NIKE-X System Manager at DA level.

17 November 1965 First guided SPRINT flight test took place at WSMR.

1966

6 May 1966 NIKE-ZEUS completed developmental flight testing of the ZEUS missile.

June 1966 Phase I of the DSCS Program initiated, with 26 satellites launched in 2-year period. Phase II began in 1971 and Phase III in October 1982.

1 September 1966 ODCSOPS Study "NIKE-X Studies for 1966 (X-66), Report to the SECDEF" concluded that: "There is adequate assurance that the probable effectiveness of NIKE-X justifies the cost of deployment at DEPEX-II."

28 September 1966 General Harold Johnson, Chief of Staff of the Army, selected the NIKE-X program for exceptional management techniques, in Memorandum 66-436.

30 September 1966 The command purchased a two-man submarine for recovery of reentry vehicles at KMR.

15 October 1966 NIKE-X System Office established as a Class II activity under the command of the NIKE-X System Manager. Located in Washington, it served as a single point of contact within DA for the coordination and direction of all activities pertinent to the NIKE-X. The NIKE-X Project Office and the NIKE-X Engineering Service Test Organization were placed under operational control of the NIKE-X System Office.

27 October 1966 The People's Republic of China announced that they had successfully test-flown a guided missile with a nuclear warhead.

10 November 1966 Secretary of Defense Robert McNamara announced that the Soviet Union had deployed an ABM system, 64 launchers deployed around Moscow.

1967

1 January 1967 To avoid confusion with the ZEUS missile, and the ZEUS DM 15X-2 missile, was renamed SPARTAN.

23 June 1967 At the Glassboro summit, President Lyndon Johnson tried to convince Soviet Premier Alexsei Kosygin that the U.S.S.R. should abandon missile defense efforts. Without this decision, Johnson stated that the U.S. "would be compelled to increase the number of warheads in its ICBM arsenal to overwhelm any defenses." Kosygin replied: "Defense is moral; offense is immoral."

15 August 1967 Created at the direction of Secretary of Defense McNamara, the DoD established the Montgomery Committee to review the Chinese threat. They released their report which "[indicated] that the NIKE-X DEMOD 1-67 constituted an adequate base for proceeding."

18 September 1967 In a speech to the UPI editors and publishers in San Francisco, Secretary of Defense McNamara announced the decision to deploy some NIKE-X components as an ABM system. The SENTINEL System, was a limited deployment production decision consistent with NIKE-X Deployment Model 1-67, designed to provide protection for urban/industrial areas against possible ICBM attacks by the People's Republic of China. It would also serve as a defense against accidental launch with an option to defend the Air Force's MINUTEMAN missile sites.

15 October 1967 U.S. Corps of Engineers Huntsville Division organized, as the U.S. Army Corps of Engineers NIKE-X Division. This was the first division organized by the Chief of Engineers in many years.

1 November 1967 The DoD announced the locations of the first ten SENTINEL sites: Boston Perimeter Acquisition Radar (PAR) and Missile Site Radar (MSR); Chicago MSR; Grand Forks AFB, ND PAR and MSR; Salt Lake City MSR; Detroit PAR and MSR; Seattle PAR and MSR; Hawaii MSR; New York MSR; and Albany, GA MSR; Sedalia, MO, and two others on 13 November 1968 (Warren AFB, WY, and Malmstrom AFB, MT). Two additional sites, Washington, D.C. and Fairbanks, AK, were never publicly announced.

15 November 1967 SENTINEL System Organization created under DA General Orders 48, replaced the NIKE-X System Office and Manager. SENTINEL System Command (SENSCOM) was established as a Class II Activity, under the direction of the SENTINEL System Manager. The System Manager position was created within the Office of the Chief of Staff.

The SENTINEL System Organization included the SENTINEL System Office in Washington, D.C., the SENTINEL System Command in Huntsville, AL, and the SENTINEL System Evaluation Agency in WSMR.

At the direction of the DoD, the SENSCOM focused on systems/operations of the SENTINEL system, while a parallel command, the BMD Research Office was created for further R&D efforts. In March 1968, the research office, also a Class II Activity, was renamed ABMDA. They reported to the Army's Chief of R&D. The two bodies, SENSCOM and ABMDA, were collocated and coordinated both in Washington and in Huntsville. The SENTINEL System Evaluation Agency, was also created as a Class II Activity located at WSMR, replacing the NIKE-X Engineering/Service Test Office. The Program Manager NIKE-X, Project Manager Kwajalein Test Site and NIKE-X Engineering/Service Test Office were discontinued as AMC activities.

15 November 1967 Secretary of the Army nominated the SENTINEL System production program to the S category of the master urgency list.

1968

March 1968 At the direction of the Secretary of Defense, the ARPA's research effort into advanced BMD concepts, Project Defender, transferred to the Army.

30 March 1968 First SPARTAN missile launched from Kwajalein. The SPARTAN "performed according to its fight plan with the flight terminating at 277.7 seconds. All test objectives were met."

10 April 1968 Ground-breaking ceremony held at SENSCOM Headquarters, for their new site at 106 Wynn Drive, in Huntsville.

16 April 1968 Kwajalein Test Site re-designated Kwajalein Missile Range (KMR).

14 May 1968 Under an MOA, the U.S. Army Engineering Division, Huntsville, was placed under operational control of the SENTINEL System Manager.

30 June 1968 The U.S. Army NIKE-X Development Office (NXDO) was established as a Class II Activity of the Chief of R&D at Huntsville, AL, under command jurisdiction of the ABMDA with responsibility for execution of the Army Advanced Ballistic Missile Defense Program. The Advanced Development Directorate of the SENSCOM and that portion of the Advanced Research Projects Division of the MICOM designated as supporting Project DEFENDER were transferred to the U.S. Army NXDO.

1 October 1968 Operational Control of PRESS Complex (less ALTAIR and ALCOR) transferred from ARPA to Kwajalein Range Directorate.

15 October 1968 Office of the Secretary of Defense approved the Sentinel Deployment Model (DEMOD 1-68 Revised).

1969

19 January 1969 The Missile Site Radar at Meck Island became fully operational.

20 January 1969 President Richard Nixon took office and initiated a DoD review of strategic offensive and defensive priorities.

6 February 1969 Secretary of Defense Melvin Laird ordered a halt in the deployment of the Sentinel system, pending the completion of a one-month review of U.S. strategic programs and other weapons systems by the new administration.

14 March 1969 President Richard Nixon redirected the BMD program. Components remained unchanged but deployment concepts were redrawn. Nixon specified three defense objectives: "Protection of our land-based retaliatory forces against a direct attack by the Soviet Union"; "Defense of the American people against the kind of nuclear attack which Communist China is likely to be able to mount within the decade"; and, "Protection against the possibility of accidental attacks from any source." The primary new focus for the program is the defense of twelve U.S. land-based ICBM sites. Authorization was subsequently given for only two MINUTEMAN bases, Grand Forks AFB, ND, and Malmstrom AFB, MT.

25 March 1969 SENTINEL System Manager became the SAFEGUARD System Manager. Similarly the SENTINEL System Organization, SENTINEL System Command, and SENTINEL System Evaluation Agency were re-designated SAFEGUARD. A separate order renamed the SENTINEL Logistics Command, the SAFEGUARD Logistics Command.

1 May 1969 The NXDO was re-designated the ABMDA, Huntsville. The mission was unchanged.

8 May 1969 The Institute of Heraldry approved the shoulder sleeve insignia for SAFEGUARD. This insignia was used until the mid-1990s.

1970

30 January 1970 President Nixon announced his decision to extend the deployment of SAFEGUARD, beyond the initial two-site Phase I program. The recommendation included a third site (Whiteman AFB, MO) and advance preparation for five additional sites (in the NE, NW, Washington, D.C., Warren AFB, WY, and in the Michigan-Ohio area). There was no deployment commitment for the latter sites.

21 July 1970 U.S. Army SAFEGUARD System Site Activation command – Grand Forks, ND, organized.

1 August 1970 The U.S. Army SAFEGUARD System Site Activation Command Malmstrom, located at Conrad, MT, organized.

12 August 1970 Forty-second and final SPRINT firing from WSMR.

8 September 1970 DoD released a revised DoD Directive 5160.32, Development of Space Systems. It assigned the following DoD Satellite Development responsibilities: (1) Each service is to conduct research and receive approval to develop the following type satellites: "unique battlefield and ocean surveillance, communication, navigation, meteorological mapping, charting and geodesy satellites." (2) The Air Force is to perform R&D, production, and developing of the following systems: launch support, launch vehicles, warning and surveillance satellites enemy nuclear capabilities, and orbital support operations; (3) the DoD Director of R&D is to serve as focal point for space technology and system to prevent unwarranted duplication minimize technical risk and cost, and ensure multiple service needs are met.

9 October 1970 The KMR Directorate SAFSCOM was organized.

23 December 1970 First live target intercept by a SPRINT missile, with intercept of an ICBM reentry vehicle, demonstrating the ability to conduct high-speed low-altitude (endo-atmospheric) intercepts Mission M1-12. The test was conducted from Meck Island.

1971

11 January 1971 The first salvo launch was made from Meck Island, in the Republic of the Marshall Islands. The test involved two SPARTAN missiles. One missile successfully intercepted an RV, the other a space point.

14 January 1971 Deputy Secretary of Defense Packard ordered the Army to proceed with a new facet in the BMD development, the Hardsite Defense (HSD) Project Office. Deployed in groups to protect Minuteman sites and each other, the concept called for a phased array radar, an interceptor, and commercial data processing equipment. Site Defense would be capable of handling a larger, more sophisticated threat than SAFEGUARD. The Project Office was under the SAFSCOM.

22 January 1971 The U.S. Army SAFEGUARD System Site Activation Command BMDC, Colorado Springs, CO, organized. In addition to directing activities of the site activation program, the BMDC, was "responsible for interface coordination of system requirements, equipment design, building configuration, logistic support and the whole spectrum of engineering and technical specifications of the integration of the BMDC with the expanded NORAD Cheyenne Mountain Complex."

April 1971 A SPARTAN missile successfully intercepted an IRBM for the first time.

6 May 1971 The SPRINT missile system intercepted its first IRBM (POLARIS MARK-2).

22 October 1971 The ARADCOM issued General Orders creating the first two units to be assigned to man the SAFEGUARD sites. Both units were organized with zero strength, but provided an organization to which newly trained personnel may be assigned. They are scheduled to be organizational 1 September 1973.

2 December 1971 Construction began on the BMD Center located in NORAD's Cheyenne Mountain Complex. It was the command and control element of SAFEGUARD.

1972

16 March 1972 A SPRINT missile successfully intercepted an ICBM. This was the first remote launch from Illeginni. The purpose was to test "toss and catch."

19 April 1972 The SAFSCOM Site Defense of Minuteman Project Office located in Huntsville, AL, was re-designated the SAFSCOM Site Defense Project Office (SAFSCOM General Orders 10, 25 April 1972). This change reflects the current Army program, previously known as the Prototype Demonstration Program for Site Defense of Minuteman" and now described as the "Prototype Demonstration Program for Site Defense."

26 May 1972 President Richard Nixon and General Secretary Leonid Brezhnev of the Soviet Union signed the ABM Treaty. Both nations agreed to a limit of two ABM sites each, one near the capital and the other near an ICBM complex. Each ABM site could have 100 missiles and 100 launchers and 15 additional launchers at test sites. In addition, the treaty regulated the type of radars for the ABM site. Finally, the treaty prevented each country from defending its entire territory, thereby negating the deterrent effect.

An interim accord, signed at the same time, set maximum limits for each country's ICBM and sea-launched ballistic missiles (SLBMs) for five years. The U.S. was allowed 1,054 ICBMs, the amount it had had since the mid-1960s, and 710 SLBMs with 44 ballistic missile submarines. The U.S.S.R. was allowed 1,618 ICBMs, 950 SLBMs and 62 submarines. The treaty was ratified by the Senate on 3 August 1972 and signed in Washington by President Nixon on 3 October.

1973

The Army created the Army Space Program Office (ASPO) and the Tactical Exploitation of National Capabilities Program (TENCAP). "TENCAP provides developmental equipment to provide the means by which national level systems can provide support to designated battlefield commanders."

3 April 1973 The Secretary of Defense issued a memorandum, which provided guidance for the DoD's BMD programs. It specified the objectives as follows: "to deploy at the Grand Forks Site a system with the objectives of providing: (1) defense of retaliatory forces and (2) a base of obtaining experience with installation, test, and operation of a deployed ABM site." A product of this memo was the development of the SAFEGUARD Operational Experience Program, designed "to provide a systematic method of assuring that the experience obtained through deployment and operation of the SAFEGUARD is available as a significant operational experience data base for use in the development and deployment of future BMD systems and other complex systems."

21 June 1973 The last SPARTAN R&D missile and the first SPARTAN production missile were successfully flight tested in a dual salvo mission.

5 August 1973 The PAR at Grand Forks, ND, tracked its first satellite.

31 August 1973 Secretary of Defense signed an Amended Program Decision which placed funding and operational constraints on the SAFEGUARD program and funding constraints on the Site Defense program.

29 November 1973 Last remote SPRINT launch from Illeginni and last planned live intercept in the SAFEGUARD Meck Test Program conducted successfully.

1974

February 1974 The last operational NIKE-ZEUS facility ceased to operate. The KMR Target Track Radar-4 had participated in a variety of test programs over the previous 13 years.

13 February 1974 First launch of the ATHENA from Wake Island, as part of the Army Special Targets Program. The seventh and final launch took place on 23 June 1974.

1 March 1974 U.S. Surveillance Battalion Grand Forks Site was reorganized at Cavalier, ND, to "provide long-range surveillance and early warning of a ballistic missile attack against the continental United States." It was assigned to the ARADCOM.

1 March 1974 SAFEGUARD Command Grand Forks Site was reorganized and assigned to the ARADCOM. Located at Langdon, ND, its mission was to "defend selected retaliatory missile sites against a ballistic missile attack."

20 May 1974 The SAFEGUARD System Organization was re-designated the Ballistic Missile Defense (BMD) Organization. Similarly, the SAFEGUARD System Manager, Program Office and the SAFSCOM became the BMD Program Manager, BMD Program Office and BMD Systems Command (BMDSCOM), respectively.

The General Orders created a new body, the BMD Advanced Technology Center (BMDATC), as a field operating agency, in Huntsville, AL, under the BMD Program Manager. The BMDATC replaced the ABMDA Huntsville, while the ABMDA Arlington, a field operating agency under the Chief of Research, Development and Acquisition was also discontinued and its personnel, etc. transferred to the BMD Program Office. The BMD Program Manger was assigned, as principal assistant and staff advisor, to the Office of the Chief of Staff. The mission is to deploy and operate the SAFEGUARD System, execute the Site Defense program, conduct research and development in advanced BMD technology, and manage KMR.

30 May 1974 Equipment Readiness Date for the SAFEGUARD BMDC at Colorado Springs, CO. Also the 721 SAFEGUARD PAR subsystem tests were completed.

21 June 1974 SAFEGUARD tactical facilities in North Dakota were designated the Stanley R. Mickelsen SAFEGUARD Complex. The word "complex" was chosen to emphasize the geographical dispersion of the MSR, PAR and four Remote SPRINT Launch sites (General Order 21, 21 June 1974).

30 June 1974 The KMR Directorate was reorganized and assigned to the BMDSCOM. The Commanding General, BMDSCOM, commanded the unit as the National Range Commander, under the direction of the BMD Program Manager.

3 July 1974 President Nixon and First Secretary Brezhnev met at Yalta and agreed to expand the 1972 ABM Treaty. The protocol limits each country to one ABM site, located at either the National Command Authority or an ICBM complex and decreased limits on the number of ABM launchers and interceptors from 200 to 100. The document was signed by President Nixon and General Secretary Brezhnev at the second Moscow summit.

27 September 1974 Acceptance ceremony held for the U.S. Army SAFEGUARD System Tactical Complex at Nekoma, ND.

1 October 1974 The Stanley R. Mickelsen SAFEGUARD Complex (SRMSC).was officially accepted by the Army and dedicated to Lieutenant General Stanley R. Mickelsen. It was the first new military installation in the U.S. since World War II. The system reached initial operating capability in April 1975 and full operational capability in September 1975. The SRMSC reached full operational capability, following the installation of the missiles 30 SPARTANs and 70 SPRINTs. Per the Secretary of Defense's direction, SAFEGUARD was used as "a base for obtaining experience with installation, test, and operation of a deployed BMD site." The SAFEGUARD Complex became the first and only ABM System in the free world. The SAFEGUARD System achieved FOC status three days ahead of schedule.

17 December 1974 The ARADCOM was disestablished.

1975

1 January 1975 The SAFEGUARD Program Office reorganized as the BMD Program Office, and remained assigned to the Office of the Chief of Staff. The new mission was to assist the PM in the following: the development of a program which insures operation of the SAFEGUARD BMD System; the execution of the Site Defense Program; conduct R&D in advanced BMD technology; and, management of the KMR as a National Range.

8 January 1975 The PAR accomplished its first tracking of two live targets.

6 February 1975 Technical Proficiency Inspection of the SAFEGUARD Complex was completed and the SAFEGUARD System certified for its nuclear mission.

1 March 1975 The BMDATC was reorganized. Its mission was to "formulate and execute approved BMD programs of exploratory and advanced development in BMD technology within the guidance and direction of the BMD Program Manager." In addition, it would "(a) provide the advanced technology foundation for improving ballistic missile defense capability; (b) provide a measure of the BMD technology art to avoid technological surprise by an adversary; and (c) assist in the development and assessment of future U.S. strategic offensive systems."

1 April 1975 The SAFEGUARD System reached initial operating capability with 28 SPRINT and 8 SPARTAN missiles deployed - and the "fully netted" system was turned over to the Commander of the Continental Air Defense Command (CONAD) for operational control.

17 April 1975 Mission M2-25 the 26th and final SPARTAN missile launch in the SAFEGUARD Meck Test Program was a success.

30 April 1975 The final SPRINT launch and final SAFEGUARD Meck Test mission. Thirty-three SPRINT missiles were launched in the Meck Test Program.

1 July 1975 The Aerospace Defense Command, a specified command, was created, superseding the previous, Air Force only, Aerospace Defense command and was tasked with space surveillance and defense responsibilities. It also assumed the past responsibilities of the CONAD, which was disestablished.

28 September 1975 The House Appropriations Committee recommended deactivation of the SAFEGUARD site by the end of the fiscal year: "Because of the improved capability of the Soviet Union's new MIRVed missiles, the limited effectiveness of the SAFEGUARD system to provide the protection it was originally intended to provide and the diminished benefits from operating the facility for only a single year."

3 October 1975 The Army formally transferred the PAR to the U.S. Air Force as the PAR Attack Characterization System. The ceremonies were conducted at the PAR site in Concrete, ND.

24 October 1975 Site Defense was changed to the Systems Technology Program.

18 November 1975 Senator Edward Kennedy introduced an amendment to the fiscal year 1976/7T Appropriations bill. The amendment read: "Provided further that funds provided in this act for the Operation and Maintenance of the ABM Facility (other than funds provided for operation and maintenance of the PAR) may be used only for the purpose of the expeditious termination and deactivation of all operation of that facility." The amendment was incorporated into the final act.

1976

10 February 1976 The Joint Chiefs of Staff directed that the deactivation of SAFEGUARD begin, as per the Congressional decision (Public Law 94-212, dated 9 February 1976). Radiation for the MSR and the missile launch capability were terminated and the warhead withdrawal commenced. Termination involved the following sites: SAFEGUARD Training Facility, Fort Bliss, TX; BMD Center, Colorado Springs, CO; SAFEGUARD Supply and Maintenance Center, Glasgow, MT; and the missile fields, MSR site, and support facilities, all located in Nekoma, ND. The SRMSC entered "caretaker status."

17 May 1976 The PAR began tracking operations against known satellites. The PAR is capable of deep space tracking.

31 August 1976 The U.S. Army SAFEGUARD Command was inactivated. Personnel and equipment transferred to BMDSCOM, with duty stations to remain in North Dakota.

1977

3 January 1977 The PAR was linked to the NORAD Combat Operations Computer. With this the NORAD Early Warning Sensor became operational under the Army. The Air Force arrived in May 1977.

22 January 1977 The BMDSCOM chartered the LoAD System.

22 August 1977 Air Force personnel assumed tactical responsibility for the PAR.

30 September 1977 Dismantling of the SRMSC facility at Nekoma, ND, completed.

October 1977 Secretary of the Army Clifford Alexander authorized the first Army Award for Project Management outside the Department of the Army Readiness Command. Brigadier General John G. Jones was recognized for his outstanding accomplishments on the BMD program.

1 October 1977 The PAR complex transferred to the Air Force as the PAR Attack Characterization System. The Spacetrack capability became operational in December. The PAR was part of the Air Force's Aerospace Defense Command (1 May 1977 to 30 November 1979);

the Strategic Air Command (1 December 1979 - 30 April 1983) and the Air Force Space Command (1 May 1983).

1978

January 1978 At the request of the Deputy Undersecretary of Defense Research and Engineering (Strategic and Space Systems), the BMD Program initiated a Minuteman Defense (MDS) II study "to define and rate the most feasible systems (or concepts) for defending the Minuteman."

16 January 1978 NASA Administrator Robert Frosch announced the selection of 35 new astronaut candidates for the Space Shuttle program. This was the first group to be selected since 1969. Major Robert Stewart, the first Army astronaut, was a mission specialist among this group of candidates.

2 February 1978 Secretary of Defense Harold Brown stated in his annual report to Congress that "An aggressive BMD R&D program is vital to this nation's interest." He noted the evolving technological base from the STP and ATP efforts "could provide, if strategic arms limitation efforts lead us in that direction, cost effective alternatives for maintaining the survivability of our strategic retaliatory elements in the ICBM threat environment," and emphasized DoD would continue the BMD R&D at a constant real program level of effort.

10 February 1978 The ALTAIR on Kwajalein concluded its test phase.

June 1978 Deputy Undersecretary of Defense Research & Engineering (Strategic and Space Systems) "directed that emphasis in the program be placed on near-term defense concepts and technologies applicable to defense of our land-based missile forces in the 1980s."

21 August 1978 In briefing the U.S. Congressional Budget Analysts, the BMDPM stated: "The restrictions on deployment previously were thought to be such that a treaty limited deployment would not be worthwhile. However, due to advancing technology, this is no longer true and a limited deployment can be useful. We are presently studying this concept."

12 December 1978 In the first of five consecutive tests, the Designating Optical Tracker (DOT) proved long-wave infrared sensors could discriminate, designate and track a reentry vehicle. The DOT set the standard for future LWIR technology.

1979

18 June 1979 President Jimmy Carter and General Secretary Brezhnev signed the SALT II agreement in Vienna. It was agreed that within 6 months each side would have only 2,250 launchers (ICBMs, SLBMs, air-to-surface ballistic missiles and heavy bombers), of these 1,200 of them could be MIRVed. There was no limit on submarine launch vehicles. The agreement included a protocol signed by Brezhnev promising to limit the range and production of the Backfire bomber and statement of principles for SALT III.

In January 1980, following the Soviet invasion of Afghanistan, President Carter removed the treaty from consideration by the Senate. However both countries agreed to observe the two SALT agreements pending clarification of the technical descriptions in SALT II.

28 September 1979 BMD Program Charter was signed.

12 October 1979 The Institute of Heraldry approved the BMD flag.

1982

4 July1982 President Ronald Reagan announced the National Space Policy. NSDD-42 superseded all previous presidential space policy directives. It included five basic commitments: "(1) To the exploration and use of space by all nations for peaceful purposes to permit activities in pursuit of national security goals. (2) To conduct international cooperative space-related activities that achieve scientific, political, economic, or national security benefits for the U.S. (3) To pursue activities in space in support of the United States' inherent right of self-defense. (4). To develop STS capabilities and capacities to meet appropriate national needs and to make the STS available to commercial and government users,.... (d) To continue to study space arms control options and to consider verifiable and equitable arms control measures that would bank, or otherwise limit, testing and deployment of specific weapons provided those measures were compatible with the U.S. national security...."

20 August 1982 DA published *FM 100-5 Operations* which launched the Army's AirLand Battle Doctrine.

1 September 1982 U.S. Air Force Space Command established "to further consolidate Air Force operational space activities."

3 September 1982 The LoAD system re-designated SENTRY.

1983

11 February 1983 The DCS unanimously recommended that the U.S. pursue a national security strategy which placed increased emphasis on strategic defenses. Their decision followed repeated meetings to determine an effective and publicly acceptable fielding program for the MX missile.

23 March 1983 President Ronald Reagan announced his SDI a shift from hardsite defense to defense of the U.S. His speech urged the exploration of the possibility of developing missile defenses as an alternative to deterring nuclear war. The address also added active defense to a primarily offensive deterrence strategy. On 25 March 1985, the President issued National Security Decision Directive 85 which implemented his plans. In July 1984, the BMDO became a part of the SDI.

18 April 1983 Guidance was issued by the President for two studies. The first, the Defense Technologies or Fletcher Report, would assess the state of missile defense technology and recommend a technology program for the new missile defense program. The second, known as the Future Security Strategy Study (or the Hoffman Report), would assess the strategic and policy implications of such a program.

July 1983 Senator Malcolm Wallop of Wyoming proposed an amendment which would give the BMD organization responsibility for all defense related development of laser technology. The amendment failed.

1 October 1983 The Naval Space Command was established at Dahlgren, VA.

1984

6 January 1984 NSDD 119 was issued authorizing the SDI, to explore the possibility of developing missile defenses as an alternative means of deterring nuclear war. The SDI program would be "focused to demonstrate the technical feasibility of enhancing deterrence and thereby reducing the risk of nuclear war through a great reliance on defensive strategic capabilities." Although non-nuclear efforts were the emphasis of the program, research work on defensive nuclear devices would continue "as a hedge against Soviet work in the same area." The directive made the Secretary of Defense responsible for the new program. Air Force Lieutenant General James Abrahamson was named the first director of the SDIO on 5 April.

23 January 1984 The Reagan administration issued its first President's Report on Soviet Non-Compliance, a series of reports on Soviet non-compliance with arms control agreements. This report deemed the Krasnoyarsk radar an outright violation of the ABM treaty.

3-11 February 1984 Lieutenant Colonel Robert L. Stewart became the first soldier to fly into space, as a Mission Specialist aboard STS-41B (*Challenger*).

2 March 1984 General John A. Wickham, Jr., Army Chief of Staff, awarded Colonel Robert Stewart his wings in a ceremony at Fort Myer, VA. Fort Myer was the site of the initial military airplane test flights in 1909.

10 June 1984 The HOE proved that it was possible to hit a bullet with a bullet, when it successfully conducted the first kinetic kill intercept of an ICBM reentry vehicle.

1 July 1984 The ERIS Project Office was established.

15 August 1984 The GPS, Ground Antenna and Monitor Station became operational.

15 August 1984 Army Space Council received charter.

1 October 1984 The Army Staff Field Element was activated at the direction of the Vice Chief of Staff of the Army. Located at the Space Command headquarters, this five-person group was to (1) "formulate Army policy pertaining to space and future participation in a Unified Space Command", (2) "exchange information pertaining to space policy, strategy and plans", (3) "monitor space-related education and training developments for Army use", (4) "represent the

Army Space Office at HQ Space Command", and (5) provide technical information to Space Command concerning Army space efforts, as appropriate and required."

4 October 1984 The Army Space Council, chaired by the Vice Chief of Staff of the Army met to discuss the Army's emerging role in space. They produced guidance for future Army efforts, the Army Space Personnel Program.

11 December 1984 Secretary of the Army approved the AOA Program Charter.

1985

January 1985 The Training and Doctrine Command directed the Combined Arms Center to establish a space directorate at Fort Leavenworth, KS. This six person unit was tasked with developing concepts, doctrine and operational requirements for the use of space to support Army operations.

26 March 1985 Secretary of Defense Caspar Weinberger invited Allied participation in the BMD program.

May 1985 The DCSOPS established the Army Space Initiative Study Group.

5 June 1985 Army published the Army Space Policy. The Office of the DCSOPS in a press release stated "that the Army will exploit space in order to enhance the capabilities of all Army elements at tactical, operational, and strategic levels of war. Of greatest interest are the abilities of space systems to provide: Reliable communications over great distances, products of observation of the earth's surface, extremely precise positioning navigation, and continuous monitoring of terrestrial environmental conditions.

1 July 1985 The USASDC was created as a field-operating agency of the office of the Army Chief of Staff. In January, the BMDATC and the BMDSCOM officially dissolved into the framework of USASDC. The position of BMD PM remained unchanged.

August 1985 The DA published Interim Space Operational Concept.

22 September 1985 The Secretary of Defense created the unified U.S. Space Command at Peterson AFB, Colorado Springs, CO. This organization is responsible for space operations, surveillance, early warning, and BMD operational planning. Concurrent with this decision, the Army Staff Field Element was re-designated the Army Space Planning Group, under the operational control of Commander, U.S. Space.

14 November 1985 Citizens of the Republic of the Marshall Islands protested the renewal of the lease by the U.S. by occupying Meck, Omelek, and Eniwetak Islands. This action began six months of demonstrations.

18 November 1985 Secretary of the Army, John O. Marsh, Jr., signed the charter for the HEDI Project Office.

December 1985 The SDIO assigned USASDC the task of developing TMD architectures.

13 December 1985 Army Space Initiative study published. The recommendation of these 30 officers, led by Brigadier General William G. Fiorentino, in conjunction with the RAND-ARROYO Study entitled "Army Master Plan for Space," reputedly led to the organization of the provisional Army Space Agency.

1986

14 January 1986 President Ronald Reagan signed the Compact of Free Association with the Republic of the Marshall Islands.

7 February 1986 The BMD Radar Project Office became the Terminal Imaging Radar Project Office.

21 February 1986 DoD Directive 5141.5 established the SDIO as an agency of the DoD. Its mission was to "manage and direct a vigorous research program, including advanced technologies, which will provide the basis for an informed decision regarding the feasibility of eliminating the threat posed by nuclear ballistic missiles of all ranges, and of increasing the contribution of defensive systems to U.S. and allied security. They were also directed to "protect the near-term deployment of limited ballistic missile defense." Programs are to be conducted in consultation and, where appropriate, participation of our allies. The SDIO program was to be conducted in compliance with existing treaties and will emphasize "non-nuclear technologies." Directive 5141.5 issued on 4 June 1997 replaced this document.

21 February 1986 General Orders 5 established the USASDC. The effective date for this transition was 1 July 1985.

27 March 1986 West Germany initiated SDI research. Italy became the second European country to undertake SDI research on 19 September 1986.

29 April 1986 The ALTAIR tracked its 100,000 deep-space satellite.

7 May 1986 The KMR became the U.S. Army Kwajalein Atoll

2 June 1986 U.S. Army Training and Doctrine Command assigned 3Y (space activities) proponency. On the same date, TRADOC established the Army Space Institute (Provisional) at Fort Leavenworth, KS, from the former CACDA Space Directorate. The institute was established as an integrating center for space-related developments. In December 1990, TRADOC downsized the program, renaming it the TRADOC Program Integration Office Space/Army Space Institute.

20 June 1986 The DCSOPS was designated the space lead within the Army staff with the creation of the Space Division, Space and Special Weapons Directorate.

30 June 1986 The Secretary of Defense directed the SDIO to "explore, the specific ways in which the U.S. led SDI research program can assist the NATO extended air defense effort in which the Europeans are taking a leading role".

July 1986 "USCINCSPACE recommended to the Chief of Staff, Army, that the Army take a more active role in space and that as initial step, the Army Component assume operational and maintenance responsibility of the DSCS GMFSC and MSQ-114 functions."

15 July 1986 The GBFEL Project Office was created at WSMR, NM.

1 August 1986 U.S. Army Space Agency (USASA), formerly the Army Space Planning Group, was provisionally activated. It was the Army component of the USCINCSPACE and a FOA of DCSOPS. August was selected because of its historical significance the first research and development Redstone missile, which first put Americans into space, was flight tested at Cape Canaveral, FL, on 20 August 1953.

18 August 1986 The Kwajalein Battlefield was dedicated as a national landmark.

October 1986 The Chief of Staff of the Army directed a joint U.S. Army Information Systems Command and USASA working group to study the feasibility of transferring the DSCS mission.

November 1986 At a meeting of the Army Space Council, the Vice Chief of the Army stated that most of the Army was not aware of space capabilities. He directed that a Space Demonstration Program be initiated with the goal of demonstrating enhanced AirLand Battle execution using space based assets to Army units.

1987

January 1987 The DSCS Command and Control Concept outlined. It includes a chain of command that runs from JCS, through U.S. Space Command to USASA, technical direction of DCA.

5 January 1987 The Multinational Programs Office initiated the TMD Architecture Study, which involved American, German, French, Italian, British, and Israeli corporations.

5 January 1987 The USASA's NASA Detachment was established at Johnson Space Center, Houston, Texas. Army astronauts and other Army personnel working at NASA are assigned to this unit. The detachment is under the operational control of the NASA with administrative control exercised by ARSPACE.

February 1987 An MOU signed by USAISC and U.S. Army Space Agency detailed operational control and transfer of GMFSC managers at Regional Satellite Support Centers (RSSCs). The MOU established the responsibilities and relationships and the two commands with regard to the transfer of operational control of the DCSOC functions worldwide.

March 1987 The DoD issued a new Space Policy, revising the 1982 version. The changes include: the addition of the SDI program; the revision of the nation's launch philosophy to include expendable launch vehicles; the successful testing of the anti-satellite system against an object in space, the formation of unified and service commands for space, the emergence of commercial space enterprises and the initiation of a manned space station program with international involvement, the increasing commitment on the part of other nations towards space exploitation and the stringent funding constraints imposed by budget limitation legislation. "Space is recognized as medium within which the conduct of military operation in support of our national security can take place, and similarly from which military space functions of space support, force enhancement, space control and force application can be performed."

4 June 1987 William Taft IV, Deputy Secretary of Defense, issued Directive No. 5141.5 re: SDIO. This document lays out the mission, organization and management, and functions and responsibilities for the SDIO, as well as relationships and authorities. While the mission remained unchanged, this document revised the overall supervision of the SDIO from the Secretary to the Deputy Secretary of Defense. Other changes can be found, for example, in the composition of the SDI Executive Committee, which provided DoD oversight and guidance for the SDI program, in the functions of the Director, etc. The document supersedes Directive 5141.5 of 21 February 1986.

August 1987 The DCSOPS approved five programs submitted by the Army Space Institute, et al, for the initial Army Space Demonstration Program. The goal of the program was "to demonstrate the capabilities of space systems to provide support to tactical units in the Army." The programs were SLGR, GPS Azimuth Determination System, WRAASE Weather Receiver, LIGHTSAT, TMD Command and Control Missile Detection.

September 1987 Secretary of Defense Cooper Weinberger approved the SDS Phase I baseline architecture and authorized six components of SDI to enter Dem/Val after a DAB recommendation. These included a SBI, a GBI, a ground-based sensor, two space-based sensors, and a battle management system.

16 September 1987 TRADOC established 3Y standards.

1 October 1987 The USASA was established under General Orders 7, dated 15 March 1987. The USASA was the Army component of the U.S. Space Command and a Field Operating Agency of the DCSOPS, DA. The USASA provided USSPACECOM an Army perspective in planning for DoD space system support to land forces and for strategic defense.

8 December 1987 American President Ronald Reagan and Soviet General Secretary Mikhail Gorbachev signed the Intermediate Range Nuclear Forces Treaty. This agreement mandated the removal of 2,611 intermediate range nuclear missiles from the European continent. The Senate ratified the treaty on 27 May 1988, by a vote of 95-5.

1988

6 January 1988 At the request of the Assistant Secretary of the Army, the U.S. Army Materiel Command established a technology manager to manage the near and possible far-term space R&D programs and to provide a developer focus both within the Army and with outside agencies the Army Space Technology and Research Office.

19 January 1988 In a speech to the Arms Control Association, Senator Sam Nunn (D-GA) called for a reorientation of the SDI program. Specifically, he advocated focusing the SDI program first on developing a "limited system for protecting against accidental and unauthorized missile launches." The long-range goal would be to develop a more comprehensive defensive system.

February 1988 The JCS approved the MILSATCOM Command and Control Concept (MJCS-11-89), which aligned the DCSC Operations Control System under Commander, U.S. Space to the Army Component and included GMFSC- RSSCs, DSCSOCs/MSQ-114, and CDOCS. The RSSCs would be collocated with the Defense Communications Agency in Washington DC, Wheeler AFB, HI, and Vaihingen, Germany.

3 February 1988 President Ronald Reagan's National Space Policy updated. This policy reaffirmed committed to exploration and addressed civil, military and commercial space had been approved by Reagan on 5 January 1988. The Presidential Directive established the following goals:

To strengthen the security of the United States;

To obtain scientific technological and economic benefits for the general population and to improve the quality of life on Earth through space related activities;

To encourage private sector investment; To promote international cooperative activities taking into account U.S. security, foreign policy, scientific and economic interests;

To cooperate with other nations in maintaining the freedom of space for all activities that enhance the security and welfare of all mankind; and,

To expand human presence and activity beyond Earth orbit into the solar system.

The following principles would guide these goals: (1) The U.S. is committed to the exploration and peaceful use of outer space for the benefit of all mankind. Peaceful use allows for activities in pursuit of national security goals. (2) The U.S. will pursue activities in support of its right of self defense and defense of its allies. (3) The U.S. rejects any claim of sovereignty over outer space or celestial bodies. (4) The U.S. considers the space system of any nation to be national property. (5) The U.S. encourages the commercial use and exploitation of space technologies. (6) The United States will conduct international cooperative space related activities that arte expected to achieve sufficient scientific, political, economic, or national security benefits for the nation.

7 April 1988 U.S. Army Space Command activated, as the Army component to the U.S. Space Command. The ARSPACE was created to provide an Army perspective in planning for DoD space system support to land forces and strategic defense operations. Responsibility for the operation of the DSCS Operations Centers transferred to ARSPACE from the Information Systems Command. The ARSPACE remained a Field Operating Agency of the DCSOPS.

The mission for the newly created organization was "As the Army component provide USSPACECOM an Army perspective in planning for DoD space system support to land forces and strategic defense operations. Ensure integration of Army requirements into USSPACECOM planning for space support. Respond to USCINCSPACE directed taskings. Conduct planning for DoD space operations in support of Army strategic, operational and tactical missions."

22 April 1988 The Institute of Heraldry approved the ARSPACE request for a shoulder sleeve insignia and a distinctive unit insignia. ARSPACE received an exception to policy, based on "the high visibility, which surrounds the Army's potential military role in space and the Army Space Command's projected growth."

1 October 1988 The ARSPACE's Ground Mobile Forces Satellite Communications MFSC managers formally activated the RSSCs planning and management cells. The planning cells support the Unified and Specified CINCs with GMF access on the DSCS.

5 October 1988 Lieutenant General Robert Hammond, Commander, USASDC, was appointed PEO for Strategic Defense. The appointment was made under the direction of President George Bush through National Security Directive 219, by Secretary of the Army Michael P.W. Stone. With this position, Lieutenant General Hammond reported directly to the Army Acquisition Executive.

14 November 1988 The GSTS Project Office was established.

1989

February 1989 The MOU signed by USAISC and ARSPACE detailed the remainder of the DSCS mission transfer.

9 February 1989 President George Bush announced in a Joint Session of Congress that he will "vigorously pursue" the SDI.

27 February 1989 The Kinetic Energy Antisatellite (KE ASAT) JPO was established in Huntsville. Brigadier General J. Morgan Jellett headed the organization.

6 March 1989 The Directed Energy portion of the Anti-Satellite Acquisition Decision Memorandum tasked the Army to develop the prime candidate for the DE ASAT weapon based upon the Army-managed, SDIO GBFEL Technology Integration Experiment. The Air Force was tasked to develop a candidate based on alternate technologies.

14 June 1989 Following a three-month general review of the U.S. national defense strategy, ordered by President George H.W. Bush, the President "concluded that the goals of the SDI program were generally sound."

July 1989 The JCS validated the USCINCSPACE Tactical Event Reporting (TERS) mission requirement, following the successful completion of tests showing the "utility of the concept." The ARSPACE role was to monitor the TERS via its constant source equipment located at the ARSPOC.

August 1989 Memorandum sent to the SDIO Director, signed by General Robert Riscassi, Vice Chief of Staff, "[prioritized] Army research needs for capabilities to protect critical assets and forces from attack by non-nuclear tactical ballistic missiles." The priorities were as follows: High Altitude Area Defense, Contingency Forces Defenses, Chemical Defense, Survivability, BM/C3I, and Launch Point Detection.

September 1989 The Army Space Demonstration Program delivered over 100 WRASSE weather receivers to units worldwide. These receivers, which receive data from U.S., Soviet, Japanese and European civil satellites, were the first product of this new program, created by Vice Chief of Staff of the Army General Maxwell Thurman in November 1986.

2 November 1989 President George Bush approved a new national space policy. "The policy reaffirmed the nation's commitment to the exploration and use of space in support of the U.S. national well being. The policy recognizes that leadership in space activities and capabilities requires preeminence in key areas It also retains the long-term goal of expanding human presence beyond Earth orbit into the Solar System." The overall goals of U.S. space activities are: (1) To strengthen the security of the U.S. (2) To obtain scientific, technological and economic benefits for the general population and to improve the quality of life on Earth. (3) To encourage continuing U.S. private sector investment in space and related activities. (4) To promote international cooperative activities. (5) To cooperate with other nations in maintaining the freedom of space for all activities that enhance the security and welfare of mankind. (6) To expand human presence and activity beyond earth's orbit into the Solar System."

December 1989 The Army is given lead in the ASAT JPO.

1990

15 March 1990 Ambassador Henry Cooper submitted the results of his independent study of the SDIO program. President George Bush had commissioned the study to examine the strategic requirements for a "'new world order.'" Ambassador Cooper "endorsed the concept of Brilliant Pebbles and spelled out the concept that became the GPALS."

26 July 1990 The Army Space Council approved the USASDC's proposed Army Tactical Surveillance Satellite (ATSS) program. The objective of the ATSS was to provide the tactical commander with a responsive, launch on demand, dedicated satellite furnishing real-time surveillance and targeting information.

9 August 1990 The ARROW had its first flight test

18 September 1990 First flight test of the Airborne Surveillance Testbed conducted.

1 October 1990 Congress directed the restructuring of the MILSTAR satellite program to emphasize its communications and support to tactical users. Accomplishing this task required the development of a smaller lightweight and more cost effective terminal. The MILSTAR is a joint service communications system.

1 October 1990 Effective date for the transfer of the High Energy Laser Systems Facility (HELSTF), from AMC to the USASDC, by the Secretary of the Army. It is to operate as a National Research and Test Facility.

1 October 1990 The ARSPACE assumed the DSCS world-wide operations and maintenance mission from USAISC.

9 November 1990 The Under Secretary of Defense for Acquisition assigned to SDIO the responsibility for the Defense Department's centrally managed TMD program.

16 November 1990 DA Memorandum 5-3, Management of Space and Special Weapons, established an ASWG. Its purpose was to support the Army Space Council. Initially, it provided feedback and concurrence to the ASEDP proposals and provided information and recommendations on space issues to the HQDA focal point for space, Director of Space and Special Weapons, ODCSOPS. In 1993, it was reorganized as the ASEWG.

1991

January 1991 All TMD functions were assigned to the USASDC. The JTMD Management Office, formerly a part of MICOM, was united with the USASDC's TMD Applications Project Office.

18 January 1991 An anti-missile missile intercepted and destroyed a ballistic missile under combat conditions on this date during the Gulf War. A Patriot air defense missile destroyed an Iraqi Scud missile that was attacking a U.S. air base in Saudi Arabia. A reporter for the *Los Angeles Times* wrote: "The age of "Star Wars" had arrived".

28 January 1991 The ERIS FTV-01 successfully intercepted a target in the exoatmosphere. This was "the first SDI experiment which successfully intercepted an exoatmospheric reentry vehicle in a countermeasures environment."

29 January 1991 President Bush announced, in his State of the Union Address, that the SDIO program would shift its focus from defense against a massive Soviet missile attack to the GPALS concept, Global Protection Against Limited Strikes. Specifically, Bush stated: "I have directed that the SDI program be refocused on providing protection from limited ballistic missile strikes, whatever their source." He added, "let us pursue an SDI program that can deal with any future threat to the U.S., to our forces overseas and to our friends and allies."

29 April 1991 Mr. Alan Sherer, HEDI Project Manager, was the first civilian to be named Project Manager of the Year.

18 June 1991 The LEAP test was successfully completed.

31 July 1991 The U.S. and the U.S.S.R. signed the START I. The Senate ratified this document on 23 May 1992. This Treaty reduced the strategic offensive arms for both the U.S. and the Soviet Union. When fully implemented the ICBMs, SLBMs and heavy bombers of the U.S. and Russia would be reduced to 1,600 with no more than a total of 6,000 attributed warheads in the arsenal of either side. Former Soviet republics signed on to the treaty with the Lisbon Protocol in May 1992.

22 August 1991 First full scale satellite lethality experiment using a high energy laser was successfully completed. This test, conducted at HELSTF, verified the effects of high energy lasers on prospective targets, permitting accurate determination of the size and power required for a DE ASAT weapon system.

5 December 1991 President George Bush signed H.R. 2100, the "National Defense Authorization Act for Fiscal Years 1992 and 1993." That portion of H.R.2100 dealing with missile defenses was known as the Missile Defense Act of 1991. This act required the DoD to "aggressively pursue the development of advanced TMD systems, with the objective of down selecting and deploying such systems by the mid-1990s." Additionally, DoD was to "develop for deployment by the earliest date allowed by the availability of appropriate technology or by fiscal year 1996 a cost effective, operationally effective, and ABM Treaty-compliant antiballistic missile system at a single site as the initial step toward deployment of an antiballistic missile system.

8 December 1991 The leaders of Russia, Ukraine and Belorussia proclaimed the Soviet Union had ceased to exist. They declared the creation of a Commonwealth of Independent States and invited other republics to join them.

1992

13 January 1992 Russia announced its succession to the Soviet Union in all treaties.

1 May 1992 Ambassador Henry Cooper, SDIO Director, concluded the MOA with the secretaries of the military services which "established the organizational structures and procedures for handling the acquisition of the GPALS system as DoD moved ahead with deploying missile defenses in accordance with instructions contained in the Missile Defense Act of 1991."

June 1992 Vice Chief of Staff of the Army and Chief of Staff of the Army approved the Army Strategic Defense realignment which produced the U. S. Army Space and Strategic Defense Command (USASSDC). Included in this proposal was the designation of the ARSPACE as the "'user' for the deployment of the ground based elements of the NMD Program."

24 August 1992 The USASDC separated into the PEO for Global Protection Against Limited Strikes (PEO-GPALS) and the USASSDC. The PEO-GPALS, was a union of the USASDC Project Offices GBI, GBR, GSTS, TMD, etc), and the PEO Air Defense from MICOM (Memorandum of Agreement, 28 July 1992). The ARSPACE, formerly a field operating agency of the office of the DCSOPS became a subordinate command of the USASSDC, a field-operating agency of the Chief of Staff.

1 October 1992 The USASSDC assumed higher headquarters funding responsibility for the ARSPACE, and expected to have this relationship expanded in the future.

2 October 1992 The Department of the Army designated ARSPACE as the responsible agent for the Milstar Network Management and Control. Duties included ensuring that the Joint MILSTAR tool is functionally designed to support Army battlefield requirements and operational conditions.

9 October 1992 The Bishkek Agreement. The Commonwealth of Independent States signed an agreement pledging to support and implement the ABM Treaty.

25 November 1992 The Army terminated the GSTS Project Office.

December 1992 The ARSPACE began to support American forces involved in Operation RESTORE HOPE (Somalia) with space based products.

1993

3 January 1993 American President George Bush and Russian President Boris Yeltsin signed the second START II, during a Moscow Summit. It was to be implemented on 1 January 2003, following the ratification by the U.S. Senate on 26 January 1996. This agreement reduced the number of attributed warheads to an actual total of 3,500, down from 6,000. It also bans land-based multiple warhead ICBMs from both arsenals. A protocol to the treaty was negotiated at the Helsinki Summit in March 1997 and later signed by both parties on 26 September 1997 in New York City. This protocol extends the implementation deadline from 31 December 2003 to the same date in 2007; adds an agreement to begin negotiations on START III to further limit warheads to 2,000-2,500 as soon as START II enters into force; and eliminates the 31 December 2003 deadline for deactivation of all delivery vehicles. The Russian Dumas adopted the Bill of Ratification for the protocol on 14 April 2000. The U.S. Senate has not ratified the Helsinki agreements, which amended both the START and ABM treaties.

28 January 1993 The first campaign of the TCMP – TMD Critical Measurements Program was completed at USAKA. The TCMP program was a product of Operation Desert Storm and the recognized need to gather data on "threat-like missiles" and improve the effectiveness of TMD systems.

13 May 1993 Secretary of Defense Les Aspin announced that with the end of the Cold War, the U.S. was no longer threatened by a massive attack from the Soviet Union. The new concern was theater ballistic missiles controlled by Third World dictators, or "hostile or irrational states that have both nuclear warheads and ballistic missile technology that could reach the U.S."

In response to these changes, the SDIO was reorganized and renamed the BMDO to reflect a new focus in DoD's missile defense program. As part of the reorganization, the BMDO will now report to the Under Secretary of Defense for Acquisition, rather than directly to the Secretary.

14 May 1993 Official opening of the DSCS – Operations Center in Fort Buckner, Okinawa, Japan.

24 May 1993 Mr. George Dausman, Army Acquisition Executive authorized the PEO-GPALS to be renamed PEO for Missile Defense.

28 May 1993 Lieutenant General William Forster, Military Deputy to the Assistant Secretary of the Army RDA, transferred the Army Space Technology Research Office (ASTRO) from the Communications-Electronics Command to the USASSDC. This transfer was made "as part of the Chief of Staff of the Army's initiative to apply sharpened focus and increased emphasis on supporting warfighters with space applications." Later in fiscal year 1993, the ASTRO became the Space Applications Technology Program.

1 July 1993 General Orders 13 designated USASSDC as the Army's focal point for space.

20 August 1993 Kwajalein converted to the west side of the international date line at midnight. As a result, Kwajalein is one calendar day ahead of Hawaii and the U.S. mainland, but is the same day as Majuro and Guam.

19 October 1993 Lieutenant General John Costello, CG USASMDC, appointed the ARSPACE Commander to the position of SMDC Deputy Commander – Space.

1994

11 February 1994 The Army System Acquisition Review Council selected the ERINT to be the missile in the Patriot PAC-3 TMD program, over the Patriot multi-mode missile. Four days later the ERINT hit a ballistic missile target vehicle in a test conducted at WSMR.

4 April 1994 Director of the Army Staff, Lieutenant General Charles E. Dominy, approved the USARSPACE Concept Plan for resourcing manpower requirements for NMD planning, Joint Tactical Ground Station (JTAGS) operations, and Contingency Space Operations now Army Space Support Teams (ARSST).

18 April 1994 The CG announced the new USASSDC Huntsville organizational structure. The principal directorates/bodies were: Executive Director, Advanced Technology, Sensors Technology, Weapons Technology, Engineering and Systems, and Targets, Test and Evaluation. The Directorates for Survivability, Lethality and Key Technologies, DEW and KEW were dissolved into the Weapons Technology Directorate. Similarly, the Battle Management/Command, Control and Communications Directorate became a part of the new Engineering and Systems organization.

24 April 1994 First flight test of the Hera, developed as a target for the THAAD.

19 May 1994 The DAB approved the PAC-3 system upgrade and validated the selection of the ERINT as the new PAC-3 missile.

June 1994 The Rapid Optical Beam Steering System (ROBS) successfully acquired and retargeted the ERINT during tests against an MQM-107D at WSMR. The ROBS is a transportable sensor system that integrates passive and active optical sensors and can track and image up to 50 targets at the same time.

14 June 1994 Deputy Secretary of Defense John Deutch issued Directive 5134.9 re: BMDO which defined the new missions. In addition, DoD oversight transferred from the Deputy Secretary of Defense to the Under Secretary of Defense for Acquisition and Technology. This Directive replaced DoD Directive 5141.5 issued on 4 June 1987.

20 June 1994 Vice Chief of Staff of the Army General J.H. Binford Peay III signed a charter making the CG of USASSDC the TMD Advocate, to serve as the Department's focal point and coordinator for operational aspects of TMD.

July 1994 Army Space Policy issued. It states, in part, that "The Army's future is inextricably tied to space."

1 July 1994 The Army Space Program Office (ASPO), a field agency of the Office of the DCSOPS, DA, transferred to USASSDC. The ASPO, created in 1973 as a Field Operating Agency of the Office of the DCSOP, executes the TENCAP in accordance with the approved ASPO Charter. This transfer was executed under General Orders 17, dated 15 December 1995.

11 July 1994 In the wake of severe flooding in the southeastern United States, the command provided support to the Federal Emergency Management Agency and the Alabama and Florida National Guards with emergency response efforts. Using the same technology used for siting missile defense systems, three teams from the Engineering and Systems Directorate ensured that maps were accurate and reflected the changing weather conditions on a daily basis. The USASSDC Disaster Relief Planning Team subsequently hosted 23 medical emergency planners from 12 former Warsaw Pact nations to address civil-military exercises for emergency planning under the NATO Partnership for Peace program.

August 1994 The ARSPACE assumed a new mission, Contingency Operations (Space) or COPS. Under this mission, they will "provide worldwide space operations support to Army forces during operations as well as other-than-war contingency missions such as floods, earthquakes, and humanitarian support. This mission would become the Army Space Support Teams. Two teams will provide on-site assistance to deployed troops or will train unit members and provide to equipment. The COPS teams resulted from an ARSPACE from the ASEDP and recognized need to make equipment available for contingency operations.

13 September 1994 The HELSTF concluded the first High Energy Laser Light Opportunity (HELLO 1). This made continuous megawatt-class laser light available and affordable for the first time ever to American researchers.

27 September 1994 In their "Contract with America" pre-congressional election platform, 350 Republican candidates for the U.S. House of Representatives pledged to deploy both ABM and TMD systems.

28 September 1994 Washington Summit "At a meeting in Washington, U.S. President Bill Clinton and Russian President Boris Yeltsin [issued] a joint statement noting that they have "agreed on the fundamental importance of preserving the viability and integrity of the ABM Treaty." The two presidents also noted that "both sides have an interest in developing and fielding effective theater missile defense systems on a cooperative basis. The presidents agreed that the two sides will conduct a joint exercise of TMD and early warning. This exercise would contribute to providing a basis for U.S. and Russian forces to operate together, for example, in peacekeeping operations."

1 October 1994 The USASSDC, as the executive agency for the BMDO, assumed custody of Wake Island. The BMDO and USASSDC have been operating on the island since 1988, when launch and support facilities were constructed for the STARBIRD program. Full transfer never took place due to issues with the environmental conditions on the island.

13 December 1994 The Director of the Army Staff, Lieutenant General Charles E. Dominy, approved the USARSPACE Concept Plan for Directed Military Overstrength Manning of the Army Theater Missile Defense Element (ATMDE), Tactical Operations Center (TOC).

1995

1 January 1995 The ARSPACE officially activated the Army Space Support Teams and teams began to deploy to the field to provide space support enhancement. Originally there were three ARSSTs, each aligned with a Combatant Command. Over the course of the year, a team was forward-deployed at Ft. Bragg to satisfy the heavy demands for support made by the XVIII Airborne Corps and special operations units.

10 January 1995 The HELSTF lased replicas of Scud missile fuel tanks to conclude a series of tests in support of the Air Force airborne laser program. The tests which began on 4 October 1994, allowed "engineers to experiment with the power of the laser and the spot-size of the beam."

16 January 1995 The USASSDC established the Missile Defense Battle Integration Center (BIC). The aim of the BIC was to connect the four elements of TMD - active defense, passive defense, attack operations, and battle management, command, control and communications.

1 February 1995 The USASSDC organization in Huntsville reorganized. The five directorates (Advanced Technology, Sensors, Systems, Targets, Test and Evaluation and

Weapons) and the Cost Analysis Office, the Program Integration Office, the PAO and the Staff Action Control Office combined to form the Missile Defense and Space Technology Center (MDSTC). This name change reflected the roles and missions of the Huntsville organization. It also underscored Huntsville's reputation as a national center of excellence for missile defense and recognized plans to expand Huntsville's role in the Army space mission. The mission for the Tech Center is "to continue as the Nation's research and development hub of space and missile defense technology excellence."

13 February 1995 The TMD Force Projection Tactical Operations Center (FP TOC) made its debut at the Pentagon. The USASSDC built the TOC to address a need envisioned by then Army Chief of Staff, General Gordon Sullivan, to provide "overarching command and control capability for the TMD fight."

21 April 1995 First flight/propulsion test of the THAAD interceptor. All test objectives were achieved to include "missile launch, booster performance, booster/kill vehicle separation, KV shroud separation, radar-to-missile communication, and flight/seeker environmental data collection."

May 1995 The Synthetic Theater of War for TMD (STOW-TMD) was first used during the Army's Roving Sands Exercises.

1 May 1995 The U.S. Army converted the MILSATCON Directorate of the ARSPACE into the 1st SATCON Battalion. The battalion plans and controls the payload of the DSCS satellites. Formed from those ARSPACE elements responsible for the DSCS, the SATCON is composed of the Fort Detrick DSCSOC Detachment now A Company, Fort Meade DSCSOC Detachment B Company; Landstuhl DSCOC Detachment C Company, Camp Roberts DSCSOC Detachment D Company and Fort Buckner DSCSOC Detachment E Company. This is the first battalion, in the history of the Army whose operational mission is directly tied to the control of space systems and capabilities.

6 May 1995 General Dennis Reimer, commander of Forces Command and the next Army Chief of Staff, visited White Sands to cut the ribbon on the first JTAGS.

11 May 1995 The TRADOC and USASSDC (the BIC) established an MOA which "[described] how TRADOC and USASSDC would jointly work together regarding materiel development, analytical and/or simulation capabilities." In addition, the USASSDC was made a voting member of the Battle Lab Board of Directors.

July 1995 The BIC tested a new long wavelength, infrared seeker the beryllium, cryogenic off-axis telescope (BeCOAT) in a radiation environment. This test, the culmination of a five-year effort, was the first demonstration of a seeker in a radiation environment. The seeker will be able to withstand the radiation effects experienced in near-outer and outer space environments.

1 July 1995 The Office of the Deputy Assistant Secretary of the Army (Procurement), Dr. Kenneth Oscar, designated the USASSDC Contracting and Acquisition Management Office as an Army "Contracting Test Bed for Acquisition Streamlining." The CAMO is the first and only Army contracting office thus designated. The office was also granted a "no protest" provision that allows offerors to voluntarily submit statements that they will refrain from protesting certain agency errors.

20 July 1995 *Field Manual 100-18* – Space Support to Army Operations published. This FM "established doctrine for the Army's use of space, enumerates current space system capabilities, and provides guidelines for the use and application of space capabilities that support Army operations."

October 1995 The USASSDC Sensors Directorate announced the development of Blue October, a user-friendly, high-technology computer simulation tool. The program permits engineering and simulation work to be accomplished at a desktop computer.

October 1995 The ARSST 1 members were deployed to Egypt to support Bright Star with satellite technologies.

November 1995 The Force Protection Tactical Operations Center (FP-TOC) made its first overseas deployment for Bright Star '95. The FP-TOC brought communications, imagery, weather, terrain analysis, intelligence and early warning systems, into a single suite for the theater commander.

1 November 1995 Proposed date for the USASSDC takeover of the War Breaker facility, developed by ARPA. With the creation of the BIC, the USASSDC required a "local center that can link via the Distributed Interactive Simulation net to the Huntsville facility and with other centers of excellence in Advanced Distributed Simulation throughout the continental United States."

10 November 1995 Secretary of the Army Togo West designated USASSDC a Reinvention Laboratory. The command was given the authority to waive DA regulations and DoD Initiatives, with justification and legal review. The purpose is to develop new and innovative business practices, streamlining the process.

14 November 1995 Army Space Executive Working Group charter issued. The charter identified the Chief, Space Integration Division, as the Chairman of the ASEWG. The Force Development Integration Center (FDIC) assumed this position upon the designation of USASMDC as the Army's specified proponent for space on 1 October 1997.

December 1995 The first units were equipped with PAC-3 Configuration 1, the first true PAC-3 system. It fields a number of improvements, especially in BMC3I and incorporates the Guidance Enhanced Missile (GEM).

14 December 1995 It was announced that a Tiger Team had been created to establish the Missile Defense and Space Technology Center as a Center of Excellence for BMDO. They had already identified five areas in which "MDSTC is the undisputed world leader in missile defense technology." These were Kinetic Energy Hit-to-Kill weapons; Lethality; Discrimination/ Phenomenology; Targets Development/Range Support; and, Radar/Ladar. As the BMDO Center of Excellence for Missile Defense, "the MDSTC would form joint product teams, evaluate service needs, and recommendation BMDO how future work should be performed." If selected a BMDO Center of Excellence, the MDSTC would serve as a clearing house in the five areas listed above assessing studies, referring proposals to related programs, permitting/denying start-up of a project. The design is to avoid duplication among the services.

1996

11 January 1996 Under Secretary of Defense for Acquisition and Technology Dr. Paul Kaminski directed the Army to form a joint program office and initiate an aerostat program. Operational control of this program was assigned to the USASSDC.

February 1996 The ASPO provided intelligence gathering support to the peace mission in Bosnia.

9 February 1996 The Nautilus program, using the Mid Infrared Advanced Chemical Laser HELSTF at WSMR, demonstrated the effectiveness of a Tactical High Energy Laser (THEL) with an intercept of a short-range rocket in flight. This was the first time that a laser had destroyed rocket in flight.

March 1996 The MDBIC Spatial Weapons System Analysis Center supported the Dayton Accord discussions with calculations to determine the impact of proposed demarcation lines.

11 May 1996 Secretary of Defense William Perry announced that the DoD had committed to work with the government of Israel to develop an Advanced Concepts Technology Demonstration for the THEL. The THEL is a joint U.S.-Israeli project to develop a tactical laser for Israeli use against enemy short-range rockets, e.g. the Katyusha rockets. This decision was based in part on the successful Nautilus test.

24 June 1996 The U.S. and Russia concluded a TMD demarcation agreement. This arrangement was described as "an initial agreement distinguishing between defenses against strategic ballistic missiles [ABM systems] … and certain defenses against non-strategic ballistic missiles, i.e., so-called 'lower-velocity' TMD. This agreement will make clear that all TMD systems with interceptor velocities up to and including 3 kilometers/second are permitted under the ABM Treaty, so long as they are not tested against target missiles with velocities above 5 kilometers/second or ranges greater than 3,500 kilometers. The sides will continue discussions on demarcation of higher-velocity TMD systems."

12 July 1996 Vice Chief of Staff of the Army General Ronald H. Griffith designated USASSDC a stand-alone Army Component Command. The HQDA Redesign Functional Area

Assessment had recommended realigning USASSDC with TRADOC. General Griffith found, however, that the USASSDC was unlike other Army organizations and its functions did not integrate well into any of the current major commands.

12 July 1996 The USASSDC was designated an Army Component Command.

18 July 1996 The Program Executive Office, Missile Defense officially became the PEO Air and Missile Defense.

20 August 1996 The ARROW-2, a two-staged missile, successfully intercepted a simulated SCUD missile. These tests completed the ARROW Continuation Experiments.

16 October 1996 Red Tigress III launched a sounding rocket experiment from Wallops Flight Facility, VA. The test included 17 experimental payloads.

1997

21 January 1997 U.S. Senate Majority Leader Trent Lott (R-MS) and 25 co-sponsors introduced the NMD Act of 1997. This act required the U.S. to deploy a NMD system by the end of the year 2003. In contrast, the Clinton administration's "3-plus-3" program required the U.S. to develop an NMD system by 2000, at which point all ballistic missile threats to the U.S. would be evaluated and a determination made as to whether or not such a system should be deployed by 2003.

8 February 1997 The Willow Dune program successfully launched a Scud ballistic missile target from the KMR, the first operation of this kind at a U.S. test range.

18 February 1997 The USASSDC signed an MOA with TRADOC, which made the command the Army Specified Proponent for Space and NMD and the overall Army integrating command for TMD. The command would now determine space requirements for TRADOC approval and lead integration of DTLOMS solutions across the Army and within appropriate joint agencies. The FDIC was created to execute these new tasks. The Battle Lab was another product of this agreement. The MOA chartered the command to establish the battle lab to plan and conduct space and missile-defense warfighting experiments.

19 February 1997 The first JTAGS unit fielded in Stuttgart, Germany.

March 1997 An MOA between the USA and the USAF signed by General Dennis Reimer and General Ronald Fogelman outlined the responsibilities of the two services with regard to NMD.

14 March 1997 The THEL Test 8A was conducted demonstrating tracking and lasing against multiple in-flight targets.

17-19 March 1997 Demonstration of the Low Earth Orbit Communications (LEOCOMM) during Gold Spear in Tampa, FL.

1 April 1997 The BMDO established the JPO-NMD. The JPO provides management oversight for NMD program elements and is responsible for the design, development, and demonstration of an NMD system to defense the U.S. from ballistic missile attack by 2003.

5 May 1997 Lieutenant General Eric Shinseki, DA DCSOPS, signed a letter of Promulgation for the Charter designating the Headquarters, USASMDC as Army Implementing Agent for the Strategic Arms Reduction Treaty (START) and START II Implementation.

27 May 1997 The second JTAGS unit was fielded at Osan AFB, Korea in support of the warfighting commander in chief. Speaking at the dedication ceremony, Lieutenant General Joseph Hurd, Osan AFB Commander, noted the joint nature of this endeavor, remarking "You are an Army element commanded by a Navy lieutenant, with half-Army half-Navy crew, operating on an Air Force Base." This first unit was fielded in Germany. These two systems replaced prototypes that had been in the field for about 36 months.

June 1997 The USASSDC established the Space Technology Integration Office (STIO) in support of the Army Vision 2010. This office was designed to "focus on space technologies and look at how USASSDC-developed technologies can be leveraged through space-related applications to meet Army requirements for the Army of the future, no just in the area of missile defense."

12 August 1997 Successful hover test of a prototype KE ASAT kill vehicle completed at the National Hover Test Facility, Air Force Systems Command's Phillips Laboratory, Edwards AFB.

29 September 1997 PAC-3 flight test conducted at WSMR. This was the first controlled test flight and data collection.

30 September 1997 The Army TMD Element FP TOC was inactivated during ceremonies at Army Space Command. The TOC was to be transferred to the AAMDC, Fort Bliss, TX, in November, and reactivated.

October 1997 The Hardware-Software Integration Center (HSIC) opened in Colorado Springs, CO. The HSIC, "provides an environment to explore, integrate, test, and evaluate space, missile defense, and related capabilities for the warfighter."

1 October 1997 Effective date of General Orders 5, dated 1 March 1998, which established the USASMDC at the MACOM level. The CG, USASMDC serves as the Army specified proponent for space and NMD and as the Army operational integrator for TMD.

1 October 1997 *SMDC Vision 2010* published. This document was the command "blueprint for reorganization." Among the goals outlined are "the integration of space support in full spectrum land operations; the creation of a global, multi-element missile defense; the cultivation of space partnerships; and, the extension of advanced space and missile defense technology for combat forces.

25 October 1997 The USASMDC successfully completed the Data Collection Exercise (DCE) at HELSTEF. In this experiment, the MIRACL and LPCL lasers successfully tracked/lased the orbiting Air Force research satellite MSTI-3. With this exercise, HELSTF "significantly improved its ability to track targets in low Earth orbit and demonstrated its ability to perform high and low power laser engagements involving on-orbit targets."

November 1997 The Laser Communications ground terminal completed its first field test. The Laser Communications is part of the Synthetic Theater of War (STOW) program.

18-21 November 1997 The ASPO fielded the Tri-Band Satellite Communications System at Fort Bragg, NC. This system is the first 6.2 meter Single Radio Frequency feed element Tri-band system that is certified in Ku band and C band.

1998

March 1998 The Ballistic Missile Targets Joint Project Office received its charter from the Army Acquisition Executive. This was the first charter for the USASMDC, which sought to centralize the requirement held by all branches of the service to develop and launch ballistic missile targets.

19 March 1998 Senator Thad Cochran (R-MS) introduced the American Missile Protection Act which established U.S. policy to deploy, as soon as technologically possible, a National Missile Defense system. The Senate passed the National Missile Defense Act of 1999 by a vote of 97 to 3 on 16 March 1999.

April 1998 The Army announced that the USASMDC Contracting and Acquisition Management Office would be recognized for their achievements, obligating more than 25% of the Army's R&D money, with Vice President Albert Gore's Hammer Award for excellence in contracting.

April 1998 The Battle Lab achieved a first by conducting training for soldiers in the Persian Gulf via simulations and a synthetic battlefield initiated in Huntsville.

April 1998 The Army announced the creation of a new officer functional area, Space Operations, or FA 40, a part of the Information Operations career field. The FDIC was responsible for this effort.

24 April 1998 - 20 May 1998 The Battle Lab introduced the new Common Operational Modeling, Planning and Simulation Strategy, or COMPASS, tool at Joint Project Optic Windmill-3. The COMPASS supported in-theater-on-site training to American and allied personnel.

24 April 1998 - 20 May 1998 The ARSPACE deployed the JTAGS to its first overseas exercise, Joint Project Optic Windmill, in The Netherlands. The airborne Surveillance Testbed and the Battle Lab also participated.

June 1998 The Army announced the selection of the Battlefield Ordnance Awareness program for the STOW. The Mosaic Array Data Compression and Processing effort became a candidate for the Army's Warfighting Rapid Acquisition Process.

10 June 1998 The Secretary of Defense selected the USASMDC to be the lead service for a joint feasibility study on the missile alert broadcast system. This is the first time that the USASMDC was selected to serve as the lead service for a joint feasibility study and a joint test and evaluation effort.

5 July 1998 The Commission to Assess the Ballistic Missile Threat to the U.S. released their report, with dissent. The report stated that "the ballistic missile threat to the U.S. is real, credible and could appear sooner than earlier intelligence predictions." Established by the 1988 Defense Authorization Act, Donald Rumsfeld chaired the commission.

September 1998 Redstone Arsenal's Army Missile Optical Range successfully tested "the world's first compact, transportable, solid-state Range Resolved Doppler Laser radar (ladar)." Part of USASMDC's Advanced Discriminating Ladar Technology Program which is designed to develop a four-dimension, solid-state imaging radar.

November 1998 The Iridium phone system developed by the Battle Lab became fully operational. The Battle Lab purchased 11 phones for warfighter demonstrations. The phones are support by a constellation of 70 satellites orbiting the globe. Captain Dwayne Dickens explained, "This is the first truly global phone system and will be invaluable to the soldier in the field." The next phase is to reduce the size of the phones to that of a cell phone.

1999

January 1999 The USASMDC published the first Directed Energy Master Plan that charts the potential uses of directed energy on future battlefields.

March 1999 The Joint Land Attack Cruise Missile, Defense Elevated Netted Sensor (JLENS) provided a link between an offshore Navy Aegis cruise and a land-based Patriot air defense system for the first time at Fort Stewart, GA.

15 March 1999 During a data collection and seeker test, the PAC-3 successfully intercepted a tactical ballistic missile at WSMR.

5 May 1999 Secretary of the Army Louis Caldera granted approval to the KMR to conduct commercial space launches.

6 May 1999 The Army announced the selection of the first FA 40 Space Operations officers. The Army's first Career Field Designation Board results listed eleven lieutenant colonels and twelve majors who were "career-field designated" into Functional Area 40 – Space Operations. Space Operations Officers will "assist in the managing, planning and integrating of space system capabilities to benefit the Force XXI and Army After Next warfighter." Lieutenant General John Costello, USASMDC Commander described the FA40 as "the pivotal position to provide the comprehensive coordination of space assets."

10 June 1999 The THAAD missile successfully intercepted its target during Flight Test 10. This test was the seventh intercept attempt.

15-27 June 1999 During Roving Sands '99, the JLENS successfully tracked multiple low altitude targets to 200 miles.

26 June 1999 The THEL Advanced Concept Technology Demonstration achieved first light at the TRW Capistrano Test Facility in California.

July 1999 The Army Vice Chief of Staff, General John Keane, signed the U.S. Army Theater Air and Missile Defense Master Plan, a significant step towards integrating present and future air and missile defense systems under a single long-term vision.

9 July 1999 DOD Space Policy announced.

23 July 1999 President William Clinton signed the National Missile Defense Act of 1999 (PL106-38) into law, saying that the legislation makes it clear that no decision on deployment has yet been made and the U.S. will continue to take its nonproliferation and arms control objectives into account.

2 August 1999 General John Abrams, CG of the U.S. Army Training and Doctrine Command, approved the charter for the NMD TRADOC Systems Manager Office. The charter authorized the new agency to act as the Army's representative, manager and integrator for the entire spectrum of doctrine, training, leader development, organizational, materiel, and soldier products (DTLOMS) associated with the land-based NMD system. The Army assigned the NMD TSM to the USASMDC.

9 September 1999 FM 40-1 JTAGS Operations published

October 1999 The first corps level Tactical Exploitation System (TES), developed by the ASPO, was fielded to the XVIII Airborne Corps.

1 October 1999 The U.S. Space Command assumed responsibility for the DoD Joint Task Force – Computer Network Defense mission. The JTF-CND is located in Arlington, VA and "orchestrates the defense of all DoD computer networks and systems." This transfer was directed by the president. The task force was originally activated on 30 December 1998 "after exercises and real-world events demonstrated the need for a single coordinating agency with authority to direct actions necessary for the defense of vital national computer networks."

2 October 1999 The first intercept test of the Ground Based Midcourse Defense Segment EKV (IFT-3), using one warhead and one decoy, was successful.

3 November 1999 The JLENS Program Office was awarded a 2000 Design and Engineering Award by *Popular Mechanics* magazine for its very clever use of existing technology to solve an extremely difficult problem."

15 November 1999 Mr. Jacques Gansler accepted the recommendation of the JROC and designated the Army as the land-based NMD system Lead Service in accordance with DoD Regulation 5000.2R. In this memo, he also supported the JROC decision to assign the Army as the User Representative for the land-based NMD system and as Operational Requirements Document (ORD) approval authority for land-based NMD system issues that are not specific Key Performance Parameter requirements. The Director BMDO remained the BMD Acquisition Executive for the NMD System.

15 December 1999 The USASMDC stood up the 1st Space Battalion "to institutionalize space within the Army by giving our soldiers a familiar structure to work with." This move brought the ARSSTs and the JTAGS under one organization. General Costello stated, "This unit is an example of the type of organization that will enable the smaller, lighter, more agile fighting forces envisioned by General Eric Shinseki, Army Chief of Staff."

2000

January 2000 The NMD User Lab, located at Army Space Command became operational.

4 January 2000 Lieutenant General Paul Kern, Director Army Acquisition Corps, and Lieutenant General Ronald Kadish, Director BMDO, issued a memorandum that ordered the streamlining of the management structure of the NMD program. As a result, all of the project managers under the direction of the Ground-Based Elements Program Office would now report to the System Program Director, NMD Joint Program.

March 2000 The JROC approved the establishment of a Single Integrated Air Picture Systems Engineer Task Force to address CINC integration and interoperability issues associated with emerging and legacy systems.

March 2000 Israel deployed the first battery of ARROW Missiles.

22 March 2000 Lieutenant General Ronald Kadish, BMDO Director, issued a memorandum in which he appointed USASMDC as the executive agent for BMD science and technology. As a result, effective 1 June 2000, the Space and Missile Defense Technology Center was realigned to establish the Center for Technology Development and the Joint Center for Technology Integration.

31 March 2000 FM 100-12 Army TMD Operations published.

4 April 2000 Forward Pass Mission #5 was the first live, over-the-horizon engagement of a cruise missile target using an elevated sensor platform. The JLENS successfully completed two CMD Forward Pass demonstrations.

28 May 2000 The command successfully conducted the demonstration flight of the Orbital/Suborbital program Target Launch Vehicle. The test occurred at Vandenberg AFB, CA.

June 2000 Lawyers from the Clinton Administration concluded that the initial work associated with the construction of an X-band tracking and discrimination radar on Shemya Island, AK, would not violate the ABM Treaty.

1 June 2000 The Office of Technology Integration and Interoperability was established as a major subordinate element of the USASMDC.

6 June 2000 The THEL demonstrator successfully tracked and destroyed a single rocket (a Katyusha) in flight for the fist time and during the fist attempt.

July 2000 The USASMDC established two deputy commanding general positions – the Deputy Commanding General (DCG) Army Space, was also the DCG for Operations, located in Colorado Springs, and the Deputy Commanding General for Acquisition, located in Huntsville, Alabama. Prior to this change there was only one DCG in the command.

6 July 2000 Ground-breaking ceremony was held for new buildings that will house the U.S. Army Space Command and the U.S. Space Command and NORAD, at Peterson AFB, Colorado. Lieutenant General John Costello, representing USARSPACE, said that the move would make them more of the team -- "It is a symbol of jointness and of working together as a joint team doing the nation's business."

10 July 2000 A ground-breaking ceremony in Stuttgart, Germany for a new combined facility for the Army Space Command-Europe and the Defense Information Systems Agency – Europe (DISA). The facility will be a one-stop-shop for communications 24/7. ARSPACE provides satellite communication support while DISA provides terrestrial communication support.

14 August 2000 In order to centralize management of the Theater Ballistic Missile Defense, PEO-AMD activated the Lower Tier Project Office. Lower Tier incorporated the Patriot, PAC-3, and Medium Extended Air Defense System (MEADS).

1 September 2000 President Bill Clinton, speaking at Georgetown University, announced: "I simply cannot conclude with the information I have today, that we have enough confidence in the technology and the operational effectiveness of the entire NMD system, to move forward to deployment. Therefore, I have decided not to authorize deployment of a NMD at this time."

October 2000 Army National Guard and Reserve personnel began to drill with the ARSPACE. Lieutenant General John Costello described the event as "a new era of cooperation between the Army and the Army National Guard and Reserve." Reserve Forces support will be provided by both individual mobilization augmentees assigned to the Army Reserve and by guardsmen assigned to the Colorado Army National Guard. These personnel will support information operations activities at the 1st Space Battalion's Mobile Technology Team as well as operations of its 1st SATCON Battalion.

1 October 2000 The U.S. Space Command assumed responsibility for the Computer Network Attack (CNA) mission for the Department of Defense. This mission was added to the existing responsibilities for Computer Network Defense and "coordinating all military space operations, to include missile warning, communications, navigation, weather and surveillance from DoD, civil and commercial satellite systems." According to the U.S. Space Command News Release, "the United States will only employ CNA after careful policy and legal review, and any use of CNA will be consistent with U.S. international obligations and the Law of Armed Conflict."

1 October 2000 Operational control of the Big Crow Program Office transferred to USASMDC. The mission of the Big Crow was to "provide projected electromagnetic environments for electromagnetic vulnerability assessments;" and "provide and operate airborne and ground-based assets for electromagnetic experiments, tests, trials and training." Big Crow, which "provides DoD's only remaining large-scale electronic warfare, high-powered, stand-off jammer capability," was assigned to ARSPACE effective 1 October 2000.

13 October 2000 The ASPO and Northrop Grumman completed fielding of the TES Main System #1 to Ft. Bragg NC. The ASPO accepted the system and handed over its ownership to C Co., 319th MI Bn, 525 MI Bde, XVIII Airborne Corps.

26 October 2000 The Army assigned ARSPACE as the single Army component commander to support U.S. SPACE CNA/CND missions. The U.S. Space Command had assumed CND for the DoD on 1 October.

November 2000 *Popular Science* magazine selected the THEL ACTD as the Grand Winner in the General Technology category for its "Best of What's New" awards for 2000.

2001

January 2001 Units from the Colorado Army National Guard and the Reserve began to drill with the ARSPACE in preparation for becoming elements of the ARSSTs and the 193rd Space Support Battalion.

11 January 2001 Space Commission Report released. Chartered by Congress to assess American National Security Space Management and Organization, the commission was headed by Mr. Donald Rumsfeld.

1 February 2001 Acting Pentagon Acquisition Chief David Oliver approved 14 new ACTDs. Included among these was the Advanced Tactical Laser.

15 February 2001 The U.S. Commission on National Security/21st Century stated that "The military cannot undertake any major operation, anywhere in the world, without relying on systems in Space."

17 May 2001 USASMDC Deputy Commanding General-Operations (DCG-O) received permission to authorize acceptance and wear of the Air Force Space and Missile Badge to members of the Army awarded this badge. The Space Badge Wear Authority for ARSPACE came via AFI 36-2923 and a PERSCOM memorandum dated 17 May 2001. The DCG-O defined two means by which an assessed FA 40 could earn the badge (1) attend and graduate from the Army or Air Force seven-week Space course or (2) have two years of service in a Space Operations position and have attended 3Y schooling. The Senior Space Badge and Master Space Badge are authorized after seven and fifteen years of space service respectively. Plans call for a separate and distinct Army space badge to be created within five years.

June 2001 At the direction of Congress, the KMR was renamed the Ronald Reagan Ballistic Missile Defense Test Site at the U.S. Army Kwajalein Atoll.

23 July 2001 A ground-breaking ceremony for the von Braun Complex, the new facility to be built for the USASMDC on Redstone Arsenal was held.

3 August 2001 A group of 14 officers graduated from the first Space Operations Officer Qualification Course the FA40 Course. This intense seven-week course was conducted in Colorado Springs, CO. The goal is to train officers to become "experts in using Space to support the warfighter." As described in *The Eagle*: Course instructions were divided into three segments beginning with 25 days of classroom instruction. A week was then devoted to off-site visits to the NRO, the National Imagery and Mapping Agency [NIMA]...This included hands-on training with the Army Space Program Office, which has developed the Tactical Exploitation of National Capabilities Space support systems used by the Army warfighters." The course also incorporates a 43-hour command post exercise designed to test each student's proficiency in 24 individual critical tasks."

9 August 2001 Record of Decision issued by BMDO to conduct initial site preparation activities for the Fort Greely, AK, portion of the Missile Defense System Test Bed.

September 2001 The USASMDC stood up the Directed Energy Center of Excellence at HELSTF.

6 September 2001 Kenneth Oscar, Acting Army Acquisition Executive, announced program realignments, which impacted the BMD Organization, PEO-AMD, U.S. Army Aviation and Missile Command and USASMDC. To this end, the Lower Tier Program left BMDO for the PEO-AMD; SHORAD transferred from AMCOM to PEO-AMD; THAAD and ARROW moved from PEO-AMD to BMDO; and, BMTJPO moved from USASMDC to BMDO. The effective date for this action was 1 October 2001.

19 September 2001 The USASMDC Tech Center chartered the Transformation Technology and Concepts Integrated Product Team.

28 September 2001 The HELSTF unveiled their new Solid State Heat Capacity Laser.

28 September 2001 Activation ceremony for the Colorado Army National Guard's 193rd Space Battalion at Peterson AFB, CO. The 193rd became the third battalion of the ARSPACE family. Another first was achieved as the Colorado unit became the first Guard unit with a space mission.

October 2001 The ASPO began to field the Grenadier Beyond line-of-sight Reporting and Tracking (BRAT) a blue-force tracking tool which allows commanders to track friendly forces in near-real time deep on the battlefield.

1 October 2001 The JLENS Program Office transferred to the PEO–AMD for formal acquisition, testing and fielding.

16 October 2001 The DCG for Space assumed the duties of Chief of Space Information Operations Element (Reach-back Element) (SIOE (RE)) for the U.S. Space Command. As the SIOE-RE, the DCG-Space "is responsible for the overall integration of Space and comprehensive IO planning into the plans of Combatant Commanders…"

26 October 2001 Effective this date, all Army acquisition programs, regardless of Acquisition Category, were to be managed by a Program/Project/Product Manager either (1) overseen by a Program Executive Officer or (2) directly reporting to the Army Acquisition Executive.

3 December 2001 At the direction of the Army, Brigadier General John Urias assumed the duties of PEO-AMD. Brigadier General Urias has also the USASMDC DCG for RD&A and Director of the USASMDC Acquisition Center.

2002

2 January 2002 Secretary of Defense Donald Rumsfeld signed a memorandum restructuring the BMDO and renaming it the Missile Defense Agency (MDA). The new MDA will report to the Under Secretary of Defense (Acquisition, Technology and Logistics). The MDA was elevated to the status of an agency in recognition of the national priority and mission emphasis on missile defense.

13 February 2002 Pentagon Acquisition Chief Pete Aldridge directed Lieutenant General Ronald Kadish, MDA Director, to "set up and carry out a single program of research and development work to develop the BMDS."

17 April 2002 Secretary of Defense Donald Rumsfeld and Chairman of the Joint Chiefs of Staff General Richard Myers announced changes to the Unified Command Plan (UCP), the document that establishes the missions and functions for combatant commanders. The new UCP established a new unified command the U.S Northern Command (NORTHCOM). Under UCP 2002, the NORAD and USSPACECOM "continue to accomplish their traditional missions and to carry out well-established actions in support of Operations NOBLE EAGLE and ENDURING FREEDOM and other U.S. military operations around the globe. There are no new mission requirements for these organizations. However, "the establishment of NORTHCOM does present future organizational implications for both NORAD and USSPACECOM, primary among them being the separation of NORAD and USSPACECOM with NORAD aligning with USNORTHCOM later this year."

24 May 2002 President George W. Bush and Russian President Vladimir Putin signed a new arms control treaty in Moscow. Under this agreement, known as the Treaty of Moscow "each party shall reduce and limit strategic nuclear warheads, so that by December 31, 2012 the aggregate number of such warheads does not exceed 1700-2200 for each Party." This treaty puts no restraint upon the number of short-range nuclear missiles held by either country. Nor is any mention made of the destruction of bombers, missiles or submarines removed from service. Once ratified, this treaty remains in effect until 31 December 2012 and may be extended. Either side may withdraw from the agreement following a three-month notice.

1 June 2002 In a speech to the graduating class at West Point, President George W. Bush suggested/outlined a new policy/doctrine of pre-emption.

13 June 2002 Following a six-month notice, the U.S. formally withdrew from the 1972 ABM Treaty.

14 June 2002 Russia formally withdrew from the START II nuclear arms treaty.

15 June 2002 Brigadier General John Holly, Program Director of the GMD JPO, oversaw the ground-breaking at Fort Greely, AK, for six underground silos, part of the GMD Testbed.

18 June 2002 The THAAD Project Office logistics team, among others, received the 2002 David Packard Excellence in Acquisition Award, the DoD's highest acquisition award. THAAD was "recognized for creating innovative logistics concepts with the potential to significantly reduce operational and support costs throughout the missile defense system's service life."

26 June 2002 During a press briefing, Secretary of Defense Donald Rumsfeld announced the merger of the U.S. Space Command and the U.S. Strategic Command (STRATCOM), with an initial operational capability of 1 October 2002. The goal of the merger is to "improve combat effectiveness and speed up information collection and assessment needed for strategic decision-making." Rumsfeld stated "the missions of SpaceCom and StratCom have evolved to the point where merging the two into a single entity will eliminate redundancies into the command structure and streamline the decision making process." This decision reflects the Bush administration's efforts "to transform the U.S. military to make it more responsive and flexible." The new command "will be responsible for both early warning of and defense against missile attack as well as long-range conventional attacks." Specifically its missions will include "control of America's nuclear forces, military space operations, computer network operations, [and] strategic warning and global planning." The new, as yet unnamed command will be located at Offutt AFB, Nebraska. Full operational capability is planned for 1 October 2003.

27 June 2002 Extended Air Defense Test Bed Product Office disestablished.

3 September 2002 The Battle Lab handed over the Advanced Warfare Environment software package to the Product Manager for Air and Missile Defense Command and Control System.

19 September 2002 Lieutenant General Ronald Kadish, MDA Director, transferred the Targets management and execution to the Air Force's Space and Missile Systems Center, headquartered at Los Angeles AFB, CA.

30 September 2002 The USASMDC Contracts and Legal Offices completed the 46th consecutive year of never having lost a protest - a feat unmatched in the U.S. Army.

October 2002 North Korea admitted that they are pursing a nuclear weapons program, in violation of their 1994 agreement with Washington.

1 October 2002 Fort Greely, AK, is officially transferred to USASMDC.

1 October 2002 The U.S. Space Command and the U.S. Strategic Command merged to create a new U.S. Strategic Command headquartered at Offutt AFB, NE. The new organization was assigned the missions of space operations, information operations, computer network operations, and strategic defense and attack missions. As the new organization stood up, officials reviewed the possibility of adding four new missions to the STRATCOM – Global Strike, Information Operations, Missile Defense, and Command, control, communications, computers, intelligence, surveillance and reconnaissance (C^4ISR).

U.S. Northern Command established; the nerve center for homeland defense. Its mission is twofold: (1) protect the nation from outside attack and (2) assist civilian agencies when attacks or natural disasters occur within the United States. Plans call for NORTHCOM to be fully operational by 1 October 2003. This is the first command of its kind since the Revolutionary War.

1 October 2002 Management of Wake Island transferred from USASMDC to the U.S. Air Force.

2 October 2002 The USASMDC made the Army Service Component Command for U.S. Strategic Command.

8 October 2002 NASA, STRATCOM, NRO, AFSPACE and the Pentagon's Director of Defense Research & Engineering established a cooperative relationship among these space-interested organizations. The goal of this relationship, outlined in an MOA, was to boost technological research and development.

9 October 2002 Opening/dedication ceremonies conducted for the new ARSPACE facility on Peterson AFB.

5 November 2002 The Mobile Tactical High Energy Laser (MTHEL) successfully tracked and intercepted an artillery projectile fired from a Howitzer. This was the first time that a laser had intercepted an artillery projectile.

17 December 2002 President George W. Bush gave the Pentagon two years to deploy a system to defend American territory, troops and allies against missile attack.

Appendix C
Army Astronaut Missions

STS Flights	Astronauts	Position	Landing Date	Ship	Mission	Key Events of Flights
41B	LTC Robert L. Stewart	Mission Specialist	11-Feb-84	Challenger	7-day 23-hours	BG Stewart was the first representative of the Department of the Army to fly into space. Two communications satellites were launched. The most significant events, however, were the first untethered space walks performed by Captain Bruce McCandless II (USN) and LTC Stewart, using manned maneuvering units. The Extra Vehicular Activity (EVA) occurred on the first and seventh days of the flight. (Note: This flight number system meant that the flight took place in 1984 - 4; the flight was launched from Kennedy Space Center -1 vs. Vandenberg AFB designated as 2; and that this was the second launch planned for that fiscal year – B.) As a BG, Stewart later served as the Deputy Commander for the U.S. Army Strategic Defense Command.
51J	COL Robert L. Stewart	Mission Specialist	10-Mar-85	Atlantis	5-day	Second mission devoted to DoD efforts: deployed 2 military satellites.
61B	COL Sherwood C. Spring	Mission Specialist	3-Dec-85	Atlantis	8-day	During the mission the crew deployed three communications satellites and performed a number of experiments. Spring was responsible for satellite deployments. Also, Spring and MAJ Jerry Ross (USAF) conducted an EVA to demonstrate the feasibility of constructing trusses in space.
28	COL James C. Adamson	Mission Specialist	13-Aug-89	Columbia	5-day	This flight was the fourth dedicated to DoD efforts.
38	LTC Charles D. Gemar	Mission Specialist	20-Nov-90	Atlantis	5-day	This mission was a DoD effort to launch a satellite, allegedly to monitor the Persian Gulf region.
43	COL James C. Adamson	Mission Specialist	11-Aug-91	Atlantis	6-day	Primary payload for this mission was the Tracking and Data Relay Satellite-5, the fourth of the TDRS cluster.
44	COL James S. Voss and CWO3 Thomas J. Hennen	Mission Specialist and Payload Specialist	1-Dec-91	Atlantis	7-day	Dedicated to DoD missions, projects included Defense Support Program satellite, Terra Scout, Military Man in Space, etc. CWO3 Hennen, the only branch office to fly in space, conducted phase one of the Terra Scout experiment, which sought to determine what an experienced imagery interpreter could observe from the Space Shuttle using the Spaceborne Direct View Optical System. This was the first time that two Army personnel flew on the same shuttle flight.

STS Flights	Astronauts	Position	Landing Date	Ship	Mission	Key Events of Flights
48	LTC Charles D. Gemar	Mission Specialist	18-Sep-91	Discovery	6-day	The primary payload was the Upper Atmosphere Research Satellite. The mission was to study the Earth's troposphere.
53	COL James S. Voss and LTC Michael R. Clifford	Mission Specialist and Payload Specialist	9-Dec-92	Discovery	7-day	The primary mission was a military payload (DoD-1), the last major military payload then planned for the shuttle fleet.
57	MAJ Nancy J.Currie Sherlock	Mission Specialist	1-Jul-93	Endeavour	9-day	The first flight of the SPACEHAB, pressurized laboratory which would more than double pressurized workspace for crew-tended experiments. The crew also retrieved the European Retrievable Carrier. This was Currie's first space flight and the first for a female Army officer.
58	COL William S. McArthur, Jr.	Mission Specialist	1-Nov-93	Columbia	14 days 12 min. 32 sec.	This mission was the second spacelab flight dedicated to life sciences research. The longest shuttle flight to date.
59	LTC Michael R. "Rich" Clifford	Mission Specialist	20-Apr-94	Endeavour	11-day	Primary payload was the Space Radar Laboratory, radar mapping of the Earth's surface to study human-induced vs. natural environmental change.
62	LTC Charles D. Gemar	Mission Specialist	18-Mar-94	Columbia	14-day	This mission was part of a series of Extended Duration Orbiter flights designed to provide information to assess the impact of long-duration space flight (10 days or more) on astronaut health. Astronauts conducted other experiments as part of the Office of Aeronautics and Space Technology 2 and U.S. Microgravity Payload.
69	COL James S. Voss	Payload Commander	18-Sep-95	Endeavour	11-day	This was the first flight during which two separate payloads were retrieved and deployed during the same mission. LTC Voss participated in a lengthy space walk (over 6 hours) to evaluate improvements made to the extravehicular activity suits and tools.
70	LTC Nancy J. Currie	Mission Specialist	22-Jul-95	Discovery	9-day	The primary mission for this flight was the deployment of TDRS, a space-based network providing communications, tracking, telemetry, data acquisition and command services essential to the Space Shuttle and other low-Earth orbital spacecraft. The crew also performed a number of scientific experiments.

STS Flights	Astronauts	Position	Landing Date	Ship	Mission	Key Events of Flights
74	COL William S. McArthur, Jr.	Mission Specialist	20-Nov-95	Atlantis	8-day	During this mission, the shuttle docked with the Russian Space Station Mir, to provide equipment and supplies. This was the second time that a space shuttle docked with the Mir, a continuation of efforts to construct an International Space Station.
76	LTC Michael R. Clifford	Mission Specialist	31-Mar-96	Atlantis	9-day	During this flight, the shuttle linked up with the Mir. LTC Clifford with Dr. Linda Godwin performed a 6-hour extravehicular activity around the two spacecraft to attach four Environmental Effects Payload experiments to the station's Docking Module.
88	LTC Nancy J. Currie	Mission Specialist	15-Dec-98	Endeavour	11-day	The first NASA mission devoted to the International Space Station (ISS). LTC Currie operated the robotic arm which connected the Zarya module to the Unity module, the first components of the ISS.
92	COL William S. McArthur, Jr.	Mission Specialist	22-Oct-00	Discovery	11-day	During this flight, the space shuttle delivered hardware components for the International Space Station. Installed two current converter units to process power.
101	COL James S. Voss and COL Jeffrey N. Williams	Mission Specialists	29-May-00	Atlantis	10-day	Williams and Voss (USA-Ret) conducted a 6 _ hour space walk to deliver materials and to work on the ISS. This was LTC Jeffery Williams' first space mission.
105	LTC Patrick G. Forrester	Mission Specialist	22-Aug-01	Discovery	12-day	The primary purpose was to rotate ISS crew members and deliver supplies using the Italian made Multipurpose Logistics Module - Leonardo. The crew also performed two spacewalks and conducted a number of scientific experiments.
109	LTC Nancy J. Currie	Mission Specialist/ Flight Engineer	12-Mar-02	Columbia	10 days 22 hrs 11 min.	Captured and maneuvered the Hubble Space Telescope, with the robot arm, to allow them to make repairs and improvements to the system.

* As of December 2002, the Army has three astronauts, LTC Timothy J. Creamer, LTC Douglas H. Wheelock, and LTC Timothy L. Kopra, who have not flown a mission.
* COL Forrester is the commander of the detachment.
* Current Army Astronauts are COL Patrick G. Forrester, COL Nancy J. Currie, LTC (P) Jeffery N. Williams; LTC Timothy J. Creamer and LTC Douglas H. Wheelock selected June 1998, and LTC Timothy L. Kopea selected July 2000.

Appendix D
Acronyms

A

AAF – Army Air Force
AAMDC – U.S. Army Air and Missile Defense Command
AAN – Army after Next
ABM – Anti-Ballistic Missile
ABMA – Army Ballistic Missile Agency
ABMDA – Advanced Ballistic Missile Defense Agency
ABMDP – Advanced Ballistic Missile Defense Program
ACES – Arrow Continuation Experiments
ACTD – Advanced Concept Technology Demonstration
ACTS – Advanced Communications Satellite
ADP – Automated Data Processing
ADS – Azimuth Determination System
AEF – American Expeditionary Force
AFB – Air Force Base
AFSAT – Air Force Satellite Communications
AFSPC – U.S. Air Force Space Command
AIAA – American Institute of Aeronautics & Astronautics
AIT – Atmospheric Interceptor Technology
ALCOR – ARPA Lincoln C-Band Observables Radar
ALTAIR – ARPA Long-Range Tracking and Instrumentation Radar
AMC – U.S. Army Materiel Command
AMOR – Army Missile Optical Range
AOA – Airborne Optical Adjunct
AOD – Army Ordnance Department
AOMC – U.S. Army Ordnance Missile Command
ARADCOM – U.S. Army Air Defense Command
ARC – Advanced Research Center
ARCENT – Army Service Component Command
ARCTIC – Advanced Research Center Telecommunications Interface Console
ARGMA – Army Rocket and Guided Missile Agency
ARL – Army Research Laboratory
ARPA – Advanced Research Project Agency
ARSPACE – U.S. Army Space Command
ARSPOC – Army Space Command Operations Center
ARSST – Army Space Support Team
ASA – Army Space Agency
ASAT – Anti-Satellite
ASC – Army Space Council

ASCC – Army Service Component Command
ASD – Assistant Secretary of Defense
ASDP – Army Space Demonstration Program
ASEDP – Army Space Exploitation Demonstration Program
ASEWG – Army Space Executive Working Group
ASI – Army Space Institute
ASIS – Army Space Initiatives Study
ASMP – Army Space Master Plan
ASPO – Army Space Program Office
AST – Airborne Surveillance Testbed
ASTRO – Army Space Technology Research Office
ASTWG – Army Science & Technology Working Group
ASWG – Army Space Working Group
ATBM – Anti Tactical Ballistic Missile
ATD – Advanced Technology Directorate
ATL – Advanced Tactical Laser
ATMDE – Army Theater Missile Defense Element
ATP – Advanced Technology Program
ATSS – Army Tactical Surveillance Satellite
AWE – Army Warfighting Experiment

B

BAMBI – Ballistic Missile Boost Intercept
BG – Brigadier General
BDU – Battle Dress Uniform
BIC – Battle Integration Center
BL – Battle Lab
BM/C^3 – Battle Management/Command, Control, and Communications
BM/C^4 – Battle Management/Command, Control, Communications, and Computers
BM/C^{4}I – Battle Management/Command, Control Communications, Computers, and Intelligence
BMD – Ballistic Missile Defense
BMDATC – Ballistic Missile Defense Advanced Technology Center
BMDC – Ballistic Missile Defense Center
BMDO – Ballistic Missile Defense Organization
BMDPO – Ballistic Missile Defense Project Office
BMDSCOM – Ballistic Missile Defense System Command
BMTJPO – Ballistic Missile Targets Joint Project Office
BN – Battalion
BOA – Battlefield Ordnance Awareness
BRAT – Beyond Line of Sight Reporting and Tracking
BSTS – Boost Surveillance and Tracking System

C

C² – Command and Control
C²/POS NAV – Command & Control/Position Navigation
C⁴I – Command, Control, Communications, Computer, and Intelligence
C⁴ISR – Command, Control, Communications, Computer, Intelligence, Surveillance and Reconnaissance
CACDA – Combined Arms Combat Development Agency
CAD – Computer-Aided Design
CALL – Center for Army Lessons Learned
CDOCS – Contingency DSCS Operational Control System
CEC – Cooperative Engagement Capability
CENTCOM – Central Command
CEP – Concept Evaluation Program
CG – Commanding General
CIS – Commonwealth of Independent States
CMD – Cruise Missile Defense
CNA – Computer Network Attack
CND – Computer Network Defense
COE – Corps of Engineers
COMPASS – Common Operational Modeling, Planning, and Simulation Strategy
COMSAT – Communications Satellite Corporation
CONAD – Continental Air Defense Command
CONOPS – Concept(s) Of Operations
CONUS – Continental United States
COPS – Contingency Operations (Space)
COTS – Commercial-off-the-shelf
CSOC – Consolidated Space Operations Center
CWO – Chief Warrant Officer

D

DA – Department of the Army
DAB – Defense Acquisition Board
DACS – Direct Altitude Control System
DARPA – Defense Advanced Research Project Agency
DCE – Data Collection Exercise
DCG – Deputy Commanding General
DCSOPS – Deputy Chief of Staff, Operations and Plans
DCSCS – Defense Satellite Communication System
DCSRDA – Deputy Chief of Staff, Research, Development and Acquisition
DE – Directed Energy
DE ASAT – Directed Energy Anti-Satellite
DEM/VAL – Demonstration/Validation
DEMP – Directed Energy Master Plan

DEW – Directed Energy Weapons
DIS – Distributed Interactive Simulation
DISA – Defense Information Systems Agency
DMSP – Defense Meteorological Satellite Program
DOCS – DSCS Operational Control System
DoD – Department of Defense
DOT – Designating Optical Tracker
DPG – Defense Planning Guidance
DR – Discrimination Radar
DRID – Defense Reform Initiative Directive
DSCS – Defense Satellite Communications System
DSCSOC – DSCS Operations Center
DSP – Defense Support Program
DSSW – Director of Space and Special Weapons
DTLOMP – Doctrine, Training, Leader Development, Organization, Materiel, Personnel
DTLOMS – Doctrine, Training, Leader Development, Organization, Materiel, and Soldier Systems
DUS – Deputy Undersecretary of Defense

E

E^2I – Endoatmospheric/Exoatmospheric Interceptor
EADSIM – Extended Air Defense Simulation
EADTB – Extended Air Defense Testbed
EFS – Enhanced Flight Screener
EIT – Exo-Interceptor Testbed
EKV – Exoatmospheric Kill Vehicle
EMD – Engineering and Manufacturing Development
ERDAS – Earth Resources Data Analysis System
ERINT – Extended Range Intercept Technology
ERIS – Exoatmospheric Reentry-vehicle Interceptor Subsystem
EVA – Extra Vehicular Activity

F

FAISS – FORSCOM Automated Intelligence Support System
FAR – Forward Acquisition Radar
FASP – Fly Away Sensor Package
FDIC – Force Development and Integration Center
FEL – Free Electron Laser
FEMA – Federal Emergency Management Administration
FLAGE – Flexible Lightweight Agile Guided Experiment
FLTSAT – Fleet Satellite Communications
FM – Field Manual
FMA – Foreign Military Acquisition

FOA – Field Operating Agency
FOC – Future Operational Capabilities
FORSCOM – U.S. Army Forces Command
FPTOC – Force Projection Tactical Operations Center
FTV – Flight Test Vehicle
FWD – Forward

G

GAMS – Ground Antenna and Monitor Station
GAO – General Accounting Office
GBEPO – Ground Base Elements Program Office
GBFEL – Ground Based Free Electron Laser
GBI – Ground Based Interceptor
GBL – Ground Based Laser
GBR – Ground Based Radar
GBR-X – Ground Based Radar-Experimental
GEM – Guidance Enhancement Missile
GMD – Ground-based Midcourse Defense
GMF – Ground Mobile Forces
GMFSC – Ground Mobile Forces Satellite Control
GPALS – Global Protection Against Limited Strikes
GPS – Global Positioning System
GSTS – Ground-based Surveillance and Tracking System

H

HALO – High Altitude Observatory
HEDI – High Endoatmospheric Defense Interceptor
HELLO – High Energy Laser Light Opportunity
HELSTF – High Energy Laser Systems Test Facility
HOE – Homing Overlay Experiment
HQDA – Headquarters Department of Army
HSD – Hardsite Defense
HSIC – Hardware/Software Integration Center

I

ICBM – Intercontinental Ballistic Missile
ICWG – Interface Control Working Group
ID – Interactive Discrimination
IFICS – In Flight Interceptor Communications
IFT – Integrated Flight Test
IGY – International Geophysical Year
IMINT – Imagery Intelligence

INF – Intermediate-range Nuclear Forces
INMARSAT – International Marine Satellite
IRBM – Intermediate Range Ballistic Missile
IRIS – Infrared Instrumentation System
ISS – International Space Station
ITB – Israeli Testbed

J

JCS – Joint Chiefs of Staff
JIOA – Joint Intelligence Objectives Agency
JLENS – Joint Land Attack Cruise Missile Defense Elevated Netted Sensor System
JPL – Jet Propulsion Laboratory
JPO – Joint Project Office
JPO NMD – Joint Project Office National Missile Defense
JROC – Joint Requirements Oversight Council
JSMB – Joint Space Management Board
JTAGS – Joint Tactical Ground Station
JTF – Joint Task Force

K

KE – Kinetic Energy
KE ASAT – Kinetic Energy Anti-Satellite
KEW – Kinetic Energy Weapons
KITE – Kinetic Kill Vehicle Integrated Technology Experiment
KMR – Kwajalein Missile Range
KREMS – Kierman Re-entry Measurements Site

L

LAM – Louisiana Maneuvers
LAR – Local Acquisition Radar
LASER – Light Amplification by Stimulated Emission or Radiation
LDS – Layered Defense System
LEAP – Lightweight Exo-Atmospheric Projectile
LEOCOMM – Low Earth Orbit Communications
LI – Light Infantry
LIGHTSAT – Lightweight Small Satellite
LoAD – Low Altitude Defense
LPCL – Low Power Chemical Laser
LRALT – Long Range Air Launched Target
LTG – Lieutenant General
LVC – Large Vacuum Chamber
LWIR – Long-Wave Infrared

M

M&S – Modeling and Simulation
MBRV – Modified Ballistic Reentry Vehicle
MACOM – Major Army Command
MAR – Multifunctional Array Radar
MASINT – Measurements and Signatures Intelligence
MASS – Mesoscale Atmospheric Simulation System
MD – Missile Defense
MDA – Missile Defense Agency
MDAP – Major Defense Acquisition Program
MDBIC – Missile Defense Battle Integration Center
MDSTC – Missile Defense and Space Technology Center
MEADS – Medium Extended Air Defense System
MG – Major General
MICOM – U.S. Army Missile Command
MILSATCON – Military Satellite Control
MILSTAR – Military Strategic Tactical and Relay
MIPS – Millions of Instruction per Second
MIRACL – Mid-Infrared Advanced Chemical Laser
MIRV – Multiple Independently-targetable Re-entry Vehicle
MOA – Memorandum of Agreement
MPRS – Mission Planning Rehearsal System
MSE – Major Subordinate Element
MSI – Multi-Spectral Imagery
MSIP – Multi Spectral Imagery Processor
MSLS – Multi Service Launch System
MSR – Missile Site Radar
MSX – Midcourse Space Experiment
MTHEL – Mobile Tactical Higher Energy Laser
MTR – Missile Track Radar
MTTV – Maneuvering Target Test Vehicle
M/V – Manpack/Vehicular Model
MX – Missile Experiment

N

NASA – National Aeronautics and Space Administration
NATO – North Atlantic Treaty Organization
NINA – National Imagery and Mapping Agency
NMD – National Missile Defense
NORAD – North American Aerospace Defense Command
NORTHCOM – U.S. Northern Command

NPB – Neutral Particle Beam
NRO – National Reconnaissance Office
NSC – National Security Council
NSD – National Security Directive
NSDD – National Security Decision Directive
NXDO – NIKE-X Development Office

O

OAMP – Optical Aircraft Measurements Program
OACSI – Office of the Assistant Chief of Staff for Intelligence
OCRD – Office of the Chief of Research and Development
ODCSOPS – Office of the Deputy Chief of Staff for Operations and Plans
ODCSR-DA – Office of the Deputy Chief of Staff for Research, Development and Acquisition
OPTEMPO – Operational Tempo
ORD – Operational Requirements Document
ORD/CIT – Ordnance Department/California Institute of Technology
OSD – Office of the Secretary of Defense
OTII – Office of Technology Integration and Interoperability

P

PAC-3 – PATRIOT Advanced Capability-3
PAR – Perimeter Acquisition Radar
PAWS – Pager Alert Warning System
PEO AMD – Program Executive Office Air and Missile Defense
PEPE – Parallel Element Processing Element
PRESS – Pacific Range Electromagnetic Signature Studies

Q, R

R&D – Research and Development
RADAR – Radio Detecting and Ranging
RAMMSO – Redstone Anti-Missile Missile Systems Office
RDA – Research, Development, and Acquisition
RDT&E – Research, Development, Test and Evaluation
RISTA – Reconnaissance, Intelligence, Surveillance and Target Acquisition
RMI – Republic of the Marshall Islands
ROBS – Rapid Optical Beam Steering
RSSC – Regional Satellite Support Center
RSTA – Reconnaissance and Target Acquisition
RV – Re-entry Vehicle

S

SAFSCOM – U.S. Army SAFEGUARD System Command
SAFLOG – U.S. Army SAFEGUARD Logistics Command
SALT – Strategic Arms Limitation Treaty
SATCOM – Satellite Communications
SATCON – Satellite Control
SBI – Space Based Interceptor
SBL – Space Based Laser
SCORE – Signal Communications by Orbiting Relay Equipment
SDC – U.S. Army Strategic Defense Command
SDI – Strategic Defense Initiative
SDIO – Strategic Defense Initiative Organization
SDS – Strategic Defense System
SECDEF – Secretary of Defense
SENSCOM – U.S. Army Sentinel System Command
SHF – Super High Frequency
SHORAD – Short Range Air Defense
SIAP – Single Integrated Air Picture
SIOE – Space Information Operations Element
SLBM – Sea Launched Ballistic Missile
SLGR – Small Lightweight Global Positioning System Receiver
SLKT – Survivability, Lethality, and Key Technologies
SMDC – U.S. Army Space and Missile Defense Command
SPECC – Space Enhanced Command and Control
SRALT – Short Range Air Launched Target
SRHIT – Small Radar Homing Intercept Technology
SRMSC – Stanley R. Mickelsen SAFEGUARD Complex
SSDC – U.S. Army Space and Strategic Defense Command
SSEB – Source Selection Evaluation Board
SSTS – Space-based Surveillance and Tracking System
STARS – Strategic Target System
START – Strategic Arms Reduction Treaty
STOW – Synthetic Theater of War
STP – System Technology Program
STR – Systems Technology Radar
STRATCOM – U.S. Strategic Command
STS – Satellite Tracking System/Space Transportation System
STTF – Systems Technology Test Facility
SWORD – Short Range Air Defense with Optimized Radar Distribution

T

TAA – Total Army Analysis
TACSAT – Tactical Communication Satellite
TAMD – Theater Air and Missile Defense

TBM – Tactical Ballistic Missile
TCMP – Theater Missile Defense Critical Measurements Program
TDRS – Tracking and Data Relay Satellite
TENCAP – Tactical Exploitation of National Space Based Capabilities Program
TERS – Tactical Event Reporting System
TES – Tactical Exploitation System
THAAD – Theater High Altitude Area Defense
THEL – Tactical High Energy Laser
TIR – Terminal Imaging Radar
TIROS – Television and Infrared Observation Satellite
TMD – Theater Missile Defense
TOC – Tactical Operations Center
TOPO – Topographic
TPIO – TRADOC Program Integration Office
TRAC^3 – Tracking, Command, Control and Communications
TRADOC – U.S. Army Training & Doctrine Command
TRT – Terrain Reconnaissance Tool
TSM – TRADOC System Manager
TTEC – Topographic Technology Exploitation Cell
TTPI – Trust Territory Pacific Islands
TTR – Target Track Radar
TTV – Test Target Vehicle

U

UAV – Unmanned Aerial Vehicle
UCP – Unified Command Plan
UHF – Ultra High Frequency
UPL – Unit Prevention Leader
USAF – U. S. Air Force
USAISC – U.S. Army Information System Command
USAKA – U.S. Army Kwajalein Atoll
USARSPACE – U. S. Army Space Command
USASA – U.S. Army Space Agency
USASDC – U.S. Army Strategic Defense Command
USASMDC – U. S. Army Space and Missile Defense Command
USASSDC – U.S. Army Space and Strategic Defense Command
USSPACECOM – U.S. Space Command
USSR – Union of the Soviet Socialist Republics

V

VCSA – Vice Chief of Staff of the Army
VDCPAD – Vehicular Data Communications and Positional Awareness Demonstration
VHF – Very High Frequency

VTC – Video Teleconference

W

WRAP – Warfighter Rapid Acquisition Program
WSMR – White Sands Missile Range (New Mexico)

X, Y, Z

XBR – X-Band Radar
ZAR – ZEUS Acquisition Radar